Modeling Neural Development

Developmental Cognitive Neuroscience
Mark Johnson and Bruce Pennington, editors

Modeling Neural Development

edited by Arjen van Ooyen

A Bradford Book
The MIT Press
Cambridge, Massachusetts
London, England

This book was set in Bembo on 3B2 by Asco Typesetters, Hong Kong, and was printed and bound in the United States of America.

Library of Congress Cataloging-in-Publication Data

Modeling neural developing / Arjen van Ooyen, editor.
    p. cm. — (Development cognitive neuroscience)
  "A Bradford book."
  Includes bibliographical references and index.
  ISBN 0-262-22066-0 (hc. : alk. paper)
  1. Developmental neurobiology—Methodology. 2. Neural networks (Neurobiology) I. Van Ooyen, Arjen. II. Series.
QP363.5.M63   2003
006.3′2—dc21   2002044450

10 9 8 7 6 5 4 3 2 1

Contents

Foreword

Looking back over the past three decades of work on neural development (approximately the period about which I have some personal knowledge), it is fair to say that in the early 1970s a book like this on modeling would have been difficult to imagine. "Modeling" was not a word often heard in the laboratories in which I was then a postdoc (Steve Kuffler's lab at Harvard and subsequently Bernard Katz and Ricardo Miledi's Department of Biophysics at University College London). In both formal and informal discussions, our mentors made it plain that the proper course for a young neurobiologist interested in development was simply to explore the relevant phenomenology, with the expectation that the important principles would in due course become clear.

Despite the obvious success of the Hodgkin-Huxley "model" of the action potential, it was understood that these equations had not preceded this remarkable body of labor-intensive work, but were rather a post hoc rationalization of a mechanism of neural signaling that had long been understood in general terms. By the same token, the "model" of chemical synaptic transmission based on Poisson statistics was understood to be a theoretical afterthought used to support the obvious experimental facts. What, then, has changed in the intervening period to make modeling not only an acceptable enterprise, but an approach to neural development that is attracting a rapidly growing number of adherents?

Several factors have no doubt combined to effect this cultural shift, all of them apparent in the present volume. First has been the gradual recognition that acquiring an understanding of development is not a problem that falls into the same category as understanding the basis of the action potential or synaptic transmission. A major difference is simply the degree of complexity. The number of causal steps required to generate an action potential or to appropriately convey information by synaptic transmission is clearly finite, as evidenced by the fact that these linkages are now understood in detail. Generating the nervous system on the other hand, involves a causal chain many orders of magnitude greater, in which no particular step provides a key to deciphering the process, and in which epigenetic factors exert a pervasive and poorly understood influence at every turn.

Given the complexity of neural development and the uncertain definition of the problems that need to be solved, it is not surprising that neurobiologists have increasingly sought to understand this process in terms of general principles that subsume a significant subset of the causal chain between the neural plate and the adult nervous system. Witness here the chapters on modeling dendritic and axonal growth and neural wiring generally, and the influence of neural activity on these events.

A second contributor to the rise of modeling as a legitimate enterprise in developmental neuroscience is the enormous success (and universal influence) of molecular genetics. The rapid increase of knowledge in this area has meant that causal linkages that only a few years ago were relegated to a "black box" can now be unraveled. As a result, a very large number of relevant causal steps are now available for incorporation into conceptual models and/or simulations of what development entails. The chapters here on early neural development and neural models of gene networks provide examples.

A third contributor is the growth of computational neuroscience. Thirty years ago, it was unusual to see neurobiologists wrestling with the awkward and relatively ineffectual computers that were then beginning to populate some laboratories. Nowadays, of course, it would be difficult to find a lab in which one of the preeminent tools is not a powerful computer system with access to databases and the ability to create simulations that were until recently beyond possibility.

It may, of course, be several more decades before any of these efforts generate the received wisdom in this field. Nevertheless, it is surely only a matter of time before generally accepted principles emerge as a result of these modeling efforts. In the meantime, the present volume serves as an important progress report in the continuing quest to understand neural development in an increasingly deep way.

Dale Purves

Introduction

Modeling Neural Development

The focus of most modeling studies in neuroscience is on information processing in the mature nervous system, at the level of ion channels, neurons, or neuronal networks. Relatively few modeling studies are directed at understanding how the nervous system develops—for example, how neurons attain their characteristic morphology, or how they are assembled into functional networks. Just as models are needed to help to understand the functioning of the nervous system, they are also needed in order to obtain a better understanding of its development. The development of the nervous system is an extremely complex dynamical process, spanning many levels of organization, from the molecular and cellular to the system level, and involving an overwhelming number of interactions and feedback loops within and between each level. To obtain insight into such a complex process, human intuition and commonsense arguments are not sufficient, and we need the guidance of appropriate mathematical and simulation models.

The present volume brings together examples from different levels of organization (from molecule to system) and from different phases of development (from neurulation to cognition) that demonstrate the power of modeling for investigating the development of the nervous system. In most cases, each chapter contains an overview of the biology of the topic in question, a brief review of the modeling efforts in the field, a discussion in more detail of some of the models, and

some perspectives on future theoretical and experimental work.

This book is intended primarily for computational and experimental neuroscientists, but it will also be of interest to anyone interested in developmental biology. We hope that it will stimulate further research in developmental neurobiology, both theoretical and experimental. Not only is development fascinating in its own right, but its study may lead to new insights into the functioning of the mature nervous system (e.g., learning and memory), since many mechanisms that operate during development remain operative in later life.

In the following section, I summarize why modeling is an integral part of neuroscience. Before describing the structure of the book in the last section, I present a brief overview of the development of the nervous system in order for the reader to be able to see where each chapter fits into the overall scheme of neural development, and thus to be better able to appreciate the choice of topics.

Modeling

All neuroscientists, not only theoretical and computational neuroscientists, make models. An experimentalist's verbal description of how a system works already constitutes a model. Such a verbal model, however, works satisfactorily only when the system under consideration is simple. When several interacting elements and feedback loops are involved, the system rapidly becomes too complex to be understood without the help of mathematical or simulation models.

Given that the nervous system is one of the most complex systems that exists, formal models are an essential research tool in neuroscience (see, e.g., Jennings and Aamodt, 2000; McCollum, 2000):

1. Formal models—in terms of mathematical equations or computer programs—provide precise and exact ways of expression. Without formal models, complex systems and their dynamics cannot be precisely described and analyzed. Constructing formal models (e.g., by translating a verbal model into a mathematical or simulation model) therefore often identifies inconsistencies, hidden assumptions, and missing pieces of empirical data. In addition, translating hypotheses and theories into formal representations makes it possible to communicate them to other researchers in an unambiguous way.

2. Models can lead to the generation of new hypotheses and give structure and meaning to empirical data. Underlying links between data can become clear, and seemingly unrelated observations or phenomena may be shown to be aspects of the same process (integration and unification). Similarly, models can highlight the overlaps among disparate fields of research or among alternative hypotheses. Finally, models may show that our intuition about the system under investigation is wrong; counterintuitive or unexpected dynamics and patterns can easily arise as the result of even simple interactions.

3. Models enable us to study how phenomena at higher levels of organization arise from processes at lower levels of organization. Even for biological systems in which the components are completely known, it is seldom understood precisely how they interact to make the system work. Using only the traditional intuitive approach, a system's collective behavior (the dynamics and the patterns it can generate, or in other words, the working of the system) is very difficult to deduce from knowledge of its constituent parts. Models can provide unique insights because they allow us to explore the consequences of postulated interactions among a system's components and to test the plausibility of hypothetical mechanisms.

4. Model studies can help guide experimental research. The insights and predictions obtained by modeling can alter our outlook and suggest new experiments, which in turn may lead to new models. A comparison between model and experiment will sometimes be qualitative and sometimes quantitative. It is important to realize that often the aim of modeling is not just to build a quantitative replica and to do "mere" computation, which would assume that the principles of how the system works are already fully known. In addition to quantification, mathematical and simulation models provide precise aids to conceptualization and to deepening structural and biological insights.

Neural Development

In order for the reader to see where each chapter fits into the overall scheme of neural development, this section briefly summarizes the development of the vertebrate nervous system. More detailed accounts can be found in, for example, Slack (1991), Cowan et al. (1997), Zigmond et al. (1999), Sanes et al. (2000), and Price and Willshaw (2000).

During the first rounds of cell division, the fertilized egg generates a ball of cells with an internal cavity (the blastula stage). Through invagination of tissues the embryo is converted into a three-layered structure with ectoderm on the outside, mesoderm in the middle, and endoderm on the inside (the gastrula stage). In response to molecules secreted by a region of the mesoderm called the organizer, a portion of the ecto-derm on the dorsal surface of the embryo becomes specified as neural tissue (neural induction) rather than as epidermis. From this flat, one cell-thick sheet of neuroectoderm cells (called the neural plate) the ner-vous system develops.

During the next phase of embryogenesis (called neurulation), the lateral edges of the neural plate ele-vate, appose each other, and later fuse at the dorsal midline to form a hollow cylinder (the neural tube) inside the embryo. Neural tube formation involves cell movements, changes in cell shape, and differential cell adhesion. As the tube forms, some cells along the edges of the neural plate (called neural crest cells) are pinched off. Guided by attractive and repulsive cues in the extracellular matrix, neural crest cells migrate away from the closing tube and eventually give rise to the peripheral nervous system. The central nervous system derives from the neural tube. Molecules ema-nating from the underlying mesoderm (and, earlier in development, from the organizer) create gradients, enabling cells in the neural tube to sense their position and express their genes differentially. As a result, the neural tube becomes specified along the rostrocaudal and dorsoventral axes into distinct domains (neural patterning), which later become the various areas of the differentiated nervous system. The neural tube develops three rostrocaudally arranged vesicles: the forebrain, midbrain, and hindbrain vesicles. The area of the tube caudal to the hindbrain is the prospective spinal cord. Later, the forebrain vesicle divides into two telencephalic vesicles and a diencephalic vesicle. At the level of the diencephalon, a bilateral evagina-tion forms the optic vesicles, which subsequently transform into cups, whose inner walls eventually give rise to the retina of each eye.

During the next stage of development, the wall of the neural tube begins to thicken as newly generated neurons migrate away from the zone of proliferation (near the inside surface of the tube) toward the outer surface of the tube, to form new zones. Neurons mi-grate radially through the wall, but also tangentially (i.e., parallel to the surface of the tube). In some regions, the movements of neurons away from the proliferation zone lead to layered structures—for ex-ample, in the cerebral cortex, which emerges from the telencephalon. In the developing retina, tangential movements of cells within their destination layer help to create regular distributions of cells (retinal mosaics).

During migration and at the site of their final desti-nation, neurons gradually become specified into many different cellular types. The determination of type involves differential gene expression and is controlled by both intrinsic and extrinsic factors. Intrinsic factors include proteins that are inherited from the neuron's precursors. Extrinsic factors are provided by other cells in the form of diffusible molecules, membrane-bound molecules, and molecules bound on the extracellular matrix (ECM); upon binding to recep-tors on the recipient cell, these molecules influence the proteins that regulate the expression of genes.

During or after migration, neurons begin to grow out by projecting many broad, sheetlike extensions, which subsequently condense into a number of small, undifferentiated neurites. Eventually one of the neu-rites increases its length and differentiates into an axon, while the remaining neurites later differentiate into dendrites. By way of the dynamic behavior of growth cones—specialized structures at the terminal ends of outgrowing neurites that mediate neurite elongation and branching—dendrites branch exten-sively and gradually form their characteristic mor-phologies. Axons continue growing out and migrate to their targets. One of the mechanisms by which they are guided is the diffusion of chemoattractant mole-cules from the target. This creates a gradient of increasing concentration, which the growth cone at the tip of a migrating axon can sense and follow. The actions of the growth cone (axon guidance, neurite elongation and branching) are influenced by the concentration of calcium within the growth cone; as a consequence, all the factors that can change this concentration, such as neuronal electrical activity, can modulate neurite outgrowth. Intracellular calcium constitutes an important signal for many aspects of neuronal development, not only for neuronal morphogenesis but also, for example, for the activity-dependent development of intrinsic and synaptic conductances.

Once arrived in their target region, axons may branch considerably before terminating to form initial synaptic connections with permanent and transient target structures. In many parts of the nervous system, large numbers of neurons die at the time that their synaptic connections are being formed, but they also die at earlier phases of development, during proliferation and migration. During the period when initial connections are being formed, the survival of individual neurons depends on anterograde signals received through their input connections as well as on retrograde, target-derived signals received through their output connections. Both types of signals involve neurotrophic factors (growth- and survival-promoting substances) and both include a component that depends on electrical activity.

Once the initial connections are well established, neuronal cell death becomes rare. Further refinement of connections occurs by retraction of axonal branches that project to the wrong targets and elaboration of branches that project to the correct targets, and thus involves both synapse elimination and continued synapse formation. In some cases, such as in the innervation of skeletal muscle fibers, retraction of connections continues until the target remains innervated by just a single axon. Remodeling of axonal arborizations, as well as axon guidance cues, also underlies the formation of topographic maps (e.g., the retinotopic map; neighboring ganglion cells in the retina project to neighboring neurons in the visual cortex) and ocular dominance stripes (i.e., alternating stripes of visual cortical cells that respond prefererentially to input from either the left or the right eye). Refinement of connectivity is influenced by patterns of neuronal electrical activity and involves competition among innervating axons for target-derived neurotrophic factors, which affect axonal arborization. In adulthood, further fine tuning of connectivity, in the form of activity-dependent changes in the signal strength of synapses, participates in learning and memory.

Structure of the Book

The order of the chapters follows loosely the chronology of development as described in the preceding section. Chapters 1 and 2 discuss the very early development of the nervous system. Chapter 1 first shows how, as a result of smooth gradients of transcription factors, sharp boundaries in gene expression can become established between cellular regions—a first step in the way in which areas become specified into distinct regions. It then shows how interactions among transcription factors, cell adhesion, cell division, and cell movement can account for neural tube formation (in vertebrates) and neural cord formation (in insects). Chapter 2 introduces a general framework for modeling molecular-level interactions—for example, gene expression—coupled to cell-cell interactions and changes in the number and properties of cells. The framework is used to examine how, in *Drosophila*, cells become specified as neural cells.

Chapters 3 through 5 cover neuronal morphogenesis and neurite outgrowth. Chapter 3 describes early morphogenesis and shows how positive feedback loops—involving calcium and active axonal transport—may underlie the formation of neuritic structures from intitally spherical cells, and the differentiation of one of the neurites into the axon. Chapter 4 then further discusses dendritic outgrowth and describes how the actions of the growth cone and the underlying cellular and molecular mechanisms give rise to characteristic branching patterns. Chapter 5 looks further at axonal outgrowth, exploring quantitative constraints on axon guidance by target-derived diffusible factors.

Chapters 6 through 8 focus on different aspects of the self-organization of neurons into networks. Chapter 6 explores the consequences for neuronal morphology and network development when neurons self-assemble into networks by means of activity-dependent neurite outgrowth. Chapter 7 examines several mechanisms by which cells in the retina can organize themselves into regular spatial patterns, or mosaics. Chapter 8 explores how single neurons and networks can self-assemble by means of activity-dependent modification of conductances to produce desired activity patterns.

Chapters 9 through 12 discuss the refinement of connectivity and the development of specific connectivity patterns, involving neuronal death (chapter 9) and remodeling of axonal arborizations and changes in synaptic strength (chapters 10 through 12). Chapter 9 describes models that show how the many neuron-to-neuron signals controlling neuronal death combine to affect development at the network level. Chapter 10 describes models of the competitive phenomena involved in the refinement of connectivity (e.g., in the visual and neuromuscular systems). Chapter 11 examines the various models that have been put forward for the generation of topographically ordered connections, or maps, between two discrete neural structures (e.g., between the retina and the visual cortex). Chapter 12 describes models for the generation of the connectivity patterns that underlie ocular dominance and orientation selectivity in the visual cortex.

While chapters 1 through 12 are concerned with how neuronal morphology and networks develop, the last two chapters of this book concentrate more on the functional implications of morphology and development. Chapter 13 focuses on the hypothesis that the development of connections and learning in the mature brain depend critically on structural plasticity at the axodendritic interface, assuming an important role for individual dendrites in computation. Chapter 14 discusses the link between developmental processes at the cellular level and those at the systems level and explores how structural brain development relates to cognitive development.

Acknowledgments

I wish to thank all who were involved in producing this book, in particular its many authors. I wish to thank Michael Rutter at MIT Press for inviting me to create this book and for help in planning it; Sara Meirowitz and Katherine Almeida assisted in all the further stages of preparing the manuscript. I am also grateful to Jaap van Pelt, Michael Corner, Geoffrey Goodhill, Stephen Eglen, David Price, and Harry Uylings for refereeing chapters and comments on the text and to Narcisa Cuceanu for assistance in proofreading.

References

Cowan, W. M., Jessell, T. M., and Zipursky, S. L. (eds.) (1997). *Molecular and Cellular Approaches to Neural Development.* Oxford: Oxford University Press.

Jennings, C., and Aamodt, S. (eds.) (2000). Computational Approaches to Brain Function. *Nat. Neurosci. Suppl.*, Suppl. Vol. 3, November 2000.

McCollum, G. (2000). Social barriers to a theoretical neuroscience. *Trends Neurosci.* 23: 334–336.

Price, D. J., and Willshaw, D. J. (2000). *Mechanisms of Cortical Development.* Oxford: Oxford University Press.

Sanes, D. H., Reh, T. A., and Harris, W. A. (2000). *Development of the Nervous System.* San Diego, Calif.: Academic Press.

Slack, J. M. W. (1991). *From Egg to Embryo. Regional Specification in Early Development.* Cambridge: Cambridge University Press.

Zigmond, M. J., Bloom, F. E., Landis, S. C., Roberts, J. L., and Squire, L. R. (eds.) (1999). *Fundamental Neuroscience.* San Diego, Calif.: Academic Press.

Molecular Models of Early Neural Development

Michel Kerszberg and Jean-Pierre Changeux

During morphogenesis, boundaries are first established among cellular territories characterized by persistent differences in gene transcription; meanwhile and afterward, cell division and movement generate embryonic form. In this chapter we present mathematical and computer models for the control of gene transcription during development. In the context of the developing muscle fiber, we first show how a single transcription factor, diffusing among nuclei and acting nonlinearly on nuclear transcriptional switches, suffices to create a sharp transcription boundary. We then study the case where several transcription factors are present, as in *Drosophila* syncytial blastoderm, and show that oligomerization of transcription factors creates combinatorial complexity, resulting in the sharp specification of multiple territories, which coincide, for example, with various ranges of concentration in a morphogen gradient. We then propose a model that describes in minimal molecular terms the early steps in the formation of the central nervous system (CNS). The formalism is based on the interaction of two transcriptional switches, a membrane receptor and its ligand, and two morphogens and incorporates cell adhesion, cell division, and cell motion. It accounts for the difference between vertebrates (which possess a neural tube) and insects (which possess a neural cord) on the basis of a few determining events in a common developmental program.

1.1 Introduction

Can the development of living forms be described mathematically? The first modern attempt to answer this question was made by Alan Turing (1952). In his landmark paper, Turing introduced a "reaction-diffusion" formalism in which at least two substances ("morphogens") were diffusing at different rates in the embryo and reacting among themselves, one of the morphogens being an "activator" and the other an "inhibitor" (see also chapters 2 and 3). He found that such a system was able to account for important aspects of morphogenesis. In particular, starting from a homogeneous configuration, a stable spatial pattern could develop, owing to the amplification of random disturbances by dynamic nonlinear instabilities. While this approach was revolutionary, at the time too little was known of the cellular and molecular bases of morphogenesis for an actual experimental demonstration to be possible.

Following in Turing's steps, Meinhardt and Gierer (1974; see also Gierer, 1981; Meinhardt, 1986; and Meinhardt and Gierer, 2000) generalized the theory and applied it to new animal models, proposing, for example, to explain the development of polarity in *Xenopus* embryos on the basis of a short-range autocatalytic activation and longer-range inhibition. The nonlinear kinetics of chemical interactions were also fundamental to the chemical "dissipative structures" defined and analyzed by Prigogine and co-workers (Prigogine and Glansdorff, 1971). However, it is fair to say that in all these instances, the precise chemical nature of the "morphogens" supposed to be involved in natural systems remained enigmatic.

Monod and Jacob (1962) were the first to draw, from the revolution of molecular biology, the important experimental and theoretical conclusion that the regulatory mechanisms operating during differentiation and development must be sought primarily (although not exclusively) at the level of gene transcription. While the nuclei of all the cells in an organism share identical DNA, not all the genes carried by this DNA are expressed in all the organism's nuclei (e.g., Alberts et al., 1999). In the nucleus of each differentiated cell, a selected subset of genes is transcribed into messenger RNA, and these RNAs are then processed and translated in the cell cytoplasm into functional proteins. Considerable experimental evidence indicates that a major cellular control point lies at the transcriptional level. The logic of transcription control was taken up in the theoretical models of Thomas and D'Ari (1990) and of Kauffman (1993), and the subject has blossomed since (for a review, see Smolen et al., 2000; see also chapter 2).

One of our goals over the past 10 years has been to achieve a synthesis of the reaction-diffusion scheme with the regulation of gene transcription. This attempt was driven by the experimental progress in molecular developmental biology, which has brought such a unified approach within reach by providing molecular candidates for previously ill-defined entities and by demonstrating their surprising degree of universality.

Thus we now realize that the reaction component of the Turing scheme is largely implemented by the molecular mechanisms that regulate gene transcription, while the diffusion component can be related to a variety of cell-cell communication systems, such as direct chemical interactions at cell contacts and intercellular transport of molecular signals, that act on cells in a concentration-dependent way just like the morphogens hypothesized by Child (1929), Morgan (1904), and Wolpert (1969). How the effects of those interactions are ultimately transduced into transcription patterns at the cell nucleus has been the subject of a vast amount of experimental work.

Experiment has revealed, in addition, that the molecular components used in morphogenesis display striking homologies that span the entire animal kingdom; i.e., they compose molecular building blocks that are used repeatedly within a given organism and that exhibit a remarkable conservation from one species to another (see also chapter 2). Notch, for instance, which is a cell membrane receptor, and Delta, a ligand able to activate it when presented on the membrane of neighboring cells (Artavanis-Tsakonas et al., 1995), were first discovered in the *Drosophila* peripheral nervous system. Today their homologs for example, are also known to be important in certain forms of Alzheimer's disease in humans. The Notch-Delta couple seems to constitute a universal "module" implementing lateral inhibition, the fundamental process that amplifies differences in genetic function among neighbors so that these neighbors are forced to adopt sharply different patterns of gene transcription.

The process of establishing boundaries in gene expression between territories is critical in embryogenesis, but it is only the first step. Discrete cellular identities specified in sharply delineated territories will underlie differential growth patterns and thus shape the embryonic tissues, mainly by regulating cell division and motion. Starting with gastrulation and neurulation, embryonic morphogenesis, at least in animals, is indeed largely a story of cells, groups of cells, and tissues moving with respect to one another in a concerted fashion (Gilbert, 1995; Jacobson, 1991).

Mechanical models of cell motion in development have been introduced (see also chapter 2), particularly with respect to the formation of the neural tube (Jacobson et al., 1986). Clearly, however, embryogenesis is ultimately powered by the coupling of mechanical effects to cell-cell communication processes and gene transcription control. What is called for, then, is a further merging of molecular genetic networks with the mechanical effects that have been extensively studied by classical embryology. Laying the foundations for such a global theoretical construct has been our second long-term goal, and in this chapter we describe some of the elements of this synthesis in their first and still very hypothetical form.

We begin with a formal model of boundary formation within the context of a relatively simple system, the muscle fiber (Kerszberg and Changeux, 1993). The muscle fiber is composed of many nuclei (it is called a syncytium), which initially are all identical from the point of view of transcription, but which mature in sharply different fashion according to their position relative to the synaptic nerve ending on the fiber.

A second model (Kerszberg and Changeux, 1994a) then deals with the question of how various boundaries can be established in a coordinated fashion, owing to the action of morphogens. We first extend the syncytial scheme to *Drosophila* syncytial blastoderm, where about a thousand nuclei sharing the same cytoplasm acquire different identities under the influence of several morphogens. We then mention briefly the patterning of a cellularized system in a gradient of retinoic acid (vitamin A), a morphogen that is common to many vertebrates and that may well play a major role in patterning the nervous system along the head-to-tail (rostrocaudal, or anteroposterior) axis (Kerszberg, 1996).

The last model we introduce (Kerszberg and Changeux, 1998) includes cell movement and cell division (coupled to gene transcription) and describes the formation and dorsoventral patterning of the CNS (a process called neurulation). To our knowledge, it is the first time that a formalism is proposed that offers a unified framework for CNS formation across evolutionary phyla. On the basis of this framework, we propose that the difference between the CNS of insects, which consists of a set of discrete ganglia organized by segments, and the CNS of vertebrates,

which consists of a continuous neural tube, arises primarily from changes in the control of cell motions. This provides an experimentally testable answer to the question of how similar gene networks may give rise to sharp morphogenetic divergences.

1.2 Transcriptional Switches and Embryonic Boundaries

As an example of boundary formation, we consider the developing muscle fiber, or myotube. At the connection between vertebrate motor nerve and skeletal muscle fiber (this connection is called the neuromuscular junction, NMJ), the neurotransmitter acetylcholine (ACh), released in the synaptic cleft by the nerve ending, binds the nicotinic acetylcholine receptor molecules (nAChR) located on the muscle, where it elicits a change in conformation (termed an allosteric transition) that leads to the opening of the associated ion channel. The receptor protein subunits are synthesized in the muscle fiber. The developing muscle fiber results from the regulated fusion of single cells, or myoblasts. Initially, nearly all nuclei in this syncytium express the genes for the various subunits composing the receptor. Progressively, however, following the first contact with the motor nerve ending, subunit gene transcription becomes restricted until in the adult it concerns only the nuclei lying directly under the stabilized motor nerve ending (Duclert and Changeux, 1995; Sanes and Lichtman, 1999). We have proposed a very simple model of such compartmentalized transcription (Kerszberg and Changeux, 1993).

1.2.1 The Model

Transcription of eukaryotic genes is under the control of so-called promoter elements, which include enhancers, short DNA sequences (often located upstream of the genes themselves) to which transcription factors (TF) bind to either enhance or inhibit the activity of RNA polymerase (the enzyme that transcribes DNA to RNA). The genes coding for TFs are, of course, also controlled by TFs. The promoters of acetylcholine receptor subunit genes have been identified (Klarsfeld et al., 1991; Schaeffer et al., 1998), and this will form the basis of our model.

In a sense, the model consists of the simplest genetic network, using a single TF that is produced in all-or-none fashion. The basic units are switches $S_i = 0, 1$ determining the OFF/ON state of gene transcription in each fiber nucleus. A simple biochemical mechanism for such switches consists of a single gene m and its protein product M, itself a TF controlling the activity of m, thus creating a positive feedback loop. If the loop gain is greater than 1, the system forms a genetic "flip-flop" with two stable states, one in which the activity of m and concentration of M are high and one in which they are low. In addition, we assume that M also turns on a genetic cascade (Britten and Davidson, 1969) leading to transcription of AChR subunit genes at the nucleus. M might stand for one or several myogenic factors, such as those of the MyoD family, which cause differentiation of mesenchyme cells into myoblasts (Weintraub et al., 1991) and AChR synthesis. The myogenic genes have indeed been shown to exhibit positive autoregulation controlled at the level of their promoters (Thayer et al., 1989).

The activity of the switches is assumed to be regulated by several effectors. Electrical activity of the nerve fiber, perhaps because of the accompanying influx of Ca^{2+}, represses transcription (Klarsfeld et al., 1989); i.e., in the present context, it reduces the efficiency of M. As a consequence, the autocatalytic effect of M is lowered and the probability of going (or staying) in the high-M state is diminished. In contrast, we postulate that anterograde factors originating presynaptically, such as agrin (Meier et al., 1998) or acetylcholine receptor-inducing activity (ARIA) (Harris et al., 1991), which are present in the immediate vicinity of the NMJ and elicit the differentiation of the postsynaptic domain, exert an enhancing effect on the switches' activity.

The myogenic factors of the MyoD family are known to diffuse within the fiber cytoplasm (Blau et al., 1983). We therefore posit that nuclei can establish communication by diffusion of M; M may thus effectively act as a cytoplasmic morphogen, synthesized near a nucleus but able to penetrate other nuclei and become trapped there as an active TF.

Two sets of equations are written to define the model, one for the states S_i of the switches i (where i denotes position along the fiber) and the other for the diffusion of the morphogen. We assume that M has concentration-dependent probabilities for occupying the a promoter site(s) and that whenever a promoter is occupied by M, transcription takes place. Thus the transition probabilities per time step for the switches are

If $S_i(t) = 1$ then $S_i(t + 1) = 1$ with probability $\mu + \alpha f(M_i)$ and $S_i(t + 1) = 0$ with probability $1 - \mu - \alpha f(M_i)$

If $S_i(t) = 0$ then $S_i(t + 1) = 1$ with probability $1 - v + \beta f(M_i)$ and $S_i(t + 1) = 0$ with probability

$$v - \beta f(M_i) \tag{1.1}$$

where μ, α, v, and β are parameters obeying restrictions that make the probabilistic interpretation of Eq. (1.1) possible, and $f(M)$ is a threshold function with threshold T. The probabilities of staying ON or going ON are both increased when $M_i \geq T$. Morphogen diffusion and synthesis obey

$$M_i(t + 1) = c_b + M_i(t)\{\tau_i - \tau_a E(t) - 2k[1 - \sigma S_i(t)]\}$$
$$+ k\{[1 - \sigma S_{i-1}(t)]M_{i-1}(t)\}$$
$$+ k\{[1 - \sigma S_{i+1}(t)]M_{i+1}(t)\}, \tag{1.2}$$

where k is the diffusion coefficient, or internuclear "hopping" probability, of M per unit of time. The first k term describes hopping away from i, the next two hopping toward i. Note that the equations need to be modified for those nuclei closest to the ends of the fiber (tendons). The parameter c_b denotes the basal level of synthesis. It is important to realize that when the nuclear switch i is turned ON, only a fraction $1 - \sigma$ of the morphogen there is available for diffusion; the rest is trapped in the nucleus, where it directs enhanced mRNA transcription. The M-enhanced transcription of the m gene yields fresh M product, at a rate described by τ_i. The latter embodies the net effect of turnover (degradation) and autocatalytic biosynthesis. It depends on whether a nerve ending is located in the immediate vicinity of nucleus i and on the state of activity S_i of this nucleus. In an isolated nucleus, $\tau_i = \tau_0 + \tau_n S_i$, while under a synaptic terminal, $\tau_i = \tau_0 + \tau_s + \tau_n S_i$. When the switch is ON, synthesis proceeds at a faster pace (at a rate determined by τ_n). Under a terminal, transcription is also boosted further by anterograde factors (τ_s). On the other hand, we introduce the depressing effect of electrical activity E through a reduction (at a rate determined by τ_a) of the net autocatalytic synthesis of M. Thus, in the absence of electrical activity, $E(t) = 0$, while with ongoing electrical activity, $E(t) = 1 - g^{n(t)}$, where $0 < g < 1$ is a constant and $n(t)$ denotes the number of subsynaptic nuclei actively transcribing the m gene at time t; we assume that $n(t)$ is indicative of the number of functional synaptic boutons and hence of the average level of electrical activity.

1.2.2   Formation of a Transcriptional Boundary

A stationary state of great interest is the one in which only a single focal nucleus ($i = 0$) per muscle fiber is actively expressing AChR genes; this is the situation most often observed in vivo. We were able to calcu-

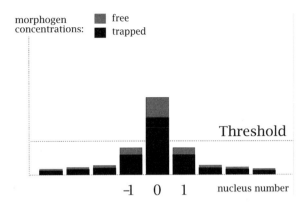

Figure 1.1
The analytical solution with one active (central) nucleus. The central nucleus is the only one where the concentration of morphogen trapped on the gene promoter is above threshold, thus driving local transcription of the morphogen and hence its synthesis. The morphogen, once synthesized, is free to diffuse away.

late an approximate analytical solution of this type (figure 1.1) (Kerszberg and Changeux, 1993). We found that the stabilized ratio of concentrations A_i to A_{i+1}, $i \neq 0$, is

$$\frac{A_i - A_\infty}{A_{i+1} - A_\infty} = e^{1/\lambda} = 1 + \frac{1 - \tau(1)}{2k(1 - \sigma S^-)}$$
$$+ \sqrt{\left[1 + \frac{1 - \tau(1)}{2k(1 - \sigma S^-)}\right]^2 - 1}. \qquad (1.3)$$

In this equation, which defines the decay length λ, S^- stands for the (low) average genetic activity in the OFF state, $\tau(1)$ is the net synthesis rate at the nonactive sites, and $A_\infty = c_b/[1 - \tau(1)]$ is the concentration distant from the active nucleus. In order for the solution to be consistent, it is imperative that A_0 (at the active nucleus) be higher than the threshold T and that A_1 (at the neighboring nucleus) be lower.

An illustrative case (Kerszberg and Changeux, 1993) yields a decay wavelength $\lambda = 7.09$, while

$A_0 = 2.97$ and $A_1 = 0.40$ (all these are under electrical stimulation). The wavelength is such that around the isolated nucleus, about seven nuclei will be inactivated on each side. The precise value depends on the threshold T above which the morphogen becomes effective. As long as the fiber is shorter than this wavelength, we may expect a single active nucleus; when it is longer, multiple active nuclei are predicted. Indeed, we have succeeded, under these conditions, in reproducing by simulation the formation of multiple, regularly spaced transcription patterns like those occurring in certain mature muscles, such as the chick anterior latissimus dorsi (Toutant et al., 1980), which are accordingly innervated at multiple, regularly spaced intervals.

Sharp boundaries may thus become spontaneously established in the fiber when electrical activity is sufficiently intense. Why are nuclei neighboring the active nucleus quiescent? According to the model, this is because the morphogen is trapped by the promoter sites of the active nucleus; its concentration in its vicinity is thus reduced, hence the low value of A_1. Computer simulations are consistent with these analytical results on focal innervation (Kerszberg and Changeux, 1993). The simulations also predicted that persistent activity may appear near the tendinous ends after the muscle fiber has grown to its mature size. While surprising at first, since no receptor seems needed where there is no synapse, this prediction was confirmed by analytical calculations and it has received experimental support (Chen et al., 1990). We also examined by simulation the progressive reinstatement of nAChR expression known (Goldman and Staple, 1989) to occur upon denervation (i.e., extinction of the electrical activity evoked by the nerve). This reexpression starts from the middle and from the tendinous ends of the fiber.

In conclusion, the example of the NMJ illustrates how the sharpening of a boundary of genetic expression may plausibly result from the operation of simple genetic circuits, namely, nonlinear transcriptional switches. These involve a positive feedback of transcription factors on the transcription of the very genes coding for them.

1.3 Hetero-Oligomeric Connections: Reading Molecular Gradients

We are now in a position to tackle more complex genetic circuits. For, as TFs interact, such interactions join independent pathways into a network. A salient feature of these interactions, shared by the vast majority of eukaryotic TFs, is that their molecular organization is oligomeric; i.e., they consist of assemblies of identical or different subunits. Examples include the *achaete-scute* gene family (Ghysen and Dambly-Chaudiere, 1989; Simpson, 1990) and the retinoic acid receptors (Laudet and Stehelin, 1992), which are involved in the anteroposterior patterning of the neural tube (Ruiz i Altaba, 1994). The combinatorial complexity that results from dimer formation is increased by the composite structure of the genomic promoter or enhancer sites to which they bind, with their variety of high- and low-affinity components (Jiang and Levine, 1993). Indeed, in addition to structural genes, TFs control transcription of their own and other genes coding for TFs (see section 1.2.1).

1.3.1 The Model

Based on these observations, we have proposed a theoretical model for gradient reading that relies crucially upon the formation of TF oligomers as well as TF self- and cross-regulation. The problem of gradient reading (i.e., the establishment of sharp transcription boundaries from a system of diffuse molecular distributions) arises, for instance, in the *Drosophila* embryo, which goes through a stage in which a thousand nuclei are present in a single syncytium, much like the skeletal muscle fiber we just considered. In this so-called syncytial blastoderm (Kerszberg and Changeux, 1994a,b), none of the genes of interest are initially active. However, these so-called zygotic genes, many of them TFs, will become expressed under the precise control of several morphogens, deposited inhomogeneously in the cytoplasm of the egg before it is laid down. We will concentrate on one of the most prominent of these shallow "maternal" gradients, namely, the anteroposterior gradient of Bicoid (Driever and Nüsslein-Volhard, 1988; Hoch et al., 1991; Struhl et al., 1989), which is a dominant factor in structuring the embryo along the head-to-tail axis.

We assume that the morphogen M (e.g., Bicoid) is a TF, or that it activates a TF. Its gene *m* is not transcribed in the embryo. It is the concentration of M that must be read. A second TF, V (e.g., Hunchback), is expressed in the embryo. V may form with M all possible homo- or heterodimers: MM, MV, and VV. The MV heterodimer establishes the crucial connection between the morphogenetic gradient and a zygotic gene product. Note that oligomer formation may be happening in the cytoplasm or in the nuclei by interaction with the DNA promoters.

V will be called the "vernier" molecule, responsible for M-gradient reading (Lawrence, 1992). We now posit that V binds to its own promoter. The sequence

v gene activity → V protein → MV, VV dimers

    → *v* gene activity

may thus compose a positive or negative self-regulation loop. It is on the existence of this loop that the model rests. As before, we formalize transcription of *v* in nuclei by a genetic switch. At first, transcrip-

tion of *v* proceeds in a statistically uniform way. Once synthesized, however, V diffuses between cytoplasmic compartments and forms the MV and VV complexes, thus regulating *v* transcription both locally and at a distance.

We study the embryo over two dimensions (i, j) and divide it into 100×25 "boxes." Five hundred of these boxes, chosen at random, contain a nucleus. M is initially distributed in an exponential gradient $M_0 e^{-i/\lambda}$, where M_0 and λ are constants. The concentrations of M, V, and MV, and optionally of MM or VV, obey straightforward discrete diffusion-reaction equations governed by mass-action laws for the formation and decay of dimers. The crucial term is the source term for V, which is present if the box (i, j) contains a nucleus, where *v* may be transcribed with a probability *P*. *P* depends on the presence of TFs at the promoter binding sites, these sites being characterized by their respective affinities and synergistic-antagonistic interactions. We assume *P* to be a sigmoid function of occupancy; for instance, if MV activates *v* transcription, one would have

$$P = \frac{1 - b - f}{2} \left[1 + \tanh \frac{\beta_0 MM + \beta_1 MV - T}{Q} \right] + b, \tag{1.4}$$

where *b* is the basal transcription rate, *f* the transcription failure rate at high promoter occupancy, and *T* and *Q* are, respectively, the threshold and width of the sigmoid. The βs are kinetic binding and activation coefficients. Note that Eq. (1.4) embodies cooperativity (Driever and Nüsslein-Volhard, 1989), owing both to the presence of the quadratic terms and to the nonlinearity of the sigmoid. The reaction scheme is described in the original publication (Kerszberg and Changeux, 1994a). We now describe the computer solutions obtained with various interaction schemes, using stochastic Monte Carlo methods.

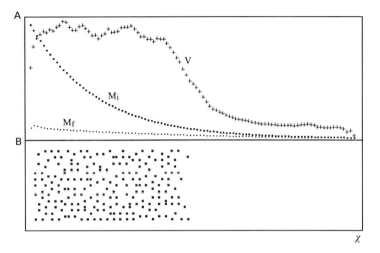

Figure 1.2

Gradient reading by a cooperatively bound morphogen. In this and subsequent figures, the upper panel (A) displays protein concentrations, with large squares denoting the initial gradient of morphogen M_i along coordinate χ of the embryo, small squares its final (late times) concentration M_f, and crosses the final vernier V distribution. The lower panel (B) represents nuclei in the simulated two-dimensional embryo. Only those nuclei are shown in which v transcription is turned on (squares) at late times. A sharp boundary is seen in the distribution of those nuclei. The protein V itself, however, shows a rather smooth distribution, which may be considered a secondary gradient.

1.3.2 *Transcription Patterns with Multiple Boundaries*

At a given M concentration [i.e., on a given column j ($\forall i$)] transcription may be ON, OFF, BISTABLE (i.e., be in a regime where transcription is either high or low, depending on the relative stability of the two states and sometimes on system history), or UNSTABLE (i.e., not settling into a stationary pattern at all). For simple cases, the existence and stability of ON and OFF can be determined analytically (Kerszberg and Changeux, 1994a). Accordingly, four distinct, generic regimes can be defined that correspond to extended ranges of values in parameter space and that yield different behaviors of the model. In what follows, we have selected parameters appropriate for obtaining representative results for three of the four regimes, which

are, respectively, MM dominated, MV dominated, and MV-MM competition-regulated (figure 1.2).

Smooth Gradient Reading by the MM Homodimer

Let us first consider the regime where the MM homodimer promotes v transcription ($\beta_0 > 0$, $\beta_1 = 0$). The results of a typical simulation are displayed in figure 1.2, where we see that the final stable situation, starting from the complete absence of V, is a graded distribution of this molecule that is steeper than that of M and in register with it. Note that in terms of transcriptional activity, the transition is sharp. Analytically, the position of this abrupt boundary is easily deduced as the location where concentration of M leads to a 50 percent chance for transcription being ON. Thus the gradient is read very precisely in terms of

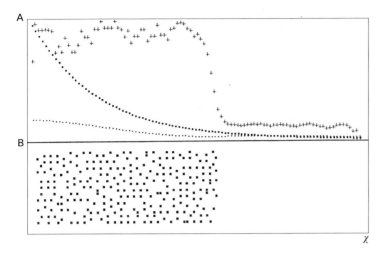

Figure 1.3
Gradient reading by the autocatalytic vernier protein V. Transcription of *v* is induced by the MV dimer, and this cooperativity gives rise to a sharp boundary in *v* expression. The boundary is at first located at the threshold $M = M_1$, the value at which the amplification factor (gain) of the V regulation loop is precisely 1. Owing to diffusion of the involved substances, it then moves somewhat toward the low-M region. When the morphogen is scarce, as here, its depletion may result in a countergradient, which reduces this later motion and stabilizes the boundary in an even stronger way.

gene activation patterns, but defines a rather smooth protein distribution suitable for playing the role of a secondary morphogenetic gradient such as that exhibited by Hunchback.

Sharp Gradient Reading by the MV Heterodimer

We now turn to the case $(\beta_0 = 0, \beta_1 > 0)$. Here the MV heterodimer controls vernier transcription positively. The morphogen gradient is now read reliably and sharply, both in concentration and in transcriptional intensity (figure 1.3). The sharpness of the boundary depends on the strength of the β_1 nonlinearity and on the sharpness of the threshold function [Eq. (1.4)]. No secondary gradient arises. The boundary occurs around a threshold M_1. In a first approximation, this is located at the point where morphogen concentration is such that the V loop

gain is 1; i.e., per unit of time, one molecule of V forms MV in such an amount that the transcription it induces will yield fresh V in just the quantity needed to balance decay (this reasoning neglects diffusion). Dynamically, the *v* gene is switched on in nuclei starting from the high-M end, and transcription is progressively established as an "ON-wave" all the way to the threshold value. Thus, a smooth M gradient leads from an initially uniform rate of basal expression of *v* to a sharp boundary.

In contrast to the situation illustrated in figure 1.2, however, a sharp boundary in V concentration is now also apparent. In the figure, one notices in addition that when M is scarce, its concentration diminishes appreciably as the MV dimer is formed in the active zone. Counterdiffusion of M toward that zone follows, stabilizing the exclusive expression of *v* there and leaving a very shallow final distribution of M. The

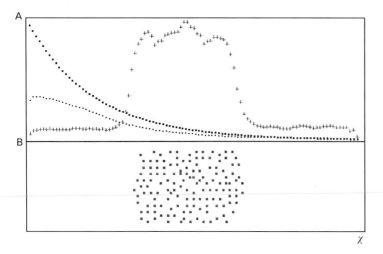

Figure 1.4
The effect of MM homodimer formation. Double gradient reading by the vernier protein V is seen to occur. Here the action of the MV dimer [see figure 1.3] is antagonized at large M concentrations by the formation of abundant MM homodimer, which competes with MV for v promoter sites, thereby reducing the V autocatalytic loop gain. At the location $M = M_2$ where MM reduces this amplification factor below 1, v transcription is turned off.

scarce morphogen situation is favorable because the ultimate location of the boundary then depends less on the fine details of the initial M distribution than on the total quantity of morphogen available on the high-M side of the embryo.

MV-MM Interaction: Reading the Gradient at Two Levels and Forming a Stripe

The final case of interest is the one where ($\beta_0 < 0$, $\beta_1 > 0$). MM dimers and MV dimers now have competing effects on transcription. Band formation may result (figure 1.4). If the competition is not too strong, the situation will be as described in the previous paragraph. However, if β_0 becomes sufficiently negative with respect to β_1, two thresholds become apparent in the system. The first is the M_1 threshold, somewhat displaced; this is the threshold above which transcription starts. But when M (and thus MM) concentration reaches a high enough value $M = M_2$, the negative

interaction will cause the V-loop gain to drop below 1; hence transcription is switched off. The figure shows that transcription is ON in a band defined by these two morphogen concentration values.

The Krüppel Band

We may summarize and illustrate our findings by an analysis of the expression of *Krüppel*, a *Drosophila* TF gene expressed in a central stripe of the syncytial blastoderm, under the control of the *bicoid* and *hunchback* genes (Kerszberg and Changeux, 1994b).

Hoch et al. (1991) showed that response elements for both Bicoid and Hunchback were present in the *Krüppel* promoter. Both Hunchback and Bicoid were known to promote *Krüppel* transcription at low concentrations and to inhibit it at high concentrations. What was not known when we performed our calculations was whether Bicoid and Hunchback were interacting. Using our formalism, we proposed that

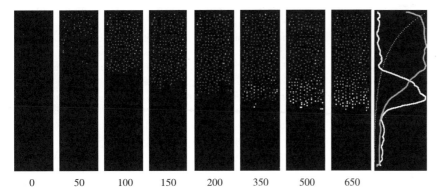

0   50   100   150   200   350   500   650

Figure 1.5

Regional patterns of transcription established in *Drosophila* syncytial blastoderm. Each rectangle represents the embryo (head upward) at a given time, which goes from 0 to 650 (left to right, in the simulation's arbitrary units). Dark dots, nuclei transcribing the *hunchback* gene; light dots, nuclei transcribing the *Krüppel* gene, which form a band at the center of the embryo. At $t = 650$, the preexisting exponential gradient of Bicoid, and the final concentrations of Hunchback and Krüppel have been represented. We observe the progress of the *hunchback* expression wave and the establishment of the Krüppel stripe in the middle of the embryo.

the experimental data could be explained by the MV scheme (Bicoid as M and Hunchback as V), i.e., by the existence of pairwise interactions between Bicoid and Hunchback (not by direct heterodimer formation, but possibly on the *Krüppel* promoter itself). Indeed, such interactions were experimentally demonstrated shortly afterward (Simpson-Brose et al., 1994).

The model offers a number of additional, specific, and testable predictions. Some of these are apparent in the results of a simulation of the full three-gene system shown in figure 1.5. For instance, anterior expression of *hunchback* is predicted by the model to start at the anterior end of the embryo and progress as a wave toward its posterior limit. This should be observable in spite of rapid kinetics. The later, observed retraction at the anterior end was predicted as well, and appears to a certain extent to be due to *Krüppel* competition. The model also explains rather simply the puzzling experimental fact that multiple doses of

Bicoid protein displace the Krüppel stripe toward the posterior end of the embryo instead of expanding it. This results from the fact that *Krüppel* expression is both enhanced by Bicoid-*Krüppel* and repressed by Bicoid-Bicoid interactions.

Cellularized Embryos

The basic model has been further extended to encompass the more general case of cellularized embryos (Kerszberg, 1996). A graded, shallow distribution of some small, diffusible molecule such as a steroid hormone or retinoic acid (RA) is now assumed. Retinoic acid has indeed been shown to intervene in the anteroposterior patterning of the neural tube (Ruiz i Altaba, 1994). The receptors for RA are nuclear proteins that function as oligomeric TFs with a rich combinatorial repertoire by binding to multiple promoter sites on the cells' DNA (see figure 1.6). It was found that such combinatorics, operating under the constraint of no intercellular receptor traffic (i.e., diffusion

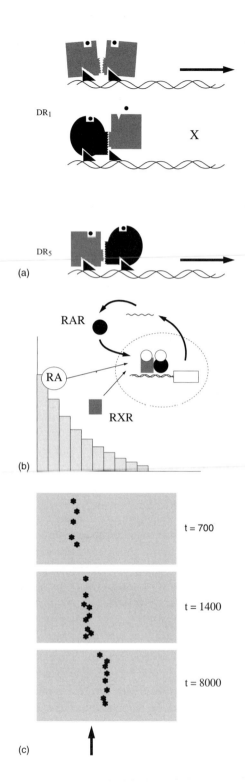

DR$_1$

X

DR$_5$

(a)

RAR

RA

RXR

(b)

t = 700

t = 1400

t = 8000

(c)

Figure 1.6

(a) Gradient reading in cellularized systems: a simplified view of the retinoic acid (RA) receptors and the response elements they bind to on DNA. Two major forms of the receptors are the RAR (black circles) and RXR (gray squares) families. They bind as dimers on two main types of response elements, consisting of directly repeated half-sites (dark triangles) spaced by one (DR$_1$) or five (DR$_5$) base pairs. Various combinations of dimers and repeats induce ($\rightarrow$) or, contrariwise, inhibit (x) transcription of downstream sequences. RA (black dots) binding to the receptors is essential for activation. In particular, the DR$_5$ element functions as an enhancer when bound by an RAR-RXR liganded heterodimer, with the RAR subunit occupying the downstream (5') position. On the DR$_1$, a homodimer RXR-RXR or a heterodimer RAR-RXR may form; in the latter case, the RAR subunit preferentially occupies the upstream position. Actually, heterodimer formation is the most likely process, but, presumably through allosteric interactions, precludes ligand binding to RXR. Hence, the homodimer is the only species that may actively promote transcription. (b) The pathway controlling the expression of the RAR receptor gene itself. In a cell nucleus (dotted line), RAR-RXR receptor dimers may activate transcription of the RAR gene if the local concentration of RA is sufficient, because of the presence of a DR$_5$ response element driving the RAR gene. (RXR is expressed throughout the embryo at a relatively constant level.) Note that the RAR pathway forms a loop of retroaction, potentially leading to many interesting morphogenetic phenomena. We assume that a second pathway (not shown), operating through another response element, controls the expression of a "reporter" gene, which is the one we use to display pattern formation in time and space; an example is shown in (c). (c) Precise gradient reading by nuclear receptors in a very narrow band of gene transcription. A gradient of retinoic acid is imposed on a cell field, with maximal value on the left, and decaying to 0.1 of this maximum at the arrow's location. The reporter gene is assumed to be driven by a DR$_1$ response element. Successive snapshots (time in units of simulation steps) display the nuclei (black) of cells where transcription of the reporter gene takes place; these cells form a finely delineated band (one cell diameter wide). Note that such precision is achieved, in the present model, in the absence of any cell-cell communication mechanism. In view of the many sources of fluctuation, this demonstrates a remarkable degree of reliability.

is now prevented), can produce gene transcription patterns of remarkable accuracy. For example, a single row of cells that is a given distance from a retinoic acid source can be assigned a unique transcriptional fate, as seen in figure 1.6. An experimental situation that this result could partly describe is the specification of rhombomere r4, a small and sharply defined portion of the developing mammalian CNS. The cells of r4 indeed assume their correct identity only when an appropriate concentration of RA is ensured and when the RA receptors of the embryo are not defective (Studer et al., 1994). Kerszberg (1996) analyzed how multiple pathways interact in reliably achieving such sharp territory formation.

1.4 Cell Adhesion, Motion, and Division: Neurogenesis

The formation of patterns of cell differentiation, as described in the preceding section, is one of the fundamental features of embryonic development. Other major aspects of morphogenesis, such as neurulation, depend on the subsequent behavior of the differentiated cells, e.g., cell adhesion, cell motion, and cell division. We have proposed a simple, plausible model (Kerszberg and Changeux, 1998) for the early development of both the neural tube (neurulation in vertebrates) and the neural cord (neuroblast delamination in insects).

It is well established that the genes involved in CNS formation in vertebrates and insects are homologous, and so is the general scheme of their interactions. We therefore posit a common genetic database for insects or vertebrates, defining in this way a minimal set of molecular components necessary for neurulation. Three distinct hierarchical levels are involved in CNS formation: (1) permanent differentiation, i.e., cell determination (see earlier discussion);

(2) cell membrane presentation of ligands and receptors and their intercellular interactions; and (3) cell responses, i.e., signal transduction pathways that dictate biochemical changes, altered gene transcription (back to level 1, for a new round of further differentiation), cellular motion, adhesion, and mitotic activity. This approach should be contrasted with previous models (Jacobson et al., 1986) in which only mechanical factors were considered.

1.4.1 Neurobiological Background

Formation of the CNS starts when the embryonic outer layer (ectoderm) differentiates into neuroectoderm, or the neural plate (which will give rise to neurons and glia), and epiderm (Gilbert, 1995). Central to the process of neuron differentiation within the neuroectoderm are the so-called proneural TFs of the *achaete-scute* (*AS-C*) complex (Ghysen and Dambly-Chaudiere, 1989; Simpson, 1990) or the homologous *XASH* genes of, for example, *Xenopus* (Zimmerman et al., 1993). Continued *AS-C* activity is sufficient for cells to become neurons. What are the upstream factors inducing *AS-C* TF activity? How is their activity later maintained? What downstream effects do the *AS-C* TFs produce on cell behavior? We address these questions in turn. The following discussion of experimental data is summarized in figure 1.7.

Morphogens and Neural Induction
Figure 1.7a is a transverse section of the vertebrate embryo's dorsal ectoderm. Early determination of neuroectoderm is characterized by the expression of a battery of genes, which is at first antagonized by endogenous repressors, the "bone morphogenetic proteins" (BMP-2 and/or BMP-4) (Smith, 1995; Liem et al., 1995). The BMPs are in turn inhibited by Chordin (Sasai et al., 1994), which is synthesized in and diffuses from the underyling dorsal mesoderm

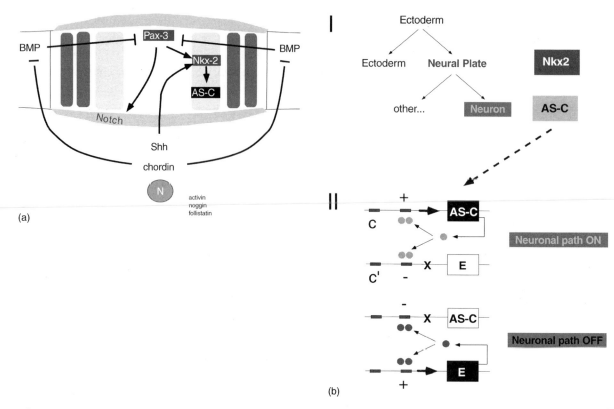

(a)

(b)

Figure 1.7

(*a*) Signaling in neural morphogenesis upstream of *achaete-scute* (*AS-C*). The *AS-C* complex of proneural genes occupies a central place in the signaling events leading to neurogenesis. Here we depict in a simplified way the part of the signaling chain lying upstream of the vertebrate homologs of *AS-C*. See the text for details. (*b*) Signaling in neural morphogenesis downstream of *AS-C*. (I) The hierarchical organization of the two genetic switches that define the neural plate (dark) and its neuronal (light) subterritories. (II) A model of the neuronal genetic switch, based on autocatalysis and mutual inhibition. See text for details. (*c*) The lateral inhibition pathway. Two cell nuclei n and n′ in which the genetic switches operate as described in (*b*) separated by cell membranes m and m′. Notch is a membrane receptor for Delta; the gene for Delta is activated downstream of *AS-C*; *Notch* is expressed over the whole neural plate (see *a*). When activated by Delta, Notch in a nearby cell signals an inhibition of the proneural genes and hence of *Delta* itself. When cells thus interacting are immersed in the two morphogen gradients (see *a*), a complex competition arises (see text).

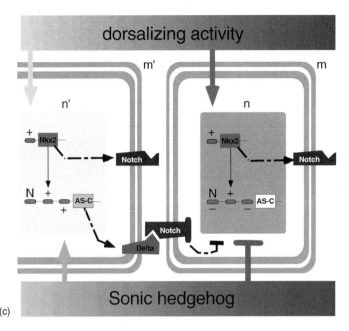

Figure 1.7 (continued)

(Spemann's organizer). This "inhibition of an inhibition" is what leads to neural induction, i.e., the formation of the neural plate. After induction, neural plate gene expression becomes autonomous.

Under the concentration-dependent control of another notochord-secreted protein, Sonic hedgehog (SHH) (Roelink et al., 1995; Fietz et al., 1995), and subject to BMP-mediated contact interactions with the epidermal ectoderm, homeobox TFs such as Msx and Nkx-2 then become expressed and, by acting on *AS-C*, play a key role in inducing the formation of neurons within the neural tissue.

The BMPs, which play a central role in neural patterning, belong to the transforming growth factor-beta (TGF-β) family. BMP is a morphogen (Nellen et al., 1996) and so is the extracellularly exported terminal of SHH. In both vertebrates and invertebrates, the net activation of the proneural TFs is thus effected by a distributed morphogen activity, modulated during a second step by another morphogen that induces the ventral neuron types.

Cell–Cell Interactions: The Lateral Inhibition Pathway

Transcription of *AS-C* starts over a large area initially and then becomes restricted by a mechanism of intercellular communication. Synthesis of AS-C activates transcription of the gene for Delta, a ligand for the receptor Notch (Artavanis-Tsakonas et al., 1995). Notch on the membrane of cell *x*, when activated by Delta present on the membrane of neighboring cell *y*, inactivates expression of *Delta* in *x*. The result is clearly lateral inhibition, implying that local differences in *AS-C* activity will tend to become amplified. The spatially graded *AS-C* activity becomes sharpened into a spatial ON/OFF distribution (Heitzler et al., 1996), with topographically separate cells undergoing separate developmental fates—neural (*AS-C* and *Delta* expression) or epithelial (no such expression).

Heterodimers and Genetic Determination

How are these different fates stabilized once the inductive signal has subsided? The key to this problem is thought to be that AS-C TFs interact with other TFs belonging to the *Enhancer-of-split* [*E*(*spl*)] family (Gigliani et al., 1996; Heitzler et al., 1996). By forming heterodimers and homodimers, the AS-C and E(spl) molecules cooperate in establishing (pro)neural identity, perhaps through a combination of autocatalysis and mutual inhibition (Nakao and Campos-Ortega, 1996).

Hypothesis

Here we hypothesize that these experiments support the idea of a bistable genetic memory system (see section 1.2.1) based on the *AS-C* and *E*(*spl*) gene families. This suggests the possibility of embryonic cell determination in the model, i.e., persistence of differentiated transcriptional states. Two major genetic switches of this type can be distinguished: (1) the proneural switch just described and (2) the neuroectodermal switch, which is responsible for the differentiation of the neural plate itself. Candidates for composing this second switch have been identified (Ruiz i Altaba, 1994).

Cell Adhesion and Motion; Mechanical Effects

Once their developmental fate is determined, neural tissues become mechanically separated from the epiderm. In insects, future neurons (neuroblasts) delaminate individually and go on to form the discrete segmental ganglia. In vertebrates, two modes of neurulation are known (Gilbert, 1995; Copp, 1993). Primary neurulation occurs as neural tissue forms a thickened neural plate, which invaginates while the surrounding epithelium folds over it. The two so-called neural folds ultimately join, closing the neural tube.

On the other hand, secondary neurulation consists in the ingression of a neuroblastic cell mass followed by cavitation. Formation of the neural tube by secondary neurulation is observed in some organisms and may sometimes coexist with primary neurulation, occurring then in those parts of the neural tube that form last. The expression of various cell adhesion molecules shows a characteristic pattern during the invagination process (Papalopulu and Kintner, 1994). During neural tube formation, neural tissue progressively ceases presenting the E-cadherin cell adhesion molecule, while beginning the production of N-cadherin instead, which plays a role in tube closure. Throughout the process, E-cadherin remains present on outer ectoderm cell membranes.

The complex movements of epithelial cells leading to neural tube formation depend on the mechanical properties of tissues. Important differences of motility may exist between the basal (inner) and apical (outer) faces of the epithelium (Viebahn et al., 1995). In addition, all cells involved are epithelial and thus form a connected sheet held together by strong intercellular junctions. An important mechanical role is also played by cell division, which undoubtedly generates complex forces, if simply by increasing the amount of tissue present.

1.4.2 The Model

The model implements many of the components described here quite accurately. Its dynamics are

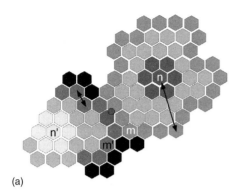

(a)

(b)

Figure 1.8
The geometric substratum of the neurulation model. (*a*) Space is divided in an underlying honeycomb array of "bins" or "pixels." Groups of pixels comprise cells, with their membranes m, m′ (in various shades of gray, indicating local ligand molecule concentration) and nuclei n, n′ (dark or light, depending on current gene transcription). Molecules diffuse and react on the membrane or are translocated to it as a result of genetic activity (arrows). Black pixels are extracellular matrix. (*b*) The bins comprising a cell can change in time (arrows); this is cell motion. Restricting this motion are strong focal attachment points among cells (circle). These may correspond to, for example, adherens junctions or to desmosomes.

studied using a Monte Carlo formalism. For details, the computer program is available from the authors.

The Geometric Substratum

For the sake of the computations, cells are assumed to exist in a restricted two-dimensional space. This is artificially divided into a honeycomb lattice (124×248) of "bins" (see figure 1.8). These are analogous to pix-

els in a digitized image; their location is fixed, but their shade of gray reflects the presence of objects, such as molecules, or parts of objects, such as cell membranes. Thus closed strings of adjacent bins compose the "membranes" of cells (see the figure). Membrane bins, as well as extracellular ones, are occupied by molecules in varying concentrations. The membranes are mobile; membrane molecules are transported along with their membrane (convection), and extracellular molecules are displaced as membranes move. In addition, membrane molecules diffuse on the membrane, and extracellular molecules may diffuse in the extracellular medium. Diffusion is described by probabilities per time step of "hopping" between adjacent bins. Reactions occur locally among the various molecular species.

The Transcriptional Switches

Genes are turned ON or OFF in a switchlike manner described by binary variables (see section 1.2.1). These switches determine cell fate and are themselves controlled by TFs. Two pairs of TFs are introduced, each composing a genetic switch. Thus, the proneural switch consists of the proneural factor *a* and an associated *E(spl)*-type factor *e*. *a* and *e* products form dimers: AA, EE, and AE. The homodimer AA enhances *a* transcription, while it represses *e*; EE does the converse, activating *e* and inhibiting *a*; AE is not active in this simple formalism. In addition, AA positively regulates the gene for a membrane ligand Δ (the equivalent of Delta). It is assumed that Δ expression is at this stage the hallmark of the neural phenotype. The second, neuroectodermal switch, is established by factors *p* (for *plate*) and *q*. These could represent two homeobox genes, for example. The same interactions are assumed to exist among the products of these genes and those of *a* and *e*, so that expression of one prevents expression of the other and conversely. The presentation on a cell membrane of N

(Notch), the receptor for Δ, depends on the product of gene p.

Morphogens

Two morphogens are assumed, B and S. Their distribution is fixed before the simulation starts and decays progressively. B activates transcription of the neuro-ectodermal gene p, increasing its ON probability. It does so by reducing a preexisting inhibition and thus operates very much like BMP when not antagonized by Chordin. The second morphogen, S, is an analog of SHH. It is responsible for the initial, relatively widespread, ventralmost region of expression of the proneural gene a, which it activates. The concentration of S reaches a peak near the midline, and is interpreted according to its value relative to two thresholds, which is in agreement with experiment (Roelink et al., 1995). At the midline (above the notochord), a high concentration of S represses the proneural gene; at intermediate levels (on both sides), activation takes place; while even farther from the midline, repression is again assumed to occur.

Lateral Inhibition

Lateral inhibition is a new feature introduced in this work. Cell membranes read at all times the state of neighboring membranes for adhesion, motion, and signal transduction purposes. Transduction occurs as each membrane bin i is examined for neighboring bins j belonging to different cells. The amounts of ligand Δ in these neighbors are added up. The sum is then multiplied by the concentration of receptor N in i; it is assumed that the local activated receptor N^* is proportional to this product. N^* is finally summed over all the bins constituting the membrane of the cell under consideration, and the resulting signal causes a reduction in the expression of the proneural gene a in this cell.

Cell Adhesion and Motion

Accounting for cell adhesion and motion is a completely novel and, as far as we know, unique aspect of our simulations. Cell motion is introduced in a simple manner. At the membrane, bins ("pixels") may cease to belong to a cell (retraction) or may be added to it. The probabilities of subtraction or addition are adjusted according to the conditions encountered locally. For the sake of illustration, consider a cell x that displays on its membrane a particular homophilic adhesion molecule. Assume now that cell x has the possibility of growing by adding a particular pixel to itself and that this addition results in cell x making contact with another cell y. If y displays on its membrane a sufficient concentration of the same adhesion molecule as displayed by x, the probability of effectively adding the pixel will be increased relative to the probability of other possible moves. The net result is an adhesion force that acts on cells and parts of cells to bring and keep them together (Edelman, 1988). In addition, epidermal cells, which do not express p, are restricted to moving in a horizontal layer. On the other hand, neural plate cells, which do express p, move more freely and movement toward their basal side is favored; i.e., the Monte Carlo moves that result in motion toward the basal side are accepted with a higher probability than those that do not result in such motion.

Cells are joined in pairs by strong focal adhesion points and form a "string of beads" much like actual epithelia when seen in transverse section. Pair relationships cannot be rearranged. The adhesion points demarcate the apical and basal faces of the cells. This distinction serves as a starting point for our modeling of cytoskeletal movement. Cell motion is in the form of a contraction of the cell cortex at the apical face, and an expansion of the basal face. By what genes is the motion controlled? In the next section we will see that the possible answers to this question form a basic prediction of our model.

Cell Division

An equally important novelty is our modeling of cell division in the context of development. In cell division, the bins previously occupied by the mother cell are split into two sets, each belonging to a daughter. A fundamental feature of cell division is anisotropy. That is, the axis defined by the two daughter nuclei is not oriented in random fashion; instead, the axis has a greater probability of lying parallel to the line joining the two focal adhesion points on the mother cell membrane. As a consequence, the new cells automatically lie in the plane of the epithelial sheet. Division itself occurs with a probability that is a function of cell size and type (expressing p and a or not expressing them). As division occurs, all molecular species present must be distributed among daughters; this is done on the basis of equipartition. Note that a rudimentary mechanism for epigenetic inheritance must be implemented during mitosis, so that each cell bequeaths to its daughters the transcriptional state of its nucleus; this is important, particularly just after division, when the complement of TFs in a daughter may not yet be sufficient to ensure continued expression or repression of the proper genes.

1.4.3   Results: Neurogenesis and its Evolution

Two main sets of simulations have been performed. They differ essentially with respect to the coupling of cell motion to the genetic switches. When cell movement is activated and modulated by the neuro-ectodermal switch, which is turned on over the relatively widespread neural plate region, we observe the analog of vertebrate primary or secondary neurulation, depending on the rate of cell division. Coupling

of cell motion to the proneural switch, on the other hand, gives rise to the delamination of discrete neuroblasts, as happens in the formation of the insect neural cord.

Computer Simulation: The Vertebrate Neural Tube

Consider the dorsal part of a vertebrate embryo in transverse section (figure 1.9). In the top left of the figure, the distribution of morphogenetic activities over such a section is shown. The light curve corresponds to B, the molecule(s) that inhibits neural plate formation. B (say, free BMP-4) is distributed in a shallow midline-centered gradient. Its concentration is high when it is far from the midline and is depressed near it by a *chordin*-type activity. The dark curve indicates the Sonic hedgehog-style gradient S. The rest of figure 1.9 shows simulated snapshots of the system, taken at successive times during a typical simulation. Cells start as a homogeneous epithelial sheet (figure 1.9a). As the morphogens act, neural plate genes are turned on. Note that while the inhibition gradient is shallow, the gene expression boundary is sharp. All neural plate cells express the N (Notch) receptor. The neural plate cells are released from the molecular forces that maintain them in a sheet; they thus grow, and the epithelium thickens as a neural plate. Apical side membrane constriction and basal side expansion also take place, the consequences of which will become clear shortly. Meanwhile (see figure 1.9b), the neurogenic gradient acts and turns on the proneural genes a (nuclei marked in white). This occurs in two patches, which are, however, very close to one another in our simulations and thus appear rapidly as a single one. A turns on $\triangle$, the ligand for N (high $\triangle$ is depicted by a light color); lateral inhibition is thus triggered, leading to a restriction of neuronal fate to isolated cells (figure 1.9c).

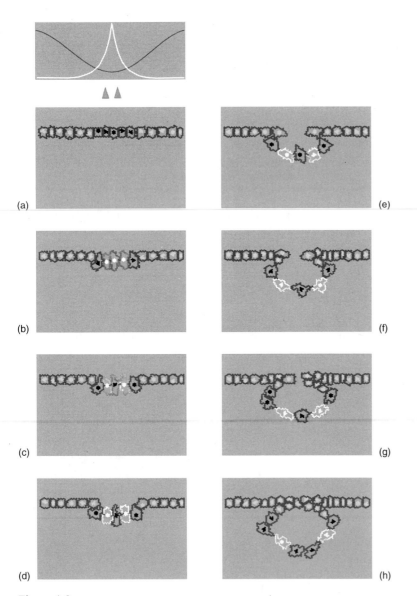

Figure 1.9
Simulated formation of a neural tube (primary neurulation). The dorsal part of a transverse embryonic section is shown at successive times from top to bottom and from left to right (the embryo's dorsal side is up). See the text for details.

Thus the model achieves accurate and reliable genetic determination of single cells at locations determined precisely by morphogenetic fields (gray arrows). The morphogen distribution has now decayed, but gene activity has become self-maintained through autocatalytic loops, and cells are committed at the genetic level. Because of differential expansion and contraction of the cell membrane, neuroepithelium ingression starts (figure 1.9d,e). Neural ingression is accompanied by a progressive further thickening near the edge of the neural plate, as ectoderm covers the plate (figure 1.9g,h).

In figure 1.9e–h, we observe how the neural folds get closer until the gap between them heals completely. Note, too, how cell shapes are affected at the folds, owing to the mechanical stresses applied there by the combined ectodermal expansion (because of cell divisions) and neural tube movements. Such shape changes have been modeled in much mechanical detail by Oster's group (Jacobson et al., 1986).

Computer Simulation: Secondary Neurulation

If one divides the spatial range of the neural inhibition activity by a factor of 2 and increases by a factor of about 5 the rate of division of the neural plate cells, the resulting behavior (figure 1.10) is in sharp contrast to our previous simulations. We now observe the ingression of a neuroblastic mass; this resembles the initial steps of secondary neurulation. The model as it stands does not encompass the full subsequent cavitation of this mass, although a tendency to cavitation can be observed in figure 1.10e,f.

Computer Simulation: The Insect Neural Cord

Let us now assume that cell motion is coupled, not to the neural plate switching genes, but to those that signal a neuronal fate (*a*). The results are remarkably different (see figure 1.11). At first, everything pro-

ceeds as in neurulation. A group of cells at the center of the system start to express the neural plate signal (figure 1.11a), and in two apposed subgroups of these, the proneural genes then become active (figure 1.11b,c). This again turns on the lateral inhibition pathway for these genes; hence, some cells progressively stop producing the neuronal signal. Ultimately, two discrete neuroblasts differentiate and their membrane motions become specialized (i.e., they perform apical contraction and basal expansion). This leads to their delamination at locations that are symmetrical with respect to the midline (figure 1.11d–f). Such behavior bears a close resemblance to the delamination of one line of ganglion precursors during neurogenesis in insects (provided the embryo is seen ventral side up).

1.5 Discussion

We started from the simplest genetic network, involving a single gene and its product. This is sufficient to obtain a major result: the establishment of sharp territorial boundaries in the embryo. We then proceeded to link different genetic pathways through the oligomerization of transcription factors involved in these pathways. Our fundamental result is that such crosstalk among cellular regulatory systems could have been a key factor in the increase of complexity in biological pattern formation. In this way, indeed, we have shown formally that coupled pathways of transcriptional control and cell-cell communication are capable of accurately reading morphogen gradients at several threshold levels, in syncytial as well as cellularized systems.

When several sharp territorial boundaries are achieved, cell adhesion, growth, and movement will drive organ morphogenesis in a direct fashion, as we have proved for the initial stages (neurulation) in the

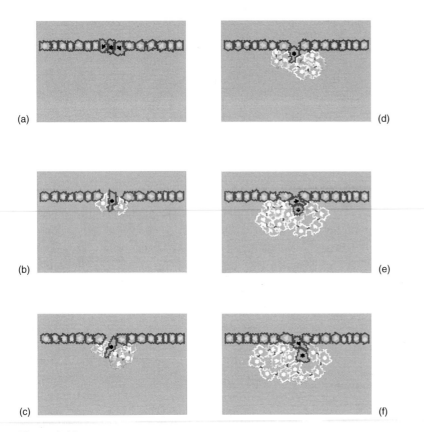

Figure 1.10
Formation of a neuroblastic cell mass (secondary neurulation). The conditions are the same as in figure 1.9 except that the spatial range over which neural plate inhibition is weak has been decreased by a factor of 2 and the rate of mitosis in neural plate tissue is five times higher.

formation of the CNS. It must be emphasized that all the results described are valid over a sizeable range of parameter values (otherwise, one may argue, we should have been unable to observe them). Our models thus demonstrate the resilience of the hypothesized developmental mechanisms. Because simulations such as those of neural tube formation were Monte Carlo processes and therefore included sources of random noise, resistance to statistical fluctuations and errors is clearly built into the system as well.

Our formalism shows that major morphological changes associated with evolutionary transitions did not require large genetic changes. Thus the genetic data (De Robertis and Sasai, 1996) and the morphogenetic capabilities of the model together suggest that no major genetic reshuffling or novelty was needed to effect the transition from the neural cord of *Drosophila* or *Octopus* to the hollow neural tube of vertebrates. Yet this was a momentous transition, since it greatly facilitated the growth and differentiation of the

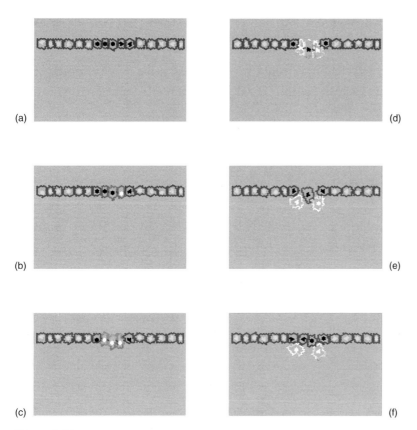

Figure 1.11

Formation of the neural cord of insects by delamination of individual neuroblasts. In this figure, the dorsal side is down. Note that the genetic differentiation events are almost exactly as in figure 1.9. It is only the coupling of the gene's activity to cell motion that generates such a different embryo. See text for details.

neural system, all the way to mammals, primates, and man.

Clearly, much work remains to be done, particularly with respect to the detailed modeling of genetic pathways and, most important, the realistic modeling of cell motion (Bray, 2001). A detailed picture of the various molecular subcomponents of the cytoskeleton will be required. This future work will also have to converge with the modeling of later stages in CNS development. Meanwhile, the connections between neural differentiation and cell motion, which have long been neglected by experimentalists, are beginning to be investigated empirically.

References

Alberts, B., Bray, D., Lewis, J., Raff, M., Roberts, K., and Watson, J. D. (1999). *Molecular Biology of the Cell*, 3rd ed. New York: Garland.

Artavanis-Tsakonas, S., Matsuno, K., and Fortini, M. E. (1995). Notch signaling. *Science* 268: 225–232.

Blau, H., Chiu, C.-P., and Webster, C. (1983). Cytoplasmic activation of human nuclear genes in stable heterocaryons. *Cell* 32: 1171–1180.

Bray, D. (2001). *Cell Movements*, 2nd ed. New York: Garland.

Britten, R., and Davidson, E. (1969). Gene regulation for higher cells: A theory. *Science* 165: 349–356.

Chen, Q., Sealock, R., and Peng, H. (1990). A protein homologous to the Torpedo postsynaptic 58K protein is present at the myotendinous junction. *J. Cell Biol.* 110: 2061–2071.

Child, C. M. (1929). The physiological gradients. *Protoplasma* 5: 447–476.

Copp, A. J. (1993). Neural tube defects. *Trends Neurosci.* 16: 381–383.

De Robertis, E. M., and Sasai, Y. (1996). A common plan for dorsoventral patterning in Bilateria. *Nature* 380: 37–40.

Driever, W., and Nüsslein-Volhard, C. (1988). The Bicoid protein determines position in the *Drosophila* embryo in a concentration-dependent manner. *Cell* 54: 95–104.

Driever, W., and Nüsslein-Volhard, C. (1989). The Bicoid protein is a positive regulator of *hunchback* transcription in the early *Drosophila* embryo. *Nature* 337: 138–143.

Duclert, A., and Changeux, J.-P. (1995). Acetylcholine receptor gene expression at the developing neuromuscular junction. *Physiol. Rev.* 75: B39–B68.

Edelman, G. M. (1988). *Topobiology*. New York: Basic Books.

Fietz, M. J., Jacinto, A., Taylor, A. M., Alexandre, C., and Ingham, P. W. (1995). Secretion of the amino-terminal fragment of the Hedgehog protein is necessary and sufficient for Hedgehog signalling in *Drosophila*. *Curr. Biol.* 5: 643–650.

Ghysen, A., and Dambly-Chaudiere, C. (1989). Genesis of the *Drosophila* peripheral nervous system. *Trends Genet.* 5: 251–256.

Gierer, A. (1981). Generation of biological patterns and form: Some physical, mathematical and logical sspects. *Prog. Biophys. Mol. Biol.* 37: 1–47.

Gigliani, F., Longo, F., Gaddini, L., and Battaglia, P. A. (1996). Interactions among the bHLH domains of the proteins encoded by the *Enhancer-of-split* and *achaete-scute* gene complexes of *Drosophila*. *Mol. Gen. Genet.* 251: 628–634.

Gilbert, S. F. (1995). *Developmental Biology*, 4th ed. Sunderland, Mass.: Sinauer.

Goldman, D., and Staple, J. (1989). Spatial and temporal expression of acetylcholine receptor RNAs in innervated and denervated rat soleus muscle. *Neuron* 3: 219–228.

Harris, D., Falls, D., Johnson, F., and Fischbach, G. (1991). A prionlike protein protein from chicken brain copurifies with an acetylcholine receptor-inducing activity. *Proc. Natl. Acad. Sci. U.S.A.* 88: 7664–7668.

Heitzler, P., Bourouis, M., Ruel, L., Carteret, C., and Simpson, P. (1996). Genes of the *Enhancer-of-split* and *achaete-scute* complexes are required for a regulatory loop between Notch and Delta during lateral signalling in *Drosophila*. *Development* 122: 161–171.

Hoch, M., Seifert, E., and Jäckle, H. (1991). Gene expression mediated by *cis*-acting sequences of the *Krüppel* gene in response to the *Drosophila* morphogens Bicoid and Hunchback. *EMBO. J.* 10: 2267–2278.

Jacobson, A. G., Oster, G. F., Odell, G. M., and Cheng, L. Y. (1986). Neurulation and the cortical tractor model for epithelial folding. *J. Embryol. Exp. Morph.* 96: 19–49.

Jacobson, M. (1991). *Developmental Neurobiology*, 3d ed. New York: Plenum.

Jiang, J., and Levine, M. (1993). Binding affinities and cooperative interactions with bHLH activators delimit threshold responses to the dorsal gradient morphogen. *Cell* 72: 741–752.

Kauffman, S. A. (1993). *The Origins of Order*. New York: Oxford University Press.

Kerszberg, M. (1996). Accurate reading of morphogen concentrations by nuclear receptors: A formal model of complex transduction pathways. *J. Theor. Biol.* 183: 95–104.

Kerszberg, M., and Changeux, J.-P. (1993). A model for motor endplate morphogenesis: Diffusible morphogens, transmembrane signaling and compartmentalized gene expression. *Neur. Comput.* 5: 341–358.

Kerszberg, M., and Changeux, J.-P. (1994a). A model for reading morphogenetic gradients: Autocatalysis and competition at the gene Level. *Proc. Natl. Acad. Sci. U.S.A.* 91: 5823–5827.

Kerszberg, M., and Changeux, J.-P. (1994b). Partners make patterns in morphogenesis. *Curr. Biol.* 4: 1046–1047.

Kerszberg, M., and Changeux, J.-P. (1998). A simple molecular model of neurulation. *BioEssays* 20: 758–770.

Klarsfeld, A., Laufer, R., Fontaine, B., Devillers-Thiéry, A., Dubreuil, C., and Changeux, J. (1989). Regulation of muscle AchR α-subunit expression by electrical activity: Involvement of protein kinase C and Ca^{++}. *Neuron* 2: 1229–1236.

Klarsfeld, A., Bessereau, J., Salmon, A., Triller, A., Babinet, C., and Changeux, J. (1991). An acetylcholine receptor α-subunit promoter conferring preferential synaptic expression in muscle of transgenic mice. *EMBO J.* 10: 625–632.

Laudet, V., and Stehelin, D. (1992). Flexible friends. *Curr. Biol.* 2: 293–295.

Lawrence, P. (1992). *The Making of a Fly.* Oxford: Blackwell.

Liem, Jr., K. F., Tremml, G., Roelink, H., and Jessell, T. M. (1995). Dorsal differentiation of neural plate cells induced by BMP-mediated signals from epidermal ectoderm. *Cell* 82: 969–979.

Meier, T., Masciulli, F., Moore, C., Schoumacher, F., Eppenberger, U., Denzer, A. J., Jones, G., and Brenner, H. R. (1998). Agrin can mediate acetylcholine receptor gene expression in muscle by aggregation of muscle-derived neuregulins. *J. Cell Biol.* 141: 715–726.

Meinhardt, H. (1986). Hierarchical inductions of cell states: A model for segmentation in *Drosophila. J. Cell Sci.* Suppl 4: 357–381.

Meinhardt, H., and Gierer, A. (1974). Applications of a theory of biological pattern formation based on lateral inhibition. *J. Cell Sci.* 15: 321–346.

Meinhardt, H., and Gierer, A. (2000). Pattern formation by local self-activation and lateral inhibition. *BioEssays* 22: 753–760.

Monod, J., and Jacob, F. (1962). General conclusions: Teleonomic mechanisms in cellular metabolism. *Cold Spring Harbor Symp. Quant. Biol.* XXVI: 389.

Morgan, T. H. (1904). An attempt to analyze the phenomena of polarity in tubularia. *J. Exp. Zool.* 1: 587–591.

Nakao, K., and Campos-Ortega, J. A. (1996). Persistent expression of genes of the *Enhancer-of-split* complex suppresses neural development in *Drosophila. Neuron* 16: 275–286.

Nellen, D., Burke, R., Struhl, G., and Basler, K. (1996). Direct and long-range action of a DPP morphogen gradient. *Cell* 85: 357–368.

Papalopulu, N., and Kintner, C. (1994). Molecular genetics of neurulation. In *Ciba Foundation Symposium*, pp. 90–99. New York: John Wiley.

Prigogine, I., and Glansdorff, P. (1971). *Thermodynamic Theory of Stability, Structure and Fluctuations.* Paris: Masson.

Roelink, H., Porter, J. A., Chiang, C., Tanabe, Y., Chang, D. T., Beachy, P. A., and Jessel, T. M. (1995). Floor plate and motor neuron induction by different concentrations of the amino-terminal cleavage product of Sonic hedgehog autoproteolysis. *Cell* 81: 445–455.

Ruiz i Altaba, A. (1994). Pattern formation in the vertebrate neural plate. *Trends Neurosci.* 17: 233–243.

Sanes, J. R., and Lichtman, J. W. (1999). Development of the vertebrate neuromuscular junction. *Annu. Rev. Neurosci.* 22: 389–442.

Sasai, Y., Lu, B., Steinbeisser, H., Geissert, D., Gont, L. K., and De Robertis, E. M. (1994). *Xenopus* Chordin: A novel dorsalizing factor activated by organizer-specific homeobox genes. *Cell* 79: 779–790.

Schaeffer, L., Duclert, N., Huchet-Dymanus, M., and Changeux, J. (1998). Implication of a multisubunit Ets-related transcription factor in synaptic expression of the nicotinic acetylcholine receptor. *EMBO J.* 17: 3078–3090.

Simpson, P. (1990). Lateral inhibition and the development of the sensory bristles of the adult peripheral nervous system of *Drosophila. Development* 109: 509–519.

Simpson-Brose, M., Treisman, J., and Desplan, C. (1994). Synergy between the Hunchback and Bicoid morphogens is required for anterior patterning in *Drosophila*. *Cell* 78: 855–865.

Smith, J. C. (1995). Mesoderm-inducing factors and mesodermal patterning. *Curr. Opin. Cell Biol.* 7: 856–861.

Smolen, P., Baxter, D. A., and Byrne, J. H. (2000). Mathematical modeling of gene networks. *Neuron* 26: 567–580.

Struhl, G., Struhl, K., and Macdonald, P. M. (1989). The gradient morphogen Bicoid is a concentration dependent transcriptional activator. *Cell* 57: 1259–1273.

Studer, M., Pöpperl, H., Marshall, H., Kuroiwa, A., and Krumlauf, R. (1994). Role of a conserved retinoic acid response element in rhombomere restriction of *Hoxb-1*. *Science* 265: 1728–1732.

Thayer, M., Davis, S. T. R., Wright, W., Lassar, A., and Weintraub, H. (1989). Positive autoregulation of the myogenic determination gene *MyoD1*. *Cell* 58: 241–248.

Thomas, M., and D'Ari, R. (1990). *Biological Feedback*. Boca Raton, Fla.: CRC Press.

Toutant, M., Bourgeois, J., Toutant, J., Renaud, D., Douarin, G. L., and Changeux, J.-P. (1980). Chronic stimulation of the spinal cord in developing chick embryo causes the differentiation of multiple clusters of acetylcholine receptor in the posterior latissimus dorsi muscle. *Dev. Biol.* 76: 384–395.

Turing, A. M. (1952). The chemical basis of morphogenesis. *Phil. Trans. Roy. Soc. London B.* 237: 37–72.

Viebahn, C., Lane, E. B., and Ramaekers, F. C. (1995). Cytoskeleton gradients in three dimensions during neurulation in the rabbit. *J. Comp. Neurol.* 363: 235–248.

Weintraub, H., Davis, R., Tapscott, S., Thayer, M., Krause, M., Benezra, R., Blackwell, T. K., Turner, D., Rupp, R., Hollenberg, S., Zhuang, Y., and Lassar, A. (1991). The *Myo-D* gene family: Nodal point during specification of muscle cell lineage. *Science* 251: 761–766.

Wolpert, L. (1969). Positional information and the spatial pattern of cellular differentiation. *J. Theor. Biol.* 25: 1–47.

Zimmerman, K., Shih, J., Bars, J., Collazo, A., and Anderson, D. J. (1993). *XASH-3*, a novel *Xenopus achaete-scute* homolog, provides an early marker of planar neural induction and position along the mediolateral axis of the neural plate. *Development* 119: 221–232.

Gene Network Models and Neural Development

George Marnellos and Eric D. Mjolsness

2

Rapid advances in molecular biology and genomics in recent years have highlighted the need for theoretical tools to analyze and integrate a flood of data. In this chapter, we review mathematical and computational models of molecular processes underlying biological development and concentrate on the role of genes and their interactions. We describe in greater detail a modeling formalism for a gene network, based on neural networks, to study gene regulatory interactions during development. We apply this framework in a computational model of early neurogenesis in *Drosophila*. Although such models are only small steps in the elucidation of how genes orchestrate the complex patterns of neural development, they could provide directions for subsequent research, and could be refined to deal with more powerful data as they become available.

2.1 Genomic Advances Document Unity of Life

From bacteria to primates, the diversity of form and function and the interdependence between the two have occupied biologists for a long time. How are morphological and behavioral patterns of individual organisms coded for and what forces produce these patterns? How is the tremendous richness of these patterns generated in nature? What rules or constraints determine their generation? What purpose might they serve? Evolutionary biology provides of course a framework for addressing questions like

these, but now with the advent of molecular techniques, we have a unique opportunity to explore at a mechanistic level how life's diversity is generated and why.

In the past 30–40 years, molecular methods for manipulating genetic material, but also other cellular components, have allowed researchers to investigate hitherto inaccessible processes both at the cell level and at higher levels of integration. Genes and entire biochemical pathways isolated in one experimental system (e.g., yeast and *Caenorhabditis elegans*) can now almost routinely be identified in many other organisms, not only providing clearer answers to questions originally investigated in individual species, but also revealing previously unimagined levels of homology.

Examples of homology abound. A particularly striking one is that of homeotic genes, which are necessary for specifying animal body patterns. Homeotic genes are transcription factors containing a conserved region that binds to DNA, and they determine regional identity. Their characteristic mutant phenotypes, which can lead to duplication of a body part or replacement of one body part by another (e.g., legs in place of antennae), had long puzzled geneticists who had studied them in *Drosophila*. In fact, in 1978, one of those researchers, E. B. Lewis, formulated a detailed hypothesis as to how homeotic genes might have contributed to the evolution of the fly body plan, thus anticipating many related discoveries in the following decade (Lewis, 1978). These genes were first cloned in *Drosophila* (McGinnis et al., 1984; Scott and Weiner, 1984), and because of the conserved region, the homeobox, their homologs were subsequently isolated in many other phyla, from cnidarians to vertebrates (Akam, 1989; Schummer et al., 1992; Krumlauf, 1994), and were shown to be organized in similar gene groupings and to be expressed in the same order along the body axis as in *Drosophila*. Such findings have prompted the reevaluation of homologies among body parts across phyla and evolutionary relationships in general, and have provided insights into the evolution of developmental mechanisms, pointing to the importance of the evolution of regulatory genes in morphological diversity (Carroll, 1995).

Other examples of genes and pathways conserved across species include, to name but a few: (1) the Ras family of G proteins, which integrate inputs from a wide range of cellular pathways and control many aspects of cell proliferation and differentiation in yeast, flies, nematodes, and mammals by transmitting signals from tyrosine kinases at the plasma membrane through a kinase cascade to the nucleus (Boguski and McCormick, 1994); (2) the mitogen-activated protein kinase (MAPK) pathway, which consists of a cascade of three classes of kinases that deliver signals from membrane to nucleus, and which was first identified in the mating response pathway of yeast and then in many metazoans (in metazoans, the MAPK cascade conveys inputs of Ras proteins, among others) (Neiman, 1993); (3) the homologous pathways that regulate fly and vertebrate limb development, as in the case of the Hedgehog family of proteins involved in the anteroposterior patterning of the limb (Fietz et al., 1994); (4) the family of nuclear receptors, such as steroid, retinoid, and thyroid hormone receptors, which mediate the effect of hormones on gene expression and which are dominant regulators of organ physiology as well as of insect morphogenesis (Beato et al., 1995; Thummel, 1995).

2.2   Need for Theoretical Tools

The list of such discoveries grows daily, supplying strong evidence about the unity of life forms and their biological building blocks, strengthening the conviction that description at the gene level will provide a unifying principle to explain diverse biological phenomena, and exposing the need for analytical methods to study these phenomena.

The situation is perhaps comparable to what happened in physics during the first decades of this century, when the structure of the atom was probed and this opened the door to the discovery of subatomic particles and a unified way of viewing matter in all its forms. However, there is a crucial difference between today's biology and the state of physics at the beginning of the century. Whereas physicists at the time had access to a formidable chest of mathematical tools, which had been the language of physics for centuries, the theoretical means currently at the disposal of biologists seem limited in comparison. There have been of course major contributions, such as the work of Lotka and Volterra (Lotka, 1925; Volterra, 1926) on population biology and predator–prey models, Fisher (1930) on gene flow dynamics in populations, Turing (1952) on the role of reaction and diffusion of substances in pattern formation during development, Thom (1972) on the topological aspects of morphogenesis, and Hodgkin and Huxley (1952) on generation and propagation of action potentials in axons. These pioneering studies have helped establish whole fields of research in ecology, population genetics, developmental biology, and electrophysiology, but the fact remains that, on the whole, in biology theory has not occupied the same place in the interpretation of data and the designing of experiments as it has in physics.

The need for theoretical methods in biology has become more apparent in the past decade as data from molecular experiments pour in at an ever-increasing rate, especially from whole genome sequencing projects. People have started responding to this realization, most notably in the area of computational molecular biology, where mathematical tools (e.g., statistical analysis and combinatorial optimization) have been used to study structure and function in genomes, genes, and gene products in order to answer specific biological questions. This work includes discovering genes in sequences, detecting gene homologies among species, building phylogenetic trees, and predicting the secondary and tertiary structure of proteins and nucleic acids from sequence information.

However, the intensity of theoretical effort in genomic research will have to be extended to the study of development and the phenotypic variation that it generates (and that selection acts upon). In particular, it will have to be directed to the study of genes and their involvement in these processes, since molecular data will contain a wealth of information on developmental and evolutionary questions.

2.3 Models of Molecular Processes and Development

2.3.1 *Reaction-Diffusion*

There has already been a considerable amount of work on mathematical models of development. A large part of this work has been in the tradition of Turing's reaction-diffusion approach. In systems such as insect segmentation, seashells, animal coats, and butterfly wings, the stable patterns that can emerge when chemical substances, morphogens (usually two in number, an activator and an inhibitor), diffuse and react with each other over a morphogenetic field

were modeled by Gierer and Meinhardt (1972), Meinhardt (1986, 1987, 1998), and Murray (1981a,b, 1993; see also chapters 1 and 3). These efforts have generally dealt with abstract quantities and have not attempted to make explicit connections between these quantities and the interactions of specific genes, although in some cases subsequent experimental work has provided candidate molecules, as in the case of Meinhardt's model of how insect leg proximodistal coordinates are set up (Meinhardt, 1983; Basler and Struhl, 1994; Diaz-Benjumea et al., 1994).

2.3.2 *Mechanical Models*

Other researchers have considered cell movements and mechanical properties of cells and tissues and modeled processes such as gastrulation and neurulation (Odell et al., 1981; see also chapter 1), cartilage condensation in limb morphogenesis and patterning of feather primordium (Murray et al., 1983; Oster et al., 1978), aggregation of *Dictyostelium* amoebae (Segel, 1984; Marée and Hogeweg, 2001), cell intercalation and sorting (Weliky and Oster, 1990; Agarwal, 1995; Graner and Sawada, 1993), and skin generation (Stekel et al., 1995).

2.3.3 *Lindenmayer Systems*

Drawing inspiration from formal languages, Lindenmayer (1968) has modeled development using sets of rules, or grammars. These rules describe cellular processes such as growth, division, and differentiation and are used to modify strings that represent organisms. These so-called Lindenmayer systems have been used to model the growth and branching patterns of plants (Prusinkiewicz and Lindenmayer, 1990). They have also been extended to include cell-cell interactions, through the use of context-sensitive grammars, as well as two-dimensional (2-D) and three-dimensional (3-D) cells that can change shape, as in some of the mechanical models mentioned earlier (Lindenmayer and Rozenberg, 1979; Lindenmayer, 1984).

2.3.4 Biochemical Kinetics

The models mentioned above do not attempt to make any connections to specific genes or biochemical pathways involved in the processes modeled, and in fact most of them do not make any reference to such factors at all. There have been models incorporating such molecular elements, however, although in this case the modeling has been restricted to processes within single cells. Savageau (1976), applying methods from chemical kinetics, has modeled biosynthetic pathways and the regulation of gene expression in prokaryotes. He analyzed the fixed points and periodic behaviors of these systems and also considered questions of optimality in the design of the pathways. Other workers have also examined dynamical features of metabolic pathways and gene expression (Goodwin, 1965; Hunding, 1974; Hastings et al., 1977; Tyson and Othmer, 1978), Ca^{2+} signaling (Dupont and Goldbeter, 1992), and the complex interactions in the progression of the cell cycle (Novak and Tyson, 1993).

In a similar framework, Bray has pointed out the similarities of biochemical signal cascades to neural networks that might be performing some kind of pattern recognition within cells. It has in fact been shown that chains of chemical reactions can be viewed as neural networks that can be reduced to Hopfield nets (Hjelmfelt et al., 1991; Hjelmfelt and Ross, 1992; Bray, 1995). Bray has modeled networks of cell-signaling reactions (Bray, 1990; Bray and Lay, 1994) and has optimized the reaction parameters to achieve a desired mode of functioning or output of a pathway. He has used this method to simulate the sig-

naling cascades involved in bacterial chemotaxis and found reaction parameter values that produce various chemotactic behaviors of known mutant phenotypes (Bray et al., 1993; Bray and Bourret, 1995).

In a biochemically more concrete look at morphogen gradients during development (see section 2.3.1), Kerszberg and Changeux (1994) examined how different assumptions about transcription factor dimers, autocatalytic feedback, and competition for regulatory binding sites by these dimers lead to different patterns of transcription and protein concentration (see also chapter 1). These authors used the results of their model to interpret morphogen gradients of Bicoid and Hunchback in the *Drosophila* blastoderm.

More recently, Von Dassow et al. (2000) have also looked at developing *Drosophila* embryos, using a fairly realistic biochemical interaction model. They examined interactions among segment polarity genes and found that the whole system is dynamically robust and can resist changes to its kinetic parameters.

2.3.5 Phage λ

In work concentrating on gene interactions, Ackers et al. (1982) and Shea and Ackers (1985) have developed a detailed quantitative model of the regulation of certain genes of bacteriophage λ that are involved in maintenance of the lysogenic state (when the prophage is integrated into the DNA of the host) and the induction of lysis (when the virus replicates). In constructing their model, the authors stayed very close to biochemical facts concerning the structure of the genes involved and their promoters, binding constants, dimerization, cooperative interactions, and so on. In a hybrid modeling approach, McAdams and Shapiro (1995) have also looked at the lysogeny–lysis switch of phage λ by integrating chemical kinetics with an electrical circuit simulation of the genes and regulatory interactions that control this switch.

2.3.6 Boolean Nets

In a different vein, abstracting from biochemical detail, Kauffman (1969) has introduced networks of elements with binary states to model gene regulatory interactions. These Boolean networks are intended to be idealizations of continuous dynamic systems with elements that behave in a sigmoidal fashion (as is the case with many cellular and biochemical processes); they are believed to capture a skeleton of the dynamical structure of such continuous systems (see chapter 5 in Kauffman, 1993, and references therein). Since Boolean networks have finitely many different states, they are guaranteed to have fixed points and state cycles, which could be viewed as corresponding to stable differentiation states and periodic behaviors of cells; such parallels are being explored in specific cases where known gene expression patterns appear to be consistent with this description (Somogyi and Sniegoski, 1996). Kauffman (1969, 1971, 1974) has explored the stability of the dynamics of these Boolean nets, which depends on the number of inputs of each element and the number of elements. He has also ascribed fitness values to different configurations of Boolean nets and investigated features of the fitness landscapes that result in such configuration spaces, such as number, similarity, and accessibility of fitness peaks (Kauffman and Levin, 1987).

2.4 Gene Net Framework

Genes being a natural module for the description of living systems, they also appear to be a natural level of abstraction for integrated biological models. Starting from this premise, Mjolsness et al. (1991) have introduced a modeling framework for the study of development, centered around genes and their interactions. This framework shares features with the models described earlier, but in a combination that is not found in any of the others. It incorporates features that allow modeling of processes at the tissue level, such as in the reaction–diffusion and mechanical models, but unlike the biochemical kinetics and phage λ models; and unlike the former but similar to the latter, it also includes a description of molecular processes, such as gene expression. The modeling framework consists of two major components:

1. The first is a neural network representation of molecular-level interactions. Gene interactions, as well as other molecular signaling and regulatory events, are modeled as a particular kind of neural net, namely, recurrent nets, with connections allowed in both directions between any pair of nodes (Hopfield, 1984; Hertz et al., 1991). In this formulation, gene product concentrations correspond to node activation levels and gene interaction strengths to connection weights.

2. The second component is a Lindenmayer-systemlike grammar of rules (Lindenmayer, 1968; Prusinkiewicz and Lindenmayer, 1990), L-grammar, which describe cell-cell interactions and changes in number, type, and state of cells.

2.4.1 Dynamics

In more detail, genes in such networks interact as nodes in a recurrent neural net. They sum up activating and inhibitory inputs from other genes in the same cell at any given time t; we represent this sum as u:

$$u_a(t) = \sum_b T_{ab} v_b(t), \qquad (2.1)$$

where genes are indexed by a and b, T_{ab} is the interaction between genes a and b, and $v_b(t)$ are gene product levels within the cell. If we include interactions with neighboring cells, this becomes

$$u_a(t) = \sum_b T_{ab} v_b(t) + \sum_i \sum_b \hat{T}_{ab} \hat{v}_b^i(t), \qquad (2.2)$$

where $\hat{T}_{ab}$ is the interaction of gene a with gene b in neighboring cells and $\hat{v}_b^i(t)$ are gene product levels in neighboring cell i. Level $v_a(t)$ of the product of gene a then changes according to

$$\frac{dv_a}{dt} = R_a g(u_a(t) + h_a) - \lambda_a v_a(t), \qquad (2.3)$$

where $u_a(t)$ is the linear sum of Eq. (2.1), R_a is the rate of production of gene a's product, λ_a is the rate of decay of gene a product, and h_a is the threshold of activation of gene a, which can be either positive or negative and could thus correspond to a constitutive positive or negative input. Function g is a monotonic, nonlinear function, usually a sigmoid, such as the following one, which we have used in gene net models and which is centered at 0 and takes values between 0 and 1:

$$g(x) = 0.5\left(1 + \frac{x}{\sqrt{(1 + x^2)}}\right). \qquad (2.4)$$

Levels of gene products should be viewed as corresponding to gene product activities in the biological system rather than to actual concentrations, and gene interactions should be viewed as corresponding more closely to genetic rather than specific biochemical (transcriptional, etc.) interactions. The form of Eq. (2.3) can be justified as follows: If we consider gene a as a producer molecule that can be either in an activated (i.e., producing) state or in an inactivated (i.e., nonproducing) one, depending on the concentrations of other gene products that can bind at its regulatory regions, then the amount of species a produced is proportional to the fraction of time that gene a spends in the activated state (or equivalently, to the fraction

of producer molecules in that state). Species a also decays at a rate λ_a that is independent of gene product concentrations. This is expressed in the following equation [which has the same form as Eq. (2.3)]:

$$\frac{dv_a}{dt} = R_a[\text{fraction activated}] - \lambda_a v_a, \qquad (2.5)$$

where [fraction activated] is the fraction of time that gene a is activated. It is this fraction, which depends on the concentrations of other gene products, that is approximated by the recurrent net formulation in Eq. (2.3) (see also figure 2.1). For a more detailed biochemical rationale of this approximation, see Mjolsness et al. (1991).

2.4.2 L-Grammar

The gene net framework allows cell transformations in the models; for instance, cells may change their state (i.e., the levels of gene products or other state

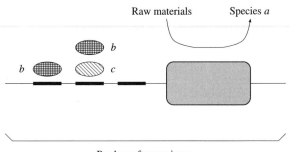

Figure 2.1
Illustration of the biochemical model incorporated in Eqs. (2.3) and (2.5). Production of species a depends on the activation of the gene for a, which is determined by the binding of gene products b and c at regulatory regions of the gene for a and by interactions of these with the transcription apparatus at the gene. (Adapted from figure 2 in Mjolsness et al., 1991.)

variables), change type, give birth to other cells, or die. These transformations are represented by a set of grammar rules, the L-grammar, as in Lindenmayer systems (Lindenmayer, 1968; Prusinkiewicz and Lindenmayer, 1990). Rules are triggered according to the internal state of each cell (or perhaps also of other cells) and are of two kinds: discrete and continuous. Transformations that happen gradually (smoothly) over time are described by continuous time rules, while processes that occur as abrupt, discontinuous changes are given by discrete time rules, which are instantaneous. Rules may involve one or more cells, representing intracellular processes and cell-cell interactions, respectively (see figure 2.2).

How rules are triggered depends on the internal state of each cell (or perhaps also of other cells). There are some constraints as to what rules may be active in a cell at any given time. Only a single continuous one-cell rule is allowed at a time, but several continuous two-cell rules may operate simultaneously. Only one birth or death rule may occur at a time, but other discrete rules that bring about changes in cell type may happen simultaneously, as long as they all transform a cell into the same cell type. A set of binary variables C keeps track of what rules are active in any particular cell at any given time. Vector $\mathbf{u}$ [see Eq. (2.2)] is therefore more accurately given, for a cell i, by

$$\mathbf{u}_i = \sum_r C_i^r \mathbf{T}_1^r \cdot \mathbf{v}_i + \sum_r C_i^r \sum_j \Lambda_{ij} \mathbf{T}_2^r \cdot \mathbf{v}_j, \qquad (2.6)$$

where $\mathbf{T}_1^r$ is the interaction strength matrix for one-cell rule r; $\mathbf{v}_i$ is the state variable (gene product level) vector for cell i; $\mathbf{T}_2^r$ is the interaction strength matrix for two-cell rule r [r, of course, in both sums of Eq. (2.6) is just a dummy variable that stands in for the actual names of the rules, which could be, for instance, mitosis, cell death, interphase, and so on]; $\mathbf{v}_j$ is the state variable vector for cell j, located in the

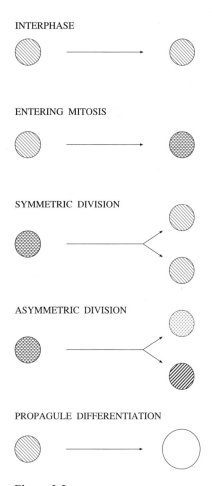

INTERPHASE

ENTERING MITOSIS

SYMMETRIC DIVISION

ASYMMETRIC DIVISION

PROPAGULE DIFFERENTIATION

Figure 2.2
Graphic representation of examples of rules that can be applied to cells in a model. Note that the asymmetrical cell division rule produces two cells (in interphase) that differ in their gene product concentrations and are depicted by different shadings, whereas the symmetrical division rule produces identical cells. In the propagule differentiation rule, the larger circle denotes a propagule with its stored reserves.

neighborhood of cell i, the type of neighborhood being specified in particular models; and Λ_{ij} is a factor that modifies the influence of cell j on cell i and depends on the geometry of cells and their positions in a model. Variables C_i^r determine which rules in Eq. (2.6) operate at any given time; if $C_i^r = 1$, then the corresponding rule is active, while if $C_i^r = 0$, the rule is inactive; these C variables encode the constraints on rule activity that were described earlier.

When two rules cannot both be active at the same time, the rule of highest "strength" at that time wins. Strength S_i^r of rule r in cell i signifies the likelihood that rule r will be triggered in that cell at that time, and depends on the internal state of the cell (and perhaps also on interactions with neighboring cells). If we consider only dependence on the internal state of the cell itself (i.e., on the state variable vector $\mathbf{v}_i$), rule strength is given by

$$S_i^r = \mathbf{v}_i \cdot \mathbf{s}_r + \theta_r, \tag{2.7}$$

where $\mathbf{s}_r$ is a vector that describes how each state variable of the cell contributes to the strength of rule r, and θ_r is the default likelihood that rule r will be triggered.

Rule strengths together with the constraints determine which rules are active at a given time and along with the parameters of state variable dynamics, $\mathbf{T}$, $\mathbf{R}$, $\mathbf{h}$, and λ of Eqs. (2.2) and (2.3), and geometry factors Λ, completely specify how the modeled system develops.

2.4.3   *Optimization*

Models using the gene net framework can be formulated as optimization tasks that seek values for the model parameters so that the model optimally fits biological data or behaves in a certain desired manner. Such requirements can be captured in a so-called

cost function (or energy function) $E(\mathbf{v})$, which depends on the state variable values $\mathbf{v}$ during development of the system; E of course ultimately depends on the values of the model parameters. A common example of a cost function is a least-squares cost function:

$$E = \sum_{i,a,t} [v_{a\text{MODEL}}^i(t) - v_{a\text{DATA}}^i(t)]^2, \tag{2.8}$$

which is the squared difference between gene product levels in the model and those in the data, summed over all cells (i) and over all gene products (a) and times (t) for which data are available.

A quadratic penalty term on these parameters, of the form

$$\text{penalty} = \sum_m w_m p_m^2, \tag{2.9}$$

where p_m are parameter values (m is an index over model parameters and there is one set of parameters for all cells in the model) and w_m are weights (usually the same for all parameters), is added to E to produce the final, objective function, which is optimized:

$$\text{objective} = w_E E[\mathbf{v}(\mathbf{p})] + w_p \text{penalty}(\mathbf{p}), \tag{2.10}$$

$\mathbf{p}$ being the vector of model parameters and w_E, w_p being weights for the energy and penalty terms of the objective function, respectively. The penalty term prevents optimized parameters from growing excessively large and hence saturating the sigmoid functions of the model or causing overflow errors in computer simulations. It effectively restricts the search space and thus may facilitate the optimization. However, if the restricted search space does not contain the optima sought or if, depending on the optimization algorithm used, parts of the parameter space are not equally accessible from all other parts of the space, this may adversely affect the optimization search. See Fonseca and Flemming (1995) for a discussion, in the context of genetic algorithms, of this and the more general problem of optimizing objective functions that have many components.

The objective functions in gene net models typically have a large number of variables, are highly nonlinear, and cannot be solved analytically or readily optimized with deterministic methods. We have therefore used numerical, stochastic techniques to optimize them: namely, simulated annealing (SA) and genetic algorithms (GA). Both of these optimization methods have a number of parameters that can affect their performance and need to be tuned for each problem. For more details on this connectionist framework and its application, see Mjolsness et al. (1991) and Marnellos (1997).

2.4.4 Overview

This combination of differential equations and grammatical structure is intended to make models computationally feasible and yet maintain a wide repertoire of behaviors at the molecular and tissue levels. Grammars can be used to summarize aspects of the intracellular and intercellular dynamics of the system being modeled, which would otherwise require a large number of extra state variables and model parameters to describe. So grammars offer a concise and computationally tractable representation. The neural net idealization representing molecular interactions is at a level of abstraction similar to that of the biochemical kinetics models. The phage λ models, in contrast, incorporate much greater biochemical detail, and it would be computationally very expensive to have that much detail in models of multicellular development.

The neural net idealization has the following additional advantage: Neural nets can be "trained" to produce desired outputs. This property of the neural net formalism has been extensively studied (Hertz et al., 1991) and there are algorithms to perform the training; for instance, in the case of sufficiently simple recurrent neural nets, there are even deterministic methods (such as those described in Pearlmutter, 1989, and Williams and Zipser, 1989) to do the training. In the gene net framework, training corresponds to fitting experimental observations, or having the simulated system behave in a desired fashion, by optimizing the adjustable parameters of the gene nets, e.g., gene interaction strengths or activation thresholds. This is similar to what Bray has done in his bacterial chemotaxis models mentioned earlier (Bray et al., 1993; Bray and Bourret, 1995), which of course are models of single cells only.

The range of biological questions that can be addressed with the gene net framework is comparable to that of Kauffman's Boolean nets, which are computationally less expensive than gene nets. However, because of the binary way in which they represent molecular events and because they do not represent cells, tissues, or such entities, Boolean nets cannot be readily used to interpret a large body of molecular and cell-level experimental observations.

A framework that is very similar to the gene net method in scope, structure, expressiveness, and level of biochemical detail has also been proposed by Fleischer and Barr (1994). It mainly differs from the gene net framework in that (1) it does not have grammar rules to represent state changes in cells, but instead uses conditional terms in the ordinary differential equations that describe how state variables change; (2) the state variable dynamics are not like in neural nets, but of a more arbitrary form that can be specified by the user; and (3) it has mainly been used to simulate artificial configurations of cells and not to interpret biological observations.

2.5   Applications of the Gene Net Framework

2.5.1   Drosophila *Blastoderm*

Reinitz et al. (1995) and Reinitz and Sharp (1995) have applied the gene net framework to an early stage of development of the *Drosophila* embryo (the blastoderm stage). They have looked at the well-characterized hierarchy of regulatory genes that control the early events of *Drosophila* embryogenesis by setting up their expression patterns along the embryo's length and dividing it into segments. This includes the products of the maternal genes *bicoid* (*bcd*) and *hunchback* (*hb*), expressed in broad gradients along the anteroposterior axis of the embryo, and products of so-called gap, pair-rule, and segment polarity genes, which end up segmenting the whole length of the embryo into stripes, each a single cell wide (Nüsslein-Volhardt and Wieschaus, 1980; Ingham, 1988; St. Johnston and Nüsslein-Volhardt, 1992). Because the expression of these genes does not vary along the dorsoventral axis of the blastoderm and because there are no separate cells in the blastoderm—the embryo is a syncytium of nuclei arranged at its surface like a shell, and cell membranes start to form only during later blastoderm stages (see Gilbert, 2000)—the authors have modeled the system as a single row of nuclei, which are the sites of gene expression and which interact with each other through diffusion of gene products.

They have investigated questions of positional specification in the blastoderm, and their model has yielded predictions and interpretations of experimental observations. The model predicted that Bicoid and Hunchback proteins cooperatively determine position in the anteroposterior axis (Reinitz et al., 1995), which has subsequently been confirmed by experiment (Simpson-Brose et al., 1994). The model also offered insights into the spatiotemporal expression pattern of the pair-rule gene *even-skipped* (*eve*), including such questions as which of the broad spatial domains of expression of the gap genes (which are expressed at the second stage of the regulatory hierarchy described above and activated by the maternal effect genes *bicoid* and *hunchback*) set the boundaries of *eve* stripes, and the timing and order of appearance of these stripes (Reinitz and Sharp, 1995). Moreover, the model provided an explanation for a cell biology observation, namely that pair-rule mRNAs and proteins (unlike those of gap genes) are apically confined in the cells forming around the surface of the blastoderm. The model showed that *eve* stripes do not form unless Eve protein has very low diffusivity, which could result from its mRNA and protein being preferentially targeted to, and retained at, the apical (facing outside the blastoderm) surface of the invaginating cell membranes, thus making it more difficult for Eve to diffuse to neighboring nuclei.

2.5.2   Drosophila *Neurogenesis*

Marnellos and Mjolsness (1998a,b) have worked on early neurogenesis in *Drosophila* and constructed models to study how neuroblasts and sensory organ precursor (SOP) cells differentiate from proneural clusters of equivalent cells. These neurogenesis models have produced predictions about the dynamics of cluster resolution and the robustness of this process (see section 2.6).

2.5.3   Xenopus *Ciliated Cells*

In a more recent model, Marnellos et al. (2000) probed lateral inhibitory signaling through the Delta-Notch pathway and its role in the emergence of *Xenopus* ciliated cells in a salt-and-pepper pattern on the initially uniform epidermis. The model reproduced the phenotypes observed experimentally in the assays tested. Statistical analysis of "genotypes" in the model suggested that the model could account for the variability of embryonic responses to the experimental assays, and highlighted a component of lateral inhibition that may be the chief source of this variability.

2.6   A Gene Net Model of Early Neurogenesis in *Drosophila*

2.6.1   *Neurobiological Background*

In *Drosophila*, neuroblasts and sensory organ precursor (SOP) cells differentiate from epithelia to give rise to the central nervous system in the fruit fly embryo and to epidermal sensory organs in the peripheral nervous system of the adult fly, respectively. Neuroblasts are neural precursor cells that divide to form neurons and glia; they segregate from the ventral neuroectoderm of the embryo in a regular segmental pattern. The SOPs appear at stereotypical positions on imaginal disks of fly third instar larvae and divide to produce a neuron and three other cells, which form *Drosophila*'s sensory organs, such as the bristles on its thorax.

The activities of two main sets of genes working in opposite directions are thought to underlie this differentiation process: one promoting neural development and the other preventing it and favoring epidermal development.

Neuroblasts and SOPs differentiate from apparently equivalent clusters of cells expressing genes of the *achaete-scute* complex, so-called proneural genes. Eventually, only one cell from each proneural cluster retains proneural gene expression and becomes a neuroblast or SOP in a process referred to as cluster resolution (see figure 2.3). Proneural genes thus promote the neuronal fate.

The other main set of genes includes a number of genes that also encode nuclear proteins, such as genes of the *Enhancer-of-split* [*E(spl)*] complex and *hairy*, as well as other genes for membrane and cytoplasmic proteins; all these tend to suppress neurogenesis and promote epidermal development. In this chapter, we refer to this set of genes as epithelial genes; in the literature they are called neurogenic genes because loss-of-function mutations of these genes lead to overproduction of neurons.

Cluster resolution and the singling out of neural precursors from within proneural clusters is brought about by inhibitory lateral signaling between adjacent cells, through the signaling pathway of receptor Notch and its ligand Delta; the neural fate is promoted in the future neuroblasts and SOPs and suppressed in other cells.

Several other genes that are involved in this specification of cell fate are also expressed in characteristic spatial and temporal patterns during the process (for reviews, see Campuzano and Modollel, 1992; Muskavitch, 1994; Artavanis-Tsakonas et al., 1995).

Despite the number of empirical observations that have been gathered, a precise characterization of lateral signaling still does not exist, and we do not understand the dynamical aspects of the system, e.g., whether and how the shape and size of proneural clusters determine how cluster resolution proceeds.

2.6.2   *Model*

In our model, cells are represented as overlapping circles in a two-dimensional hexagonal lattice; the extent of overlap determines the strength of interaction between neighboring cells (but note that in the examples presented here we have only used identical overlaps between cells). Cells in the model express a small number of genes corresponding to genes that are involved in neuroblast and SOP differentiation.

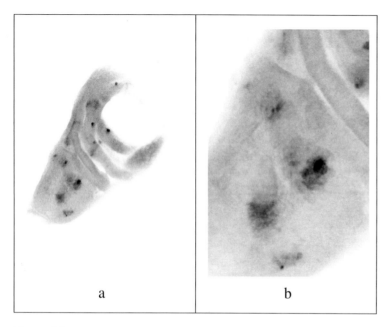

a   b

Figure 2.3
(*a*) Proneural gene expression in clusters in a *Drosophila* wing disk (the appendage of the fruit fly larva that gives rise to the wing and the back of the adult). The *lacZ* reporter gene indicates *achaete* expression (*achaete* is one of the proneural genes). (*b*) Detail of (*a*). Note the cluster on the lower left that has not yet resolved; other clusters appear to be at a more advanced stage of resolution.

In the work presented here, we have used networks with four genes (one corresponding to the proneural group; another for the epithelial group; and two for the ligand and receptor, respectively, that mediate cell-cell signaling).

Genes interact as nodes in recurrent neural nets, as described in Eqs. (2.2) and (2.3). The matrix **T** of gene interactions has the structure depicted in table 2.1. This table shows that we have allowed only proneural and epithelial gene products to directly regulate the expression of other genes (themselves included), since these two genes correspond to transcription factors in the real biological system.

We have modeled lateral interactions between cells by the binding of ligand to the receptor in the neigh-

Table 2.1
Intracellular interactions. The structure of matrix **T** of gene interactions

	Proneural	Epithelial	Receptor	Ligand
Proneural	●	●		
Epithelial	●	●		
Receptor	●	●		
Ligand	●	●		

Notes: Columns are for input genes, and rows are for genes affected. Empty boxes signify zero interaction strength, i.e., no interaction.

boring cell and subsequent regulation of the epithelial gene by the active ligand–receptor complex; this corresponds to the signal relayed from the activated Notch receptor to epithelial gene *E(spl)*. In more detail, the ligand-receptor reaction is assumed to be governed by mass action-type kinetics:

$$L + R \rightleftarrows L \circ R, \tag{2.11}$$

where L is the ligand (on one cell), R is the receptor (on a neighboring cell), and $L \circ R$ is the active receptor–ligand complex; the rate of the reaction to the right is, say, k_1 and to the left k_2. If v_L is the ligand concentration, v_R is the receptor concentration, and $[L \circ R]$ is the concentration of the receptor–ligand active complex, then we have

$$\frac{d[L \circ R]}{dt} = k_1 v_L v_R - k_2 [L \circ R], \tag{2.12}$$

$$\frac{dv_L}{dt} = \frac{dv_R}{dt} = -k_1 v_L v_R + k_2 [L \circ R]. \tag{2.13}$$

This reaction is assumed to take place at a much faster time scale than gene expression and to have reached a steady state before influencing gene expression. Thus the epithelial gene in a cell receives input from receptor–ligand complexes activated by ligand in the six surrounding cells (the lattice is hexagonal).

A more recent model for the summation of gene inputs that extends Eq. (2.2) and could cover receptor-ligand signaling between cells, as described earlier, is presented in Eq. (2.14) (Jonsson et al., 2002):

$$u_a(t) = \sum_b T_{ab} v_b(t) + \sum_i \sum_b \hat{T}_{ab} \hat{v}_b^i(t)$$

$$+ \sum_i \sum_b \sum_c \tilde{T}_{ac}^{(1)} \tilde{T}_{cb}^{(2)} v_c(t) \hat{v}_b^i(t) \tag{2.14}$$

where the first two terms are as in Eq. (2.2) and where

in the last term, v_c are receptor concentrations, $\hat{v}_b^i$ are ligand concentrations excreted by neighboring cell i, $\tilde{T}_{cb}^{(2)}$ is the connection strength for excitation of receptor c that is due to ligand b, and $\tilde{T}_{ac}^{(1)}$ is the connection strength for production of protein a via receptor c activation.

We optimize on gene interaction strengths in order to fit gene expression patterns described in the literature; the cost function optimized is a least-squares one, as in Eq. (2.8). We have used a stochastic algorithm, namely simulated annealing, for this optimization. For more details on the model and the optimization method used, see Marnellos (1997) and Marnellos and Mjolsness (1998b).

The gene expression data sets we optimize on (i.e., the training data sets) are adapted from schematic results described in the experimental literature (Cubas et al., 1991; Skeath and Carroll, 1992; Jennings et al., 1994). These specify the initial pattern of concentrations of the gene products (i.e., the proneural clusters), the desired intermediate pattern, and the desired final pattern when the proneural clusters have resolved into single cells expressing the proneural gene at high levels (see figure 2.4). It is left to the optimization to find the right model parameters so that the system develops from the initial state through the intermediate one to the desired final one. The initial concentrations of receptor and ligand are uniform for all cells, and their subsequent concentrations are not constrained by the data set (in this respect they are comparable to hidden units in neural nets).

2.6.3 Model Results and Predictions

We have tried to limit the number of parameters we optimize on in order to avoid overfitting our rather small data sets. The optimization procedure used (simulated annealing) has produced very good and consistent fits to the training data sets. Successful

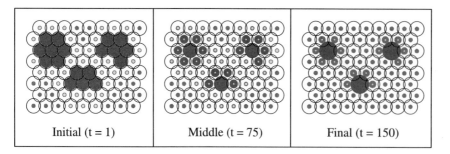

Initial (t = 1)	Middle (t = 75)	Final (t = 150)

Figure 2.4

Cells are modeled as circles on a hexagonal lattice. Gene expression is represented by disks; proneural expression is in dark gray, epithelial in medium gray, and where the two overlap, it is in light gray; the disk radius is proportional to the level of expression. This figure shows the training data set. The left panel shows the initial concentrations of the gene products; there is proneural gene expression in three clusters. The middle panel shows the desired intermediate pattern of expression. The right panel shows the desired final pattern of gene expression; proneural expression is retained only in the central cell of each cluster, the future neuroblast or sensory organ precursor, whereas all other cells express the epithelial gene. Times (t) indicate the points in the run when the desired expression pattern was compared with the actual one [see Eq. (2.8)]; at $t = 1$ there is of course only initialization and no comparison. Initial concentrations of ligand and receptor are not shown.

optimization runs have yielded solutions that not only perform well on the training data set shown in figure 2.4 but also work for other data sets with greater numbers of similar-sized or larger clusters in various spatial arrangements. This indicates that optimization does not just find parameter values that only work for the specific size and cluster arrangement of the training data set, but produces solutions incorporating "rules" for cluster resolution.

With parameter values derived by optimization on the data set of figure 2.4, the model makes predictions about how the interplay of factors such as proneural cluster shape and size, gene expression levels, and strength of cell-cell signaling determines the timing and position of neuroblasts and SOP cells as well as the robustness of this process and the effects of perturbations in gene product levels on cell differentiation.

In figure 2.5, for instance, we have the same optimization solution parameters in both rows, but in the run in the top row, initial concentrations of proneural and epithelial gene products are identical for all cells in a cluster, while in the bottom row initial proneural concentrations vary and differ among cells by about 10–15 percent. Despite this and despite the fact that in this particular example the future neural precursors start out with lower proneural concentrations than other cluster cells (even the lowest in the cluster), the pattern of cluster resolution remains identical, as the end result shows (compare the right panels of the top and bottom rows of figure 2.5). So optimization solutions are robust to small changes in initial conditions. Such robustness is a feature that a biological system would need during development.

Proneural clusters in test sets resolve well apart from the small, diamond-shaped four-cell clusters; this is probably because four-cell clusters do not have a cell that is much more encircled than the others (as the five-, six-, and seven-cell clusters of figure 2.6 do), but all cells are almost equally exposed. The optimization solution also resolves the large cluster in figure 2.6, for which it was not optimized; this is another aspect of the robustness of the solution. As

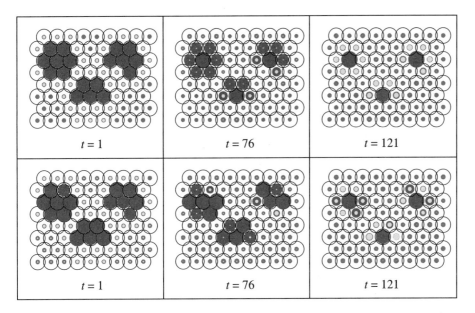

Figure 2.5
Computer simulation of neural precursor differentiation with parameter values in the model derived by optimization on the data set of figure 2.4. From left to right, different time frames of the evolution of gene product concentrations. (*Top*) A run with identical initial gene product concentrations for all cells in each proneural cluster. (*Bottom*) Initial proneural concentrations vary by about 10–15 percent between cells in each cluster. In both runs, the clusters resolve in the same way, as the comparison of the two panels at $t = 121$ shows; in the bottom run, the clusters take slightly longer to resolve. This illustrates the robustness of cluster resolution to small changes in initial gene expression levels in proneural clusters. Conventions are as in figure 2.4. (From Marnellos and Mjolsness, 1998b.)

in the previous example, variation in initial concentrations in large clusters usually does not alter the final outcome of cluster resolution, but in rare cases large clusters do not resolve to a single cell, but to two or three cells. This is consistent with experimental observations (Huang et al., 1991) and illustrates the role of the interplay between position in cluster and level of proneural expression in determining whether a cell becomes a neural precursor.

Since lateral interactions are crucial for cluster resolution, we have varied their strength to see the effects on the dynamics of the process. Stronger lateral interaction increases the speed of cluster resolution, the ef-

fect being much more pronounced in large clusters, which take longer to resolve in our model. With even stronger lateral interaction, clusters start to fail to resolve and proneural expression is extinguished. When lateral interactions are abolished, clusters do not resolve, but all cells in them retain proneural gene expression. This parallels the effect of the neurogenic mutations in a real biological system; these mutations disrupt lateral communication between cells and lead to overproduction of neurons (Poulson, 1940; Skeath and Carroll, 1992). Thus, a variation in the value of a single or a few parameters controlling the strength of lateral interactions can produce a "heterochronic"

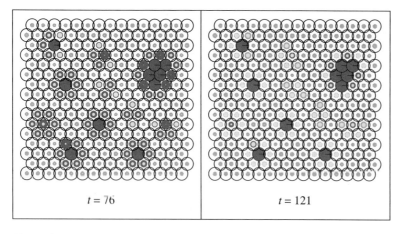

$t = 76$ $t = 121$

Figure 2.6

Simulation with perturbations of gene expression in individual cells of two symmetrical, seven-cell clusters. The clusters are the two in the lower left corner of the data set. At $t = 60$ (not shown), the level of epithelial expression in the central cell of the upper one of these two clusters is instantaneously increased, while in the lower cluster proneural expression is increased in a peripheral cell. Both perturbations can be detected in the left panel ($t = 76$). The first perturbation abolishes cluster resolution, while the second has no effect on resolution, as can be seen in the right panel ($t = 121$).

change in the process of cluster resolution or even prevent neural precursor differentiation. This is an interesting and testable prediction of the model.

As a final illustration of how one can probe the dynamics of cluster resolution in our model, we perturbed the levels of expression of proneural and epithelial genes in specific cells during a run, as illustrated in figure 2.6. In this simulation, we increased instantaneously at $t = 60$ the level of epithelial expression in the central cell of a symmetrical, seven-cell cluster and also the level of proneural expression in a peripheral cell of a different symmetrical cluster. Whereas the first perturbation prevents normal resolution of the cluster involved (as can be observed at $t = 121$ for instance), the second one has no effect on resolution and the cluster involved resolves normally. The effects of such perturbations will vary, depending on what time and in which cell they are carried out, and on

whether they occur singly as in the two examples of figure 2.6, or in various combinations. Such manipulations are therefore a rich source of predictions of the model and most likely can be tested in *Drosophila* (Halfon et al., 1997).

2.7   Discussion

In this chapter we have presented an overview of computational models of gene regulation—including fairly detailed models of biochemical reactions and their dynamics as well as more abstract reaction-diffusion and Boolean network models—that have been applied to questions in biological development. In particular, we have concentrated on a gene network model based on the framework of Mjolsness et al. (1991) as it has been applied to early neurogenesis in *Drosophila*.

Although different in details, all such models offer insights into the processes under study by posing biological questions in more concrete terms and testing the logical consistency and inferences of underlying assumptions. The gene network approach has the advantages (1) that it can represent, through its grammar structure, a large spectrum of molecular and tissue-level processes while at the same time being computationally tractable; and (2) because of its neural net dynamics, its adjustable parameters can be trained on experimental data or optimized to produce other desired behaviors. However, as with other models using parameters fitted on experimental data, the available regression methods (such as optimization through simulated annealing, used in the neurogenesis model presented here) are often inadequate for fitting the parameters of gene network models; this could limit the scalability of such models.

Future Experimental Studies

Neurogenesis, which we have examined in this chapter using our gene network model, encompasses general questions of cell differentiation in epithelia and tissue-level dynamic interactions, which are relatively simple to formulate and yet complex enough to be of theoretical and experimental interest since such processes occur repeatedly in metazoan development. We have been able to extract predictions about the dynamics of proneural cluster resolution and draw conclusions about mechanisms that may be sufficient or necessary for neuroblast and sensory organ precursor differentiation. We have also investigated how robustly various genotypes lead to cluster resolution under various perturbations. The predictions of the model and the model's robustness under perturbations can be tested experimentally.

The kinds of questions that can be posed with the model described here are not only of relevance to

neurogenesis in *Drosophila* but are common to many developing organisms, especially in view of the fact that homologs to genes involved in *Drosophila* neurogenesis have been isolated in many species, from worms to mammals, and participate in a variety of developmental processes. In vertebrate neurogenesis, such homologs act in ways similar to those of the *Drosophila* genes in regulating the number of neurons generated (Chitnis et al., 1995; Lewis, 1996). One would therefore expect that a theoretical and empirical understanding of *Drosophila* neurogenesis would provide insights into neurogenesis in higher vertebrates, for example, into questions surrounding neuronal proliferation in the developing mammalian cortex (Caviness et al., 1995; Rakic, 1995).

Future Modeling Studies

In view of this, it would be interesting to see the neurogenesis model presented here extended to mammalian neurogenesis now that molecular data are becoming increasingly available in this area also.

Considering ways to move ahead with developmental modeling approaches more generally, one has to observe that the models described in this review have been primarily concerned with the dynamics of development, i.e., how cells develop in space and time. Such models can, of course, be constrained by the large amounts of genomic data becoming available, but still, they cannot incorporate all the available data without becoming intractable. A good portion of these data is of a qualitative nature and therefore cannot be readily mapped to precise interaction strengths between the components of a model. Other methods are needed to achieve a more comprehensive representation of biological knowledge.

Researchers have recently experimented with graphical models and pathway-model databases. Graphical models (Gifford, 2001) attempt to systematically describe relationships (edges) between elements (nodes) in biological systems, bringing together data on mRNA expression, protein interactions, environmental conditions, etc. These descriptions are probabilistic, with probabilities conditioned on existing data. They allow inferences from these data and point to areas where more data are needed. Pathway-model databases (Karp, 2001) describe metabolic and gene-regulatory networks, enzymes, and other proteins and try to present a more global picture of many interacting processes within an organism. They are based on an ontology, which is a database structure or schema that captures important features of the underlying system and precisely defines their relationships, and include theories about how the organism works that can be derived from the ontology. For such databases to be able to extract knowledge from large data sets, methods for efficiently generating mathematical models, storing them in the databases, comparing them with each other, and validating them against existing data will also be needed. In connection with this last point, see, for instance, Cellerator (Shapiro and Mjolsness, 2001), a program that allows users to specify a set of biochemical reactions, group them in a hierarchical graph structure that corresponds to the process modeled, and translate them into ordinary differential equations and solve them.

Both graphical models and pathway-model databases appear to be attractive vehicles for storing the more detailed kinds of dynamical models presented in this review, for comparing them with each other and with available data, and for assessing how well they fit in the more global ontology of an organism.

References

Ackers, G. K., Johnson, A. D., and Shea, M. A. (1982). Quantitative model for gene regulation by λ phage repressor. *Proc. Natl. Acad. Sci. U.S.A.* 79: 1129–1133.

Agarwal, P. (1995). Cellular segregation and engulfment simulations using the cell programming language. *J. Theor. Biol.* 176: 79–89.

Akam, M. (1989). Hox and HOM: Homologous gene clusters in insects and vertebrates. *Cell* 57: 347–349.

Artavanis-Tsakonas, S., Matsuno, K., and Fortini, M. E. (1995). Notch signaling. *Science* 268: 225–232.

Basler, K., and Struhl, G. (1994). Compartment boundaries and the control of *Drosophila* limb pattern by Hedgehog protein. *Nature* 368: 208–214.

Beato, M., Herrlich, P., and Schütz, G. (1995). Steroid hormone receptors: Many actors in search of a plot. *Cell* 83: 851–857.

Boguski, M. S., and McCormick, F. (1994). Proteins regulating Ras and its relatives. *Nature* 366: 643–654.

Bray, D. (1990). Intracellular signalling as a parallel distributed process. *J. Theor. Biol.* 143: 215–231.

Bray, D. (1995). Protein molecules as computational elements in living cells. *Nature* 376: 307–312.

Bray, D., and Bourret, R. B. (1995). Computer analysis of the binding reactions leading to a transmembrane receptor-linked multiprotein complex involved in bacterial chemotaxis. *Mol. Biol. Cell* 6: 1367–1380.

Bray, D., and Lay, S. (1994). Computer simulated evolution of a network of cell-signaling molecules. *Biophys. J.* 66: 972–977.

Bray, D., Bourret, R. B., and Simon, M. I. (1993). Computer simulation of the phosphorylation cascade controlling bacterial chemotaxis. *Mol. Biol. Cell* 4: 469–482.

Campuzano, S., and Modollel, J. (1992). Patterning of the *Drosophila* nervous system—the *achaete-scute* gene complex. *Trends Genet.* 8: 202–208.

Carroll, S. B. (1995). Homeotic genes and the evolution of arthropods and chordates. *Nature* 376: 479–485.

Caviness, V. S., Takahashi, T., and Nowakowski, R. S. (1995). Numbers, time and neocortical neuronogenesis: A general developmental and evolutionary model. *Trends Neurosci.* 18: 379–383.

Chitnis, A., Henrique, D., Lewis, J., Ish-Horowicz, D., and Kintner, C. (1995). Primary neurogenesis in *Xenopus* embryos regulated by a homologue of the *Drosophila* neurogenic gene *delta*. *Nature* 375: 761–766.

Cubas, P., De Celis, J.-F., Campuzano, S., and Modolell, J. (1991). Proneural clusters of *achaete-scute* expression and the generation of sensory organs in the *Drosophila* imaginal wing disc. *Genes Dev.* 5: 996–1008.

Diaz-Benjumea, F. J., Cohen, B., and Cohen, S. M. (1994). Cell interaction between compartments establishes the proximal-distal axis of *Drosophila* legs. *Nature* 372: 175–179.

Dupont, G., and Goldbeter, A. (1992). Oscillations and waves of cytosolic calcium: Insights from theoretical models. *BioEssays* 14: 485–493.

Fietz, M. J., Concordet, J.-P., Barbosa, R., Johnson, R., Krauss, S., McMahon, A. P., Tabin, C., and Ingham, P. W. (1994). The *hedgehog* gene family in *Drosophila* and vertebrate development. *Development*. Suppl. 43–51.

Fisher, R. A. (1930). *The Genetical Theory of Natural Selection.* Oxford: Clarendon Press.

Fleischer, K., and Barr, A. H. (1994) A simulation testbed for the study of multicellular development: The multiple mechanisms of morphogenesis. In *Artificial Life III: Proceedings of the Workshop on Artificial Life*, C. G. Langton, ed. Reading, Mass.: pp. 389–416. Addison-Wesley.

Fonseca, C. M., and Fleming, P. J. (1995). An overview of evolutionary algorithms in multiobjective optimization. *Evolut. Comput.* 3: 1–16.

Gierer, A., and Meinhardt, H. (1972). A theory of biological pattern formation. *Kybernetik* 12: 30–39.

Gifford, D. K. (2001). Blazing pathways through genetic mountains. *Science* 293: 2049–2051.

Gilbert, S. F. (2000). *Developmental Biology*, 6th ed. Sunderland, Mass.: Sinauer.

Goodwin, B. C. (1965). Oscillatory behaviour in enzymatic control systems. *Adv. Enzyme Reg.* 5: 425–438.

Graner, F., and Sawada, Y. (1993). Can surface adhesion drive cell rearrangement? Part II: A geometrical model. *J. Theor. Biol.* 164: 477–506.

Halfon, M., Kose, H., Chiba, A., and Keshishian, H. (1997). Targeted gene expression without a tissue-specific pro-

moter: Creating mosaic embryos using laser-induced single-cell heat shock. *Proc. Natl. Acad. Sci. U.S.A.* 94: 6255–6260.

Hastings, S. P., Tyson, J. J., and Webster, D. (1977). Existence of periodic solutions for negative feedback control systems. *J. Diff. Eqs.* 25: 39–64.

Hertz, J. A., Palmer, R. G., and Krogh, A. S. (1991). *Introduction to the Theory of Neural Computation.* Reading, Mass.: Addison-Wesley.

Hjelmfelt, A., and Ross, J. (1992). Chemical implementation and thermodynamics of collective neural networks. *Proc. Natl. Acad. Sci. U.S.A.* 89: 388–391.

Hjelmfelt, A., Weinberger, E. D., and Ross, J. (1991). Chemical implementation of neural networks and Turing machines. *Proc. Natl. Acad. Sci. U.S.A.* 88: 10983–10987.

Hodgkin, A. L., and Huxley, A. F. (1952). A quantitative description of membrane current and its application to conduction and excitation in nerve. *J. Physiol. (London)* 117: 500–544.

Hopfield, J. J. (1984). Neurons with graded response have collective computational properties like those of two-state neurons. *Proc. Natl. Acad. Sci. U.S.A.* 81: 3088–3092.

Huang, F., Dambly-Chaudiere, C., and Ghysen, A. (1991). The emergence of sensory organs in the wing disc of *Drosophila*. *Development* 111: 1087–1095.

Hunding, A. (1974). Limit-cycles in enzyme systems with nonlinear negative feedback. *Biophys. Struct. Mech.* 1: 47–54.

Ingham, P. W. (1988). The molecular genetics of embryonic pattern formation in *Drosophila*. *Nature* 335: 25–34.

Jennings, B., Preiss, A., Delidakis, C., and Bray, S. (1994). The Notch signalling pathway is required for *Enhancer-of-split* bHLH protein expression during neurogenesis in the *Drosophila* embryo. *Development* 120: 3537–3548.

Jonsson, H., Mjolsness, E., and Shapiro, B. (2002). Resources and signaling in multicellular models of plant development. *Proceedings of the Third International Conference on Systems Biology* (ICSB 2002), Karolinska Institutet, p. 127.

Karp, P. D. (2001). Pathway databases: A case study in computational symbolic theories. *Science* 293: 2040–2044.

Kauffman, S. A. (1969). Metabolic stability and epigenesis in randomly connected nets. *J. Theor. Biol.* 22: 437–467.

Kauffman, S. A. (1971). Differentiation of malignant to benign cells. *J. Theor. Biol.* 31: 429–451.

Kauffman, S. A. (1974). The large-scale structure and dynamics of gene control circuits: An ensemble approach. *J. Theor. Biol.* 44: 167–190.

Kauffman, S. A. (1993). *The Origins of Order.* Oxford: Oxford University Press.

Kauffman, S. A., and Levin, S. (1987). Towards a general theory of adaptive walks on rugged landscapes. *J. Theor. Biol.* 128: 11–45.

Kerszberg, M., and Changeux, J. (1994). A model for reading morphogenetic gradients: Autocatalysis and competition at the gene level. *Proc. Natl. Acad. Sci. U.S.A.* 91: 5823–5827.

Krumlauf, R. (1994). *Hox* genes in vertebrate development. *Cell* 78: 191–201.

Lewis, E. B. (1978). A gene complex controlling segmentation in *Drosophila. Nature* 276: 565–570.

Lewis, J. (1996). Neurogenic genes and vertebrate neurogenesis. *Curr. Opin. Neurobiol.* 6: 3–10.

Lindenmayer, A. (1968). Mathematical models for cellular interaction in development, parts I and II. *J. Theor. Biol.* 18: 280–315.

Lindenmayer, A. (1984). Models for plant tissue development with cell division orientation regulated by preprophase bands of microtubules. *Differentiation* 26: 1–10.

Lindenmayer, A., and Rozenberg, G. (1979). Parallel generation of maps: Developmental systems for cell layers. In *Graph Grammars and Their Application to Computer Science; First International Workshop,* Lecture Notes in Computer Science, Vol. 73, V. Claus, H. Ehrig, and G. Rozenberg, eds. pp. 301–316. Berlin: Springer-Verlag.

Lotka, A. J. (1925). *Elements of Physical Biology.* Baltimore, Md.: Williams and Wilkins.

Marée, A. F. M., and Hogeweg, P. (2001). How amoeboids self-organize into a fruiting body: Multicellular coordination in *Dictyostelium discoideum. Proc. Natl. Acad. Sci. U.S.A.* 98: 3879–3883.

Marnellos, G. (1997). "Gene Network Models Applied to Questions in Development and Evolution." PhD thesis, Yale University, New Haven, Conn.

Marnellos, G., and Mjolsness, E. (1998a). A gene network approach to modeling early neurogenesis in *Drosophila.* In *Pacific Symposium on Biocomputing,* Vol. 3, pp. 30–41. Singapore: World Scientific.

Marnellos, G., and Mjolsness, E. (1998b). Probing the dynamics of cell differentiation in a model of *Drosophila* neurogenesis. In *Artificial Life VI, Proceedings of the Sixth International Conference on Artificial Life,* Vol. 6, pp. 161–170 Cambridge, Mass.: MIT Press.

Marnellos, G., Deblandre, G. A., Mjolsness, E., and Kintner, C. (2000). Delta-Notch lateral inhibitory patterning in the emergence of ciliated cells in *Xenopus*: Experimental observations and a gene-network model. In *Pacific Symposium on Biocomputing,* Vol. 5, pp. 329–340. Singapore: World Scientific.

McAdams, H. H., and Shapiro, L. (1995). Circuit simulation of genetic networks. *Science* 269: 650–656.

McGinnis, W., Garber, R. L., Wirz, A., Kuroiwa, A., and Gehring, W. (1984). A homologous protein-coding sequence in *Drosophila* homeotic genes and its conservation in other metazoans. *Cell* 37: 403–408.

Meinhardt, H. (1983). Cell determination boundaries as organizing regions for secondary embryonic fields. *Dev. Biol.* 96: 375–385.

Meinhardt, H. (1986). Hierarchical inductions of cell states: A model for segmentation in *Drosophila. J. Cell Sci.* Suppl. 4: 357–381.

Meinhardt, H. (1987). A model for pattern generation on the shells of molluscs. *J. Theor. Biol.* 126: 63–89.

Meinhardt, H. (1998). *The Algorithmic Beauty of Sea Shells.* Berlin: Springer-Verlag.

Mjolsness, E., Sharp, D. H., and Reinitz, J. (1991). A connectionist model of development. *J. Theor. Biol.* 152: 429–453.

Murray, J. D. (1981a). On pattern formation mechanisms for lepidopteran wing patterns and mammalian coat markings. *Phil. Trans. Roy. Soc. London B.* 295: 473–496.

Murray, J. D. (1981b). A pre-pattern formation mechanism for animal coat markings. *J. Theor. Biol.* 88: 161–199.

Murray, J. D. (1993). *Mathematical Biology.* Berlin: Springer-Verlag.

Murray, J. D., Oster, G. F., Harris, A. K. (1983). A mechanical model for mesenchymal morphogenesis. *J. Math. Biol.* 17: 125–129.

Muskavitch, M. A. T. (1994). Delta-Notch signalling and *Drosophila* cell fate choice. *Dev. Biol.* 166: 415–430.

Neiman, A. (1993). Conservation and reiteration of a kinase cascade. *Trends Genet.* 9: 390–394.

Novak, B., and Tyson, J. J. (1993). Modeling the cell division cycle: m-Phase trigger, oscillations and size control. *J. Theor. Biol.* 165: 101–134.

Nüsslein-Volhardt, C., and Wieschaus, E. (1980). Mutations affecting segment number and polarity in *Drosophila*. *Nature* 287: 795–801.

Odell, G. M., Oster, G., Alberch, P., and Burnside, B. (1981). The mechanical basis of morphogenesis. I. Epithelial folding and invagination. *Dev. Biol.* 85: 446–462.

Oster, G. F., Murray, J. D., and Harris, A. K. (1978). Mechanical aspects of mesenchymal morphogenesis. *J. Embryol. Exp. Morphol.* 78: 83–125.

Pearlmutter, B. A. (1989). Learning state space trajectories in recurrent neural networks. *Neur. Comput.* 1: 263–269.

Poulson, D. (1940). The effect of certain X-chromosome deficiencies on the embryonic development of *Drosophila melanogaster*. *J. Exp. Zool.* 83: 271–318.

Prusinkiewicz, P., and Lindenmayer, A. (1990). *The Algorithmic Beauty of Plants.* New York: Springer-Verlag.

Rakic, P. (1995). A small step for the cell, a giant leap for mankind: A hypothesis of neocortical expansion during evolution. *Trends Neurosci.* 18: 383–388.

Reinitz, J., and Sharp, D. H. (1995). Mechanism of *eve* stripe formation. *Mech. Dev.* 49: 133–158.

Reinitz, J., Mjolsness, E., and Sharp, D. H. (1995). Model for cooperative control of positional information in *Drosophila* by Bicoid and maternal Hunchback. *J. Exp. Zool.* 271: 47–56.

Savageau, M. A. (1976). *Biochemical Systems Analysis.* Reading, Mass.: Addison-Wesley.

Schummer, M., Scheurlen, I., Schaller, C., and Galliot, B. (1992). HOM/HOX homeobox genes are present in Hydra (*Chlorohydra viridissima*) and are differently expressed during regeneration. *EMBO J.* 11: 1815–1823.

Scott, M. P., and Weiner, A. J. (1984). Structural relationships among genes that control development: Sequence homology between the *Antennapedia*, *Ultrabithorax* and *fushi-tarazu* loci in *Drosophila*. *Proc. Natl. Acad. Sci. U.S.A.* 81: 4115–4119.

Segel, L. A. (1984). *Modeling Dynamic Phenomena in Molecular and Cellular Biology.* Cambridge: Cambridge University Press.

Shapiro, B., and Mjolsness, E. (2001). Developmental simulations with Cellerator. In *Second International Conference on Systems Biology (ICSB)*, pp. 342–351. California Institute of Technology.

Shea, M. A., and Ackers, G. K. (1985). The O_R control system of bacteriophage lambda: A physical-chemical model for gene regulation. *J. Mol. Biol.* 181: 211–230.

Simpson-Brose, M., Treisman, J., and Desplan, C. (1994). Synergy between the Hunchback and Bicoid morphogens is required for anterior patterning in *Drosophila*. *Cell* 78: 855–865.

Skeath, J. B., and Carroll, S. B. (1992). Regulation of proneural gene expression and cell fate during neuroblast segregation in the *Drosophila* embryo. *Development* 114: 939–946.

Somogyi, R., and Sniegoski, C. A. (1996). Modeling the complexity of genetic networks: Understanding multigenic and pleiotropic regulation. *Complexity* 1: 45–63.

Stekel, D., Rashbass, J., and Williams, E. D. (1995). A computer graphic simulation of squamous epithelium. *J. Theor. Biol.* 175: 283–293.

St. Johnston, D., and Nüsslein-Volhardt, C. (1992). The origin of pattern and polarity in the *Drosophila* embryo. *Cell* 68: 201–219.

Thom, R. (1972). *Stabilité Structurelle et Morphogenèse: Essai d'une Théorie Générale des Modèles.* Reading, Mass.: W. A. Benjamin.

Thummel, C. (1995). From embryogenesis to metamorphosis: The regulation and function of *Drosophila* nuclear receptor superfamily members. *Cell* 83: 871–877.

Turing, A. M. (1952). The chemical basis of morphogenesis. *Phil. Trans. Roy. Soc. London B.* 237: 37–72.

Tyson, J. J., and Othmer, H. G. (1978). The dynamics of feedback control circuits in biochemical pathways. *Prog. Theor. Biol.* 5: 1–62.

Volterra, V. (1926). Variazioni e fluttuazioni del numero d'individui in specie animali conviventi. In *Memorie (Accademia Nazionale dei Lincei; Classe di Scienze Fisiche, Matematiche e Naturali)*, Ser. 6, Vol. 2, pp. 31–113.

Von Dassow, G., Meir, E., Munro, E. M., and Odell, G. M. (2000). The segment polarity network is a robust developmental module. *Nature* 406: 188–192.

Weliky, M., and Oster, G. (1990). The mechanical basis of cell rearrangement: I. Epithelial morphogenesis during *Fundulus* epiboly. *Development* 109: 373–386.

Williams, R. J., and Zipser, D. (1989). A learning algorithm for continually running fully recurrent neural networks. *Neur. Comput.* 1: 270–280.

Early Dendritic and Axonal Morphogenesis

H. G. E. Hentschel and A. Fine

3

In this chapter we review the neurobiology of early dendritic and axonal morphogenesis and the underlying physical mechanisms that may control the growth and form of axons and dendrites. An increasing body of experimental and modeling work suggests that several physicochemical instabilities may underlie the generation of dendritic and axonal forms. Lamellipodia formation and the early stages of dendritic arborization may be the consequence of an instability involving the control of growth by the local calcium concentration, while axonal differentiation may be due to an instability involving actin depolymerization and microtubule formation in the developing neuron. In this chapter we describe models of neurite formation and axonal differentiation that incorporate these instabilities and discuss their implications for neuronal development.

3.1 Neuronal Morphogenesis as Pattern Formation

The diversity of neuronal morphologies has been recognized since the time of Ramon y Cajal (1911; see also chapter 4). The mechanisms determining neuronal morphology, however, remain obscure. From the observation that many neurons can recreate their characteristic dendritic branching patterns when grown in cell culture, we may infer that dendritic morphology is genetically determined. At the same time, however, dendritic growth and the resulting morphology can be influenced by various extrinsic (epigenetic) factors, such as external electric fields

(Jaffe and Poo, 1979), glial- or target-derived substances (Tessier-Lavigne et al., 1988), electrical activity (Schilling et al., 1991), neurotransmitters (Mattson, 1988), cyclic adenosine monophosphate (cAMP) (Lohof et al., 1992), and neurotrophins (Ruit and Snider, 1991).

Despite their diversity, many neurons share common aspects of growth and form. Cells, typically, exhibit a set of well-defined growth stages involving the development of both axons and dendrites (see figure 3.1A). The initially spherical cell begins to grow by projecting many broad, short, wavelike extensions called lamellipodia. This period, typically lasting a few hours, has been referred to by Dotti et al. (1988) as stage 1 of outgrowth. The lamellipodia subsequently condense into a number of small neurites of approximately equal length, which undergo a period (roughly half a day) of growth and contraction (stage 2). Eventually (typically by the second day), one of the neurites rapidly increases its growth rate, becoming differentiated as the axon (stage 3), while the growth rates of the other neurites are profoundly reduced. Over the next several days, the remaining neurites begin to grow again and acquire the full dendritic characteristics (stage 4). Finally, the neuron develops into its mature form (stage 5). In this chapter we concentrate on modeling the early stages (stages 1–3) of neuronal development.

What are the biological mechanisms underlying neuronal development? One possibility is that a detailed specification of the total structure of individual neurons is genetically encoded. A problem with this suggestion, however, is that too many genes

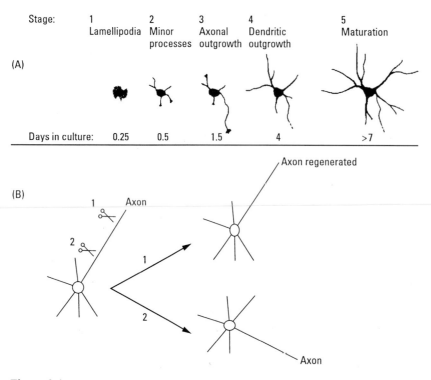

Figure 3.1

(*A*) A schematic diagram of stages 1–5 in the development of hippocampal neurons in vitro. See the text for details. (*B*) A schematic diagram showing axon regeneration after axotomy. If the axon is transected so that its remaining length is still greater than that of the other neurites, it regrows as the axon (1). If after transection the axon is no longer the longest neurite, the longest remaining neurite becomes the axon (2). (Redrawn from Andersen and Bi, 2000.)

would be required. An alternative possibility is that only rules are encoded. At first sight, this possibility appears attractive. The amount of information that would have to be stored is very small compared with detailed encoding, and in addition, evidence for a degree of self-similarity in the structure of some dendritic branching patterns (Montague and Friedlander, 1989; Caserta et al., 1990; Takeda et al., 1992) suggests that underlying the observed shape there possibly is a recursive algorithm, repeating some simple rule at several scales (e.g., Pellionisz, 1989; see also chapter 4). A problem with rule encoding, however, is its

arbitrariness; there is no underlying justification for the choice of one set of rules over another except for a posteriori agreement with observed structures. An interesting alternative is that a large fraction of the observed variation among classes of cells is controlled by epigenetic factors such as variations in their local environments (see McAdams and Arkin, 1997). In reality, we would expect both genetic and epigenetic factors to play a role in neuronal development.

As first suggested by Turing (1952), much of neuronal morphogenesis may result from physicochemical self-organization arising naturally from reaction-diffusion processes in the cell. Turing (1952) showed that in domains of a simple geometry containing reaction-diffusion processes, the homogeneous state can be unstable to stripe formation. With different initial and boundary conditions, more complex spatiotemporal patterns can be generated, such as evolving and colliding whorl-like patterns (see also Gierer and Meinhardt, 1972; Meinhardt and Gierer, 2000). These are examples of self-organization in the sense that the specific pattern has much more structure than the reaction-diffusion equations that created it.

Such reaction-diffusion processes have also been applied to cellular morphogenesis by coupling the resulting spatiotemporal patterns of a morphogen with membrane growth. For example, this approach has been used to explain the whorls of hair formation at the growth tips of single-celled algae such as *Acetabularia* (Goodwin and Trainor, 1988), and an explicit morphogen has been suggested in membrane-bound calcium (Harrison et al., 1988). The free boundary problem in which the pattern depends on the shape of the enveloping membrane (through its boundary conditions) while at the same time the shape of the membrane changes in response to this pattern is a highly complex nonlinear problem requiring the kinematics of curved surface evolution. Such methods have been applied to algal tip growth by Pelce and Pocheau (1992), who assumed the elongation rate of the cell wall to be proportional to the local osmotic pressure. Mechanical strain fields can also affect cellular morphogenesis, and their influence has been studied by several groups (Odell et al., 1981; Goodwin and Trainor, 1988).

A dominant aspect of neuronal form is its dendritic arborization (see chapter 4). An important class of physical processes that can generate such dendritic structures are regulated by diffusion (Witten and Sander, 1981) and based on the Mullins-Sekerka in-

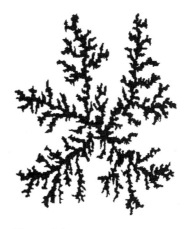

Figure 3.2
A typical dendritic structure grown by diffusion-limited aggregation (DLA). The branching was allowed to continue to approximately the same degree as a neuronal arbor. If allowed to proceed indefinitely, a fractal structure emerges with a fractal dimension $d_f \approx 1.72$.

stability (Mullins and Sekerka, 1963). Qualitatively, the basic principle is that morphogen diffusion gives rise to spatial gradients in morphogen concentration at curved interfaces. These concentration gradients increase at the tips of curved surfaces. If the interface then grows at a rate proportional to this gradient, a positive feedback loop is created in which the tips of curved interfaces grow at the greatest rates, giving rise to even larger gradients. It has been found that numerous different physical processes controlled by the Mullins-Sekerka instability all give rise to similar dendritic patterns and are in the same universality class as diffusion-limited aggregation (DLA) (see figure 3.2). They generate fractal structures at large scales, with a fractal dimension of $d_f = 1.718\ldots$ in two dimensions. The dendritic arbors of neurons also display statistical self-similarity (although only over a small range of scales). For example, fractal dimensions d_f close to those associated with two-dimensional DLA have been found for a number of approximately flat neuronal arbors, e.g., for cat retinal ganglion cell dendrites (Caserta et al., 1990; Montague and Friedlander, 1989), $d_f \approx 1.68-1.73$; and for mouse cerebellar Purkinje cell dendrites (Takeda et al., 1992), $d_f \approx 1.71$. Is this a coincidence or does this observation provide a clue to how neurons actually grow? We argue that the observation is not coincidental and that calcium ion may be the key morphogen (Hentschel and Fine, 1994, 1996).

Of course, even if diffusion-limited processes control aspects of dendritic arborization, many other dynamic mechanisms may be involved in neuronal growth. Not only can retraction and growth occur during dendritic morphogenesis—a feature absent in most diffusion-limited growth processes—but a crucial feature of neuronal development is axonal differentiation, or neuronal polarization (Dotti and Banker, 1987; Goslin and Banker, 1989). Generally, the longest neurite becomes the axon. Moreover (see figure 3.1B), if the developing axon is transected so that its remaining length is still greater than that of the other neurites, it will usually regrow as the axon; however, if it is transected close to the soma, a change in polarity usually occurs, with the longest remaining neurite becoming the new axon (Dotti and Banker, 1987; Craig and Banker, 1994). Positive and negative feedback loops leading to dynamic instability may underlie such behavior (Samuels et al., 1996; Anderson and Bi, 2000). Neuronal polarization may result from competition for a diffusible, anterogradely transported "determinant chemical" (produced in the soma) whose concentration at the neurite tip affects the rate of outgrowth. In section 3.3.3, we show that active transport mechanisms coupled to diffusion inside the neurites may indeed lead to such a "winner-take-all" instability in growth (Samuels et al., 1996) and thus to neuronal polarization.

3.2 Neurobiological Background

In this section, we summarize some key phenomena of early dendritic and axonal morphogenesis.

3.2.1 Influence of Calcium on Neuronal Outgrowth

Many factors have been implicated in the appearance of lamellipodia and filopodia and their subsequent transformation into branched neurites. These factors include neurotransmitters, cell adhesion molecules (CAMs), and growth factors. An important common link, however, is that they may work, at least partially, by changing local internal calcium concentrations within the neuron (Kater et al., 1988; Cohan et al., 1987; Mattson et al., 1990; Grinvald and Farber, 1981).

Electrical stimulation and gradients of various chemicals can change the neuron's internal concentration of calcium via local calcium influx through voltage- or ligand-gated calcium channels (Lohof et al., 1992; Zheng et al., 1994; Grinvald and Farber, 1981) and further augmentation by calcium-induced calcium release from internal stores (Holliday et al., 1991); in consequence they can modulate the growth and form of neurons. The modulation of neurite outgrowth by calcium in growth cones appears to be bell shaped (Kater et al., 1988). Outgrowth is enhanced by increased calcium concentration up to an optimum, whereas very large elevations of calcium can suppress outgrowth (Cohan et al., 1987; see also chapter 6). Accordingly, preventing such large elevations in calcium by a blockade of electrical activity (Schilling et al., 1991) or depletion of internal stores (Holliday et al., 1991) can increase outgrowth.

In accordance with the bell-shaped modulation of outgrowth by calcium, imaging studies have demonstrated that calcium levels are higher in actively growing growth cones (but not quiescent ones) than in the cell body (Kater et al., 1988; Cohan et al., 1987). Asymmetrical elevations in calcium on one side of a growth cone may precede turning toward that side (Zheng et al., 1994). Focal elevations in internal calcium concentration (Meberg et al., 1999; Lau et al., 1999) lead to spatially restricted formation of filopodia, which in turn give rise to dendritic or axonal branches. The precise molecular mechanisms by which calcium exerts these effects on neurite growth are not known, but are likely to involve calcium-regulated processes such as cytoskeletal actin dynamics and the exocytotic addition of plasma membrane (Kater et al., 1988; Forscher, 1989).

3.2.2 Influence of Other Signaling Molecules on Neuronal Outgrowth

Neuronal outgrowth can be influenced by molecules secreted by other cells, such as nerve growth factor (NGF) (e.g., Levi-Montalcini, 1976; Campenot, 1977; Ruit and Snider, 1991; see also chapter 10), semaphorins, netrins, and Slit (see chapter 5). These molecules may be diffusible or bound to the extracellular matrix. They may function via changes in levels of second messengers, including calcium and cyclic nucleotides, or by modulating the activity of enzymes (Song and Poo, 1999; Polleux et al., 2000). Nerve growth factor, in particular, may exert its effects via modulation of guanosine triphosphatases (GTPases) of the Rho family (Davies, 2000; Li et al., 2000). These enzymes influence cellular polarity and lamellipodia and filopodia formation (Bradke and Dotti, 1999; Nobes and Hall, 1999; see also chapter 4), at least in part by regulating cytoskeletal actin (Tapon and Hall, 1997). Among the diverse targets of these GTPases are several phosphatidyl inositol kinases (Kaibuchi et al., 1999), which may mediate some of these effects. In addition, through its influence on membrane

cycling (Stenmark, 2000) and modulation by growth-associated protein 43 (GAP-43) and related protein kinase C substrates (Laux et al., 2000), phosphatidyl inositol phosphate metabolism may complement calcium in constituting a common pathway uniting a variety of outgrowth-regulating pathways.

3.2.3 Axonal Differentiation

The onset of rapid growth in the presumptive axon is the first known sign of neuronal polarization (see figure 3.1A). It is preceded in stage 2 by the enlargement and increased dynamics of growth cones (Bradke and Dotti, 1997) and an increased intracellular vesicular and microtubule transport (Bass, 1999); it is followed by a slowing of the other neurites' growth (Esch et al., 1999). Also at this stage, certain cell constituents begin to be segregated to either the axonal or the somato-dendritic compartments (Craig and Banker, 1994). For example, only microtubules with plus ends oriented distally to the cell body are found in axons (Burton and Paige, 1981), whereas in dendrites comparable numbers of plus and minus ends occur distally (Burton, 1988). This is not the result of the exclusion of minus-end distal microtubules from axons; rather, the appearance of minus-end distal microtubules in dendrites requires the presence of a specific motor protein, CHO1/MKLP (Sharp et al., 1997). In addition, GAP-43 and microtubule-associated protein (MAP) tau are conspicuous components of axons but not of dendrites (Goslin et al., 1988), whereas MAP2 is largely restricted to dendrites (Matus, 1988). The mechanism of their sorting is unknown.

Axotomy experiments (see figure 3.1B) have shown that all neurites have the potential to develop into an axon (Dotti and Banker, 1987; Goslin and Banker, 1989) when the axon is cut during developmental stages 2 and 3. The origin of the new axon is determined primarily by the length of the neurites, with the longest one most often becoming the axon (Goslin and Banker, 1989). If the two longest neurites are roughly of equal length, then a latency period occurs before a decision is made as to which neurite will form the new axon. It appears, therefore, that identical growth processes occur in every neurite and that an instability occurs that accelerates the growth rate of a single neurite, favoring the longest and inhibiting the growth of all others. External cues, such as contact with a novel permissive substrate, can influence this process (Esch et al., 1999).

Several mechanisms have been suggested for axonal differentiation. One possibility is that preexisting internal cellular asymmetries bias one neurite to become the axon. Dotti and Banker (1991), however, found no evidence of correlations between site of axon initiation and asymmetry or position of various cell organelles. Certain molecules, such as the membrane-associated protein GAP-43 (Goslin et al., 1988) and the microtubule-associated protein tau (Binder et al., 1984), are segregated to the axon soon after the axon can first be identified on morphological grounds; on this basis, it has been suggested that these molecules could determine the axon. While such segregated molecules may participate in the formation, stabilization, or function of the axon, the mechanisms responsible for their segregation—and thus for the uniqueness of the axon—remain unknown. Thus, such "prepattern" hypotheses do not provide a full account of neuronal polarization. Moreover, prepattern mechanisms are inconsistent with the observed outcome of axotomy experiments, as described earlier.

To avoid this inconsistency, Caceres and Kosik (1990) hypothesized that a phosphatase-mediated cooperative dephosphorylation of microtubule-associated proteins (e.g., tau), stochastically acting within each neuron to increase microtubule stability and elongation, could account for the emergence of the fastest growing neurite as the axon. Evidence for a distally increasing gradient in the ratio of dephosphorylated to phosphorylated tau has recently been obtained (Rebhan et al., 1995). Such a mechanism may indeed participate in axon differentiation, but cannot by itself account for the inhibition of the growth of the other neurites.

Another mechanism that may be involved in neuronal polarization is actin depolymerization at the growth cone (Bradke and Dotti, 1999). When randomly selected neurites of hippocampal neurons at stage 2 were perfused for 15–30 min with cytochalasin D (an actin polymerization inhibitor), the perfused neurite formed the axon within a day (Bradke and Dotti, 1999; Forscher and Smith, 1988). It is not certain how actin depolarization results in axon formation, but it may involve the exposure of more barbed ends of actin filaments, thus ultimately increasing net actin polymerization or leading to enhanced microtubule formation, which results in neurite extension (Waterman-Storer and Salmon, 1999).

3.2.4 *Active Transport and Diffusion in Axonal Differentiation*

Axon formation must involve enhanced transport of vesicles and microtubule elements, and may also involve reduced transport of other elements. Transport of such "determining factors" must occur either by active mechanisms or by diffusion. In addition, after selective delivery to the axon, some membrane proteins appear to be retained by association with cytoskeletal elements, constituting a diffusion barrier at the axon's initial segment (Dotti and Simons, 1990; Winckler et al., 1999). Such polarization may depend on complexes between scaffold proteins such as protease-activated receptor (PAR) 3 and PAR-6, which can bind to, and modulate, enzymes such as

Rho GTPases and protein kinase C, that are involved in cell polarity (Lin et al., 2000).

Evidence of intracellular sorting during axonal differentiation suggests that active transport is involved. During axonal differentiation, cytoplasmic flow into the incipient axon is associated with morphological changes, including the development of a larger growth cone and the accumulation of organelles such as vesicles and ribosomes (Bradke and Dotti, 1997).

Active transport of substances along neurites is analogous to transport along a series of highways. The highways themselves are formed by microtubules, and transport can occur both from the soma to the growth cones (anterograde transport) and from the growth cone to the soma (retrograde transport). Such active transport is mediated by motor proteins such as dynein and kinesin (Vallee and Bloom, 1991), which, utilizing adenosine triphosphate (ATP) as an energy source, move along the microtubules, transporting vesicles and other organelles to which they bind. One place where such active transport against diffusive gradients is likely to have a great influence is in the development of the axon, which can in principle grow many centimeters or even meters in length.

Proteins competing unsuccessfully with axonally segregated substances for cytoskeletal binding would be excluded from the axon. Tau proteins segregate to the axon, and it has been suggested that they could displace MAP2 from axonal microtubular binding sites (Schoenfeld and Obar, 1994). Also, if substances that selectively cap microtubule minus ends, preventing their elongation, segregate to the axon, this could account for the differential orientation of microtubules in axons and dendrites (see section 3.2.3). Axons thus contain only (the faster growing) plus-end distal microtubules. In dendrites, in the absence of such selective caps, the initial preponderance of the faster growing plus-end distal microtubules would

in time equilibrate with slower growing minus-end distal microtubules, so that both orientations occur. Cellular elements would thus be transported from the soma along minus-end distal microtubules (e.g., Stebbings and Hunt, 1983) into the dendrites and would be excluded from the axon if they selectively bound dynein (which is responsible for transport toward the minus end) rather than kinesin (which is responsible for transport toward the plus end).

Why could active transport have an important influence on axonal differentiation? The rate of anterograde transport could be coupled to growth rate by at least two mechanisms. The growing tip exerts tension on the neurite (Lamoureux et al., 1989), and increased growth may lead hydrostatically to increased flow of cytosol into the proximal neurite. This increased proximal inflow will allow increased loading, and thus transport, of a chemical that determines axonal differentiation. Alternatively, the capacity or velocity of the transport mechanism could be increased by the growth-associated increase in neurite tension if the increased strain leads to conformational changes (Heidemann and Buxbaum, 1994) in transport proteins.

Whereas attention has generally focused on special characteristics of the nascent axon, complementary mechanisms for neuronal polarization may involve special characteristics of the nascent dendrites. Depletion of the motor protein CHO1/MKLP1, required for transport of minus-end distal microtubules into dendrites, causes the dendrites to become longer, thinner, and, ultimately, indistinguishable from axons (Yu et al., 2000).

Finally, diffusion is also likely to be of major importance because it may be the source of inhibiting signals during neuronal polarization. Diffusion of determinant chemical back from the neurite tips opposes active anterograde transport, and may be responsible for the slowing of growth in "losing" neurites as their tips come into equilibrium with the soma.

3.3 Models of Early Dendritic and Axonal Morphogenesis

The neurobiology described in the preceding sections provides a framework in which models for the early stages of neuronal morphological development need to fit. In this section, we describe possible mechanisms by which calcium (and other signaling molecules), active transport and diffusion may give rise to dynamic mechanisms underlying the emergence of dendritic form and axonal differentiation. Models that study the implications of calcium (or electrical activity)-dependent neurite outgrowth for a later developmental stage (namely, the development of networks of synaptically connected cells) are described in chapters 6 and 7.

3.3.1 Stage 1 Growth: The Development of Lamellipodia

How does a spherical neuron begin to develop lamellipodia, the wavelike protrusions of its membrane? If the internal structure of the cell also had spherical symmetry, some sort of symmetry-breaking instability would have to be involved. In reality, of course, even if the cell appears spherical on a coarse scale, its internal structure is already highly asymmetrical in the distribution of organelles and proteins. Thus, even at the earliest stages of development, asymmetrical signals could exist that initiate stage 1 of development. Nevertheless, it is reasonable to ask whether such asymmetry is a necessary precondition for stage 1 of development to occur. The answer is no, and by the time a cell reaches a critical size R_{crit}, lamellipodia are likely to appear as a result of a dynamic instability

even in a truly spherical cell in the presence of a diffusive morphogen such as calcium.

The development of a cell whose growth depends upon the local concentration of calcium near the inner surface of the cell membrane, $[Ca^{2+}]_{in}$, (in a growing, complex-shaped cell, this concentration varies with both time and position) can be modeled by solving the diffusion equation for the calcium concentration (in the presence of voltage-dependent ionic fluxes through the membrane) and by taking the growth rate $V([Ca^{2+}]_{in})$ to be locally determined according to the empirically observed bell-shaped dependence of growth upon calcium (see section 3.2.1). For an initially spherical cell, such growth can be shown analytically (Hentschel and Fine, 1994) to be unstable, i.e., to generate lamellipodia rather than a smoothly advancing circular border. This growth exhibits important differences in cells with excitable versus passive membranes, particularly with regard to the dependence of dendritic morphogenesis on cell size $R_0(t)$ and calcium diffusivity D_{Ca}. For cells with excitable membranes, instability results mainly from a positive feedback between calcium influx and submembrane calcium concentration. This instability will be enhanced as the cell increases in size, and also by any reduction in internal morphogen diffusivity (Hentschel and Fine, 1994). In contrast, for a passive membrane—where the unstable growth is mainly due to membrane convexities, generating regions of higher intracellular calcium concentration—the instability will be strengthened by a decrease in cell size and by an increase in calcium diffusivity.

To understand these results intuitively, consider a spherical cell from which a dendrite protrudes spontaneously. Since the calcium concentration outside the cell is much higher than that inside the cell, an inward flux through the protrusion into the cell will occur. The magnitude of this flux will depend both on the geometric properties of the protruding dendrite and on whether the membrane is passive or excitable. The internal calcium concentration, and especially the submembrane calcium concentration $[\text{Ca}^{2+}]_{\text{in}}$, will be affected by this influx. The spontaneous protrusion is likely to be unstable if $[\text{Ca}^{2+}]_{\text{in}}$ in this protrusion increases sufficiently above its submembrane value elsewhere in the soma to cause increased growth and consequently an even more dominant protusion. From the flux boundary condition at the membrane surface, $-D_{\text{Ca}}\vec{\nabla}[\text{Ca}^{2+}]_{\text{in}} = \vec{J}_{\text{Ca}}$, we can estimate that for spherical cells with a passive membrane of permeability κ, the local submembrane calcium concentration scales as

$$[\text{Ca}^{2+}]_{\text{in}} \sim \frac{\kappa[\text{Ca}^{2+}]_{\text{out}}}{D_{\text{Ca}}/R + \kappa}, \qquad (3.1)$$

where R is the cell radius. From Eq. (3.1), we see that for small cells $R \ll D_{\text{Ca}}/\kappa$, protrusions are likely to lead to significant variations in the submembrane calcium concentration with variations in curvature, while the actual magnitude of the submembrane concentration will be greater in larger cells $R \gg D_{\text{Ca}}/\kappa$. Thus we might expect that for passive membranes there will be large variations in the calcium concentration in small cells with large internal calcium diffusivities, encouraging lamellipodia formation; while as the actual magnitude of $[\text{Ca}^{2+}]_{\text{in}}$ is likely to be greater in larger cells, there is a greater likelihood that this will open channels in excitable membranes, leading to an even greater calcium influx and consequently to lamellipodia formation by this new positive feedback mechanism available only to active membranes.

These intuitive ideas can be developed in a more rigorous manner. The nature of the instability for cells with excitable membranes will depend upon the details of the membrane's ionic conductances. For example, for an excitable membrane whose conductivity is modulated mainly by the morphogen cal-

cium, growth instability increases with cell radius $R_0(t)$, rising rapidly when the cell grows beyond a size

$$R_{\text{crit}} \sim D_{\text{Ca}} \bigg/ \left(\frac{\partial J_{\text{Ca}}}{\partial [\text{Ca}^{2+}]_{\text{in}}} \right), \qquad (3.2)$$

where J_{Ca} is the calcium flux into the cell. Note that for a cell with a passive membrane, this positive feedback caused by channel opening with increased submembrane calcium, $\partial J_{\text{Ca}}/\partial [\text{Ca}^{2+}]_{\text{in}} > 0$, does not exist, and indeed $\partial J_{\text{Ca}}/\partial [\text{Ca}^{2+}]_{\text{in}} = -\kappa$ is negative.

If, alternatively, the membrane conductivities are set by a sodium- and potassium-dependent transmembrane potential ϕ, the instability increases rapidly as the radius approaches a value

$$R_{\text{crit}} \sim \frac{\frac{e}{kT}([\text{K}^+]_{\text{in}} + [\text{Na}^+]_{\text{in}})}{D_{\text{Na}}^{-1}\frac{\partial J_{\text{Na}}}{\partial \phi} + D_{\text{K}}^{-1}\frac{\partial J_{\text{K}}}{\partial \phi}}, \qquad (3.3)$$

where J_{Na} and J_{K} are, respectively, the sodium and potassium fluxes into the cell; D_{Na} and D_{K} are the internal sodium and potassium diffusivities; and $[\text{Na}^+]_{\text{in}}$ and $[\text{K}^+]_{\text{in}}$ are the internal sodium and potassium concentrations. The conditions for these kinetic instabilities resemble those leading to endogenous ionic currents through spherical cells (Pelce, 1993), and indeed the two phenomena appear to be related; if the ions behave as morphogens, lamellipodia will result.

Equations (3.2) and (3.3) come from a linear stability analysis of spherical cell growth (Hentschel and Fine, 1994, 1996). Linear stability analysis investigates whether infinitesimal geometric perturbations of the cell membrane vanish or grow with time. If we observe that the perturbations grow, then a spherical cell of radius $R_0(t)$ is unstable, and we can investigate what shape takes its place. Any observed instability will place an upper bound on the real cell size that can remain spherical, because any internal cell variations will only enhance the observed dynamic instability.

Consider a circular cell taken to grow as $dR_0/dt = V([Ca^{2+}]_{in}(t))$, where $V([Ca^{2+}]_{in}(t))$ is the rate of cell growth, which is dependent in a bell-shaped manner on $[Ca^{2+}]_{in}(t)$. In the presence of surface perturbations, the cell will ruffle and can be parametrized by its now variable radius $R(\theta, t)$, where θ is the angle. $R(\theta, t)$ has the Fourier expansion $R(\theta, t) = R_0(t) + \sum_{m=1}^{\infty} \delta_m(t) \cos(m\theta)$, where $R_0(t)$ is the radius of the growing unperturbed cell and $\delta_m(t)$ is the magnitude of a surface fluctuation, which varies with angle θ as $\cos(m\theta)$. Owing to this surface perturbation, the calcium flux through the membrane, and consequently the calcium concentration in the cell, will change. The submembrane calcium concentration $[Ca^{2+}]_{in}(\theta, t)$ will gain an angular dependence, and this spatial dependence will in turn modulate the growth, resulting in an equation of motion for the interface perturbations (obtained from linear stability analysis and valid for both excitable and passive membranes):

$$\frac{d\delta_m}{dt} = \delta_m \left[\frac{m^2(1 - m^2)\gamma}{R_0^4} + X_m \right], \qquad (3.4)$$

where γ is the membrane rigidity and

$$X_m \equiv \frac{\partial C}{\partial R_0} \frac{\partial V}{\partial [Ca^{2+}]_{in}}$$

$$\times \left\{ 1 + \frac{1 + (R_0/D_{Ca}) \dfrac{\partial J_{Ca}}{\partial [Ca^{2+}]_{in}}}{m - (R_0/D_{Ca}) \dfrac{\partial J_{Ca}}{\partial [Ca^{2+}]_{in}}} \right\}. \qquad (3.5)$$

In Eq. (3.5), $\partial C/\partial R_0 \equiv \partial C(r, t)/\partial r|_{r=R_0(t)}$, where $C(r, t)$ is the intracellular morphogen concentration at a point r and time t during unperturbed growth, while $[Ca^{2+}]_{in} \equiv C[R_0(t), t]$.

An analysis of Eqs. (3.4) and (3.5) suggests the possibility of unstable growth, i.e., $d\delta_m/dt > 0$; therefore, dendritic morphogenesis is in principle possible for both passive and excitable membranes provided that calcium stimulates growth, i.e., $\partial V/\partial [Ca^{2+}]_{in} > 0$. But we can see that excitable and passive membranes show interesting differences with regard to their dependence on cell size $R_0(t)$ and diffusivity D_{Ca}, which is in agreement with the intuitive arguments presented earlier.

For an excitable membrane, $\partial J_{Ca}/\partial [Ca^{2+}]_{in} > 0$, and the cell surface can in principle be unstable however small the cell is, provided membrane rigidity γ is weak. As the cell grows, the instability is enhanced and by the time the cell reaches a size $R_0(t) \approx D_{Ca}/(\partial J_{Ca}/\partial [Ca^{2+}]_{in})$, nonlinear terms need be added to Eq. (3.4) for a complete analysis; however, it is clear that the cell surface will ruffle, as suggested by Eq. (3.2). Note also that as the morphogen diffusivity D_{Ca} decreases, instability is enhanced for excitable membranes because the submembrane calcium is larger and therefore more efficient at opening calcium channels.

For a passive membrane, $\partial J_{Ca}/\partial [Ca^{2+}]_{in} = -\kappa$, where κ is the membrane permeability. In consequence, $X_m = (\partial C/\partial R_0)(\partial V/\partial [Ca^{2+}]_{in})(m + 1)/(m + \kappa R_0/D)$, and therefore, although unstable growth may still occur in this case, the instability will be strongly damped by membrane rigidity γ and will decrease with cell size $R_0(t)$. For a passive membrane, increases in the diffusivity D_{Ca} will enhance unstable growth.

3.3.2 Stage 2 Growth: Lamellipodia Condensation into Neurites

The strong conclusion from these analytical investigations is that stage 1 of development can be understood as a dynamic instability involving calcium as a morphogen. To investigate neurite formation in a biologically plausible manner, we describe here simulations of neuronal growth that incorporate the sodium field as the major modulator of the transmembrane potential. That is, the transmembrane potential is set mainly by the potassium field, but this resting potential is modulated mainly by variations in the submembrane sodium concentration. The calcium permeability is active in the sense that the calcium channels are voltage dependent; thus, calcium channels open in response to local depolarization, allowing an inward flux of calcium, which then acts as the morphogenic field.

We are interested here in the dynamics of growth on a time scale of hours—which is significantly larger than the diffusive time scale $T_{\text{diff}} \approx l^2/D_{\text{Ca}} \approx 10$ sec, where $l \approx 100$ μm is the neurite length scale and $D_{\text{Ca}} \approx 10^{-7}$ cm^2/sec is the calcium diffusion constant in the presence of buffering. Therefore, in our simulations we made the adiabatic approximation of solving Laplace's equation $\nabla^2 C = 0$ for quasi-equilibrium ionic concentrations, using relaxational methods rather than the full diffusion equation.

In our simulations (which are two-dimensional), the cell interior is divided into a number of discrete compartments corresponding to about 1 μm^2 of cytoplasm, while the surface is divided into segments about 1 μm in length. The relaxational procedure yielded the concentrations of sodium and morphogen (calcium) at each internal compartment, or pixel. The cell surface was treated as a discrete linked list of elements, referred to as membrane pixels, which contain information not only on transmembrane potential and ionic conductivities at each interfacial site s_i but also on the local submembrane concentration of relevant ions, especially calcium $C_i \equiv [\text{Ca}^{2+}]_{\text{in}}(s_i)$ and sodium $\text{Na}_i \equiv [\text{Na}^+]_{\text{in}}(s_i)$.

The solution of these relaxational equations for the ionic concentrations requires knowledge of the explicit boundary conditions. These boundary conditions are found by equating the flux just internal to the surface with the total flux by all mechanisms through the membrane. This flux boundary condition then fixes the submembrane ionic gradients. Thus, the total morphogen flux through the cell membrane at point s_i is the sum of its diffusive influx and active extrusion by pumping, i.e., $J(s_i) = J_{\text{diff}}(s_i) - J_{\text{pump}}(s_i)$. For each ionic species, we then have the boundary condition

$$-D\vec{\nabla} C_i = [J_{\text{diff}}(s_i) - J_{\text{pump}}(s_i)]\vec{n}(s_i), \tag{3.6}$$

where $\vec{n}(s_i)$ is the unit normal vector into the cell at point s_i on the surface. We also need explicit expressions for the diffusive and pumping fluxes. For example, we employed for the diffusive calcium flux,

$$J_{\text{Ca, diff}}[\phi(s_i)] = g_{\text{Ca}}[\phi(s_i)]\frac{\phi_{\text{Ca, Ne}}(s_i) - \phi(s_i)}{2e}, \tag{3.7}$$

where $g_{\text{Ca}}[\phi(s_i)]$ is the voltage-gated calcium conductivity at site s_i and $\phi_{\text{Ca, Ne}}(s_i) = (kT/2e)\log([\text{Ca}^{2+}]_{\text{out}}/C_i)$ is the local Nernst potential for calcium in terms of its charge $2e$ and concentration C_i inside and concentration $[\text{Ca}^{2+}]_{\text{out}}$ outside the cell. (We took the outside of the cell to be well mixed, an assumption that is in accord with the observation that normal development occurs in neurons grown in vitro in the presence of well-mixed media.)

To calculate the transmembrane potential employed in Eq. (3.7), we assume that the principal voltage-dependent ionic conductance modulating the potential is sodium. Thus, membrane potential is solved from the nonlinear approximation:

$$\phi(s_i) = \frac{g_{\text{leak}}\phi_{\text{rest}} + g_{\text{Na}}[\phi(s_i)]\phi_{\text{Na, Ne}}(s_i)}{g_{\text{leak}} + g_{\text{Na}}[\phi(s_i)]}, \tag{3.8}$$

where g_{leak} and ϕ_{rest} are, respectively, a lumped conductance and its associated Nernst resting potential, terms that represent the aggregate contribution of the other main permeant ions (e.g., K^+, Cl^-), which are assumed to have a constant spatial distribution in the cell. The Nernst potential for sodium, $\phi_{Na, Ne}(s_i)$, which is a function of the distribution of sodium, is determined by solving the diffusion and flux equations for sodium in the same manner as for calcium.

Active extrusion of sodium and calcium is assumed to occur via saturable mechanisms, i.e., to depend sigmoidally upon their internal concentrations near the membrane, with pumping negligible at low concentrations but above some concentration increasing significantly to a maximum (Blaustein, 1988). For example, for calcium we used

$$J_{Ca, pump}(C_i) = \frac{J_{Ca, max}}{1 + e^{(C_{crit} - C_i)/\Delta C}}, \qquad (3.9)$$

where ΔC is the width of this sigmoidal dependence. For $C_i \approx C_{crit}$, the pumping becomes significant; this occurs shortly after the voltage-gated channels open. The effect can be seen in simulations as a slowing down of the total inward flux at $C_i \approx 1000$ nM.

Membrane parameters (ionic conductivities, pump fluxes, etc.) were chosen as far as possible to be consistent with known biology (Blaustein, 1988). For example, the calcium pumping and diffusive fluxes were typically found to vary from 0.1 to 10 pmol/cm^2 sec at different points on the cell surface. With the external calcium concentration fixed at $[Ca^{2+}]_{out} = 1$ mM, the observed internal calcium concentrations were found to range from 100 nM to 400 nM.

We incorporated the empirically observed bell-shaped dependence of outgrowth upon calcium (see section 3.2.1) by parameterizing growth velocity normal to the surface as

$$V(s_i) = k_1 C_i^\alpha - k_2 C_i^\beta + W(t), \qquad (3.10)$$

where k_1 and k_2 are lumped rate constants for the growth and retraction mechanisms, respectively, and α and β reflect the cooperativity of the dependence on calcium. The global rate of cell growth is taken to be constant, reflecting a constant rate of synthesis of new cytoplasm; the constraining velocity term $W(t)$, so defined to ensure this, is therefore included. A problem arises as to how membrane stiffness and short-range membrane repulsion are to be included (growing membrane segments cannot intersect). In addition, Eq. (3.10) implies deterministic growth unaffected by noise.

How do we treat time in our simulations? Since we assume that cytoplasmic material is manufactured at a constant rate in the cell, time is divided into a series of intervals in which M new elements of cytoplasmic material are added to the surface of the growing cell. If in this time interval N_+ new compartments are added to the surface and N_- compartments disappear as a result of retraction, then clearly $M = N_+ - N_-$, but there is no unique algorithm specifying how this is to be achieved. Each interfacial site s_i has a probability $P_{g,i}$ of growing, a probability $P_{r,i}$ of retracting, and a probability $P_{s,i}$ of remaining stationary $(P_{g,i} + P_{r,i} + P_{s,i} = 1)$; these probabilities depend on C_i and are not uniquely defined. Indeed, different neuronal morphologies could be created by specifying different growth rules. We have employed a plausible but non-unique, two-stage stochastic growth rule.

First, a calcium-stimulated growth step is attempted, with the probability of growth at a membrane pixel s_i given by $P_{g,i} = P_{max}[2(C_i/C_{max}) - (C_i/C_{max})^2]$, where P_{max} is the maximal growth probability (which occurs when the submembrane calcium concentration $C_i = C_{max}$, the peak of the bell-shaped curve). This functional form is equivalent to taking $\alpha = 1$ and $\beta = 2$ in Eq. (3.10). In general in our simulations, $P_{g,i} > 0$, since the calcium concentration C_i was typically less than C_{max}; very rarely, at very high calcium concentrations (if $C_i > 2C_{max}$),

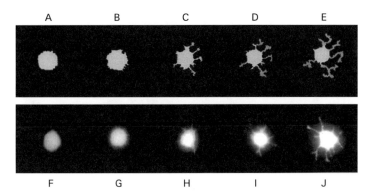

Figure 3.3
Typical growth of a model neuron (*A–E, top*) and of real retinal neurons (*F–J, bottom*), from 0.5 to 4 days in vitro. See the text for details. (From Hentschel and Fine, 1996.)

$P_{g,i} < 0$, and the surface was retracted. Growth is attempted at each membrane pixel by comparing $P_{g,i}$ with the output of a random number generator. In this manner, a total number of N_+ sites are grown at each discrete time step.

Second, if $N_+ > M$, we then attempted to retract each nongrowing membrane pixel with a probability P_r of retraction, whose value was fixed by demanding that a fixed amount M of cytoplasm be added to the whole interface per unit of time $[P_r = (N_+ - M)/(N - N_+)$, where N is the number of membrane pixels at any time]. Thus the probability of retraction at a particular membrane pixel s_i is given by $P_{r,i} = P_r(1 - P_{g,i})$, and the probability of its remaining stationary is therefore $P_{s,i} = 1 - P_{r,i} - P_{g,i}$. Retraction or stationarity was then assigned by use of a random number generator.

All sites in the domain were flagged as either inside the growing cell (internal), belonging to the membrane (one pixel thick, although in reality the interface is much thinner on the scale of a single pixel), or outside the cell (external). Whether growth or retraction occurred was further constrained by topology and rigidity (no pinching off of membrane, curvature not too large). Finally, the intersection of growing dendrites was discouraged. The approach of processes closer than the diameter of a filopod-bearing growth cone was disallowed (thus implementing contact-mediated inhibition). Growth or retraction was achieved by replacing the membrane pixel in question by an internal (growth) or external (retraction) domain point, while preserving membrane contiguity.

Growth and Form
Simulations of the growth of cells according to the model described here (figure 3.3A–E) display emergent properties that resemble the dynamic behavior of actual growing neurons (figure 3.3F–J).

In both model neurons (figure 3.3A,B) and natural neurons (figure 3.3F,G), initial outgrowth consists of broad, irregular extensions (lamellipodia) and short, very fine extensions (filopodia) of the cell membrane. Distinct processes (neurites) emerge only subsequently in both model and natural neurons (figure 3.3C,H); in both cases, the processes spontaneously form enlargements (growth cones) at their actively growing tips (figure 3.3C–D,H–I), which in turn give rise to branches (figure 3.3D–E,I–J). In both model and natural neurons, small processes often retract, and the extension of large processes may be punctuated by episodes of stasis or retraction.

Influence of Biological Parameters on Morphology

An important emergent property of the model is the ability of changes in membrane excitability and calcium conductivity to affect dendritic thickness, branching, and neurite length. Figure 3.4A–D shows the effects of changes in membrane electrical excitability (changes in excitability influence calcium permeability via voltage-gated conductances). The growth of model neurons with reduced electrical excitability (figure 3.4B) results in longer, thinner neurites than equivalent growth of control model neurons (figure 3.4A). Similar changes are seen in cerebellar Purkinje neurons grown in culture in the presence of 1 mM tetrodotoxin (TTX) (which blocks voltage-dependent sodium channels) (figure 3.4D) compared with those grown in standard medium (figure 3.4C) (Schilling et al., 1991). Figure 3.4E–H shows the effects of changes in calcium permeability. Increasing calcium permeability in model neurons (figure 3.4F) leads to the formation of more compact dendrites with broader growth cones than equivalent growth under standard conditions (figure 3.4E). Comparison with the growth of hippocampal neurons in standard culture medium (figure 3.4G) and in the presence of the calcium ionophore A23187 (500 nM) (figure 3.4H) reveals similar effects (Mattson et al., 1990).

Ionic Gradients

The spatiotemporal distributions of ionic concentrations inside growing cells can be extracted from our simulations (see figure 3.5). These findings are in accord with the basic mechanisms of unstable growth. The existence of voltage-gated sodium channels leads to focal depolarizations and elevated sodium concen-

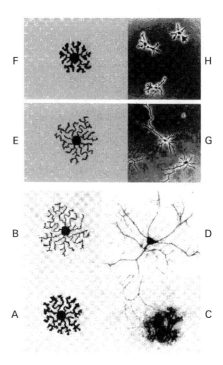

Figure 3.4
Effects of changes in membrane electrical excitability (*A–D*) and calcium permeability (*E–H*). (*C* and *D* are taken from Schilling et al., 1991, with permission; *G* and *H* are taken from Mattson et al., 1990, with permission.) See the text for details.

trations whose tendency to diffuse away is exceeded by an influx that is due to increased permeability. Calcium follows through voltage-gated calcium channels, causing increased growth and neurite formation. These internal sodium and calcium gradients and irregularities of contour emerge spontaneously at early stages of growth and lead to the observed instability through positive feedback, as the emerging neurites lead to even greater focal membrane depolarization. The appearance of calcium gradients along neurites early in the growth of a model neuron can be seen in figure 3.5B. In figure 3.5C, similar intra-

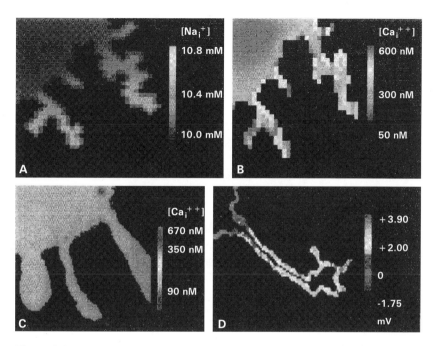

Figure 3.5
Gradients of intracellular concentration in neuronal outgrowth. (*A*) Sodium concentration in a model neuron increases distally along neurites in concert with membrane depolarization. (*B*) Calcium concentration in the same model neuron, influenced by membrane potential, is also higher along the neurite than within the cell body. (*C*) Calcium concentration in a mollusk neuron. (From Cohan et al., 1987 with permission.) (*D*) The transmembrane potential variation about the resting potential in the model neuron.

cellular calcium gradients along growing neurites of a real (molluscan) neuron in culture (Cohan et al., 1987), monitored with fura-2, can be seen.

Membrane Potentials

The sodium gradients are enhanced by, and contribute to, local depolarization of the cell membrane along the neurite. Gradients of transmembrane potential, progressively more depolarized toward the tips of growing processes, are a robust emergent property of our simulations. Recent observations indicate that similar gradients exist along growing processes of real neurons (Bedlack et al., 1994); to our knowledge, gradients of sodium along neurites have not been

investigated in real cells. Our simulations (see figure 3.5D) show relative depolarization (lighter pixels) at the tips of growing neurites, an emergent property in keeping with observations of such gradients in real neurons (Bedlack et al., 1994).

External Electric Fields

It is known that in the presence of external electric fields, neurites grow preferentially toward the cathode (Jaffe and Poo, 1979), and it has been suggested that intracellular calcium gradients generated by these fields may be responsible for this (Robinson, 1985). This suggestion is supported and clarified by the behavior of the model, where external electric field effects were incorporated as an additive contribution to the transmembrane potential. The external fields induced slight depolarization of cathode-facing membranes and hyperpolarization of anode-facing membranes. Preferential opening of voltage-dependent calcium channels on cathode-facing membranes results in higher local calcium influx and consequently in a tendency for neurites to extend toward the cathode.

3.3.3   Stage 3 Growth: Axonal Differentiation

We have shown (Samuels et al., 1996) that a "winner-take-all" dynamic instability is in all likelihood involved in axonal differentiation (see also Andersen and Bi, 2000) and that a great deal of the data about axonal differentiation can be understood by assuming that the formation, motion, and consumption of a single chemical determines the rate of growth of the neurites. The determinant chemical in this model should therefore be interpreted as the rate-limiting substance among all of those necessary for early neurite growth. At present, the identity of this determinant chemical is not known.

We assume that the determinant chemical is produced in the soma of the neuron and transported to the tips of the growing neurites, where it is consumed by the growth process. Depending on the identity of this chemical, this consumption may be viewed as, for example, the polymerization of cytoskeletal monomers, the addition of microtubule-associated proteins to microtubules, the consumption of energy or metabolites in cytoskeleton assembly, or the addition of new membrane. These processes are believed to occur primarily at the distal ends of growing neurites (Craig et al., 1995).

So far in this chapter we have employed two main mathematical methods to study early neuronal devel-

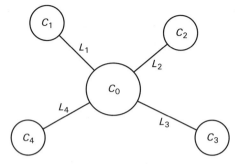

Figure 3.6
A schematic diagram of the simulation for a neuron with four neurites. The simulation variables are the concentration C_0 of the determinant chemical in the soma, the concentrations C_1 to C_4 at the four neurite tips, and the neurite lengths L_1 through L_4.

opment: (1) linear stability analysis to understand the stage 1 appearance of lamellipodia, and (2) detailed computer simulations to understand aspects of stage 2 neurite formation and its dependence on membrane parameters. To study axonal differentiation, we use a third approach: dynamic systems theory. We treat the developing neuron at a very coarse scale and compartmentalize the complex neuronal architecture to the level necessary for developing the essence of the proposed dynamic instability. In particular, we segment the neuron into its soma and a set of N neurites of different lengths.

The variables we consider are the concentration C_0 of the determinant chemical in the soma, the concentrations C_1 to C_N at the N neurite tips, and the neurite lengths L_1 through L_N (see figure 3.6). The development of the concentration in the soma is given by

$$\frac{dC_0}{dt} = \frac{1}{V_{\text{soma}}} \left(S - \sum_{i=1}^{N} T_i \right), \qquad (3.11)$$

where S is the rate of production of the determinant chemical by the cell, the i subscript labels each of the N neurites, T_i is the net transfer rate of the chemical from the soma to the neurite tip, and V_{soma} is the volume of the soma. We take S to be constant; in reality, the value of S is likely to change with time as the cell enters different stages of development, and it may also be dependent on C_0 so that production decreases when the soma concentration becomes too high.

The net transfer of the determinant chemical will result from two processes, diffusion and anterograde transport (more precisely, the net result of combined retrograde and anterograde transport). These two processes will tend to counteract each other, since anterograde transport acts to concentrate material at the neurite tip, while diffusion acts to equilibrate concentrations at the tip and soma. If a unique axon is to form, the net transfer rate must increase for the axon—at the expense of the other neurites—in order to supply adequate amounts of determinant chemical. Diffusion becomes less effective with increased neurite length, but the required instability only emerges if anterograde transport increases with either the length or the growth rate of the neurite. It is difficult to imagine how neurite length per se could affect the transfer rate, but an effect due to the growth rate is plausible (see also section 3.2.4). The transfer rate is a product of two factors: the speed of transport along the neurite and the rate at which substances enter the transport process from the soma. Then if, for example, neurite outgrowth results from traction exerted by the growing tip (Lamoureux et al., 1989), an increased growth rate would, as noted earlier, pull more cytosol—and thus more of the determinant chemical —into the neurite's proximal end, where there is access to the transport mechanism. The rate at which any determinant chemical is used during growth— whether it is the addition of microtubule monomers,

the addition of new material to the growth cone membrane, or the use of some important protein in microtubule stabilazation—is likely to be proportional to the rate of axon growth. With such a picture in mind, we model the anterograde transport as proportional to the growth rate of the neurite. The net transfer rate from the soma to the neurites is then

$$T_i = DA\frac{C_0 - C_i}{L_i} + F\frac{dL_i}{dt}C_0, \qquad (3.12)$$

where D is the diffusion constant of the determinant chemical, A is the cross-sectional area of the neurite, and F is a growth-dependent active transport parameter.

Because all neurites have the potential to form the axon (Dotti and Banker, 1987), we model all the neurites identically. The development of the chemical concentrations C_i and the neurite lengths L_i are given by

$$\frac{dC_i}{dt} = \frac{1}{V_{tip}}\left(T_i - G\frac{dL_i}{dt}\right)$$

$$\frac{dL_i}{dt} = \alpha C_i, \qquad (3.13)$$

where G and α are growth parameters and V_{tip} is the volume of the neurite tip, where the growth processes occur. The term $G(dL_i/dt)$ represents the consumption of the determinant chemical by the growth process at a rate proportional to the growth rate. The growth rate of the neurite is assumed to be proportional to the local concentration C_i of the determinant chemical. More complex models for this growth rate, as well as retraction and noise, can easily be incorporated, but Eq. (3.13) is sufficient to show the dynamic instability.

There are nine parameters in Eqs. (3.11)–(3.13), but only special combinations of them influence form.

The others simply influence the temporal and spatial scales at which growth occurs, and we can extract the important combinations of parameters controlling form by converting Eqs. (3.11)–(3.13) into a nondimensional form. To do this, we must choose scales for the concentration, time, and length. We choose the scales

$$C_{\text{scale}} = \frac{S}{\alpha G}; \quad t_{\text{scale}} = \frac{V_{\text{soma}}}{\alpha G}; \quad L_{\text{scale}} = \frac{V_{\text{soma}} S}{\alpha G^2}.$$

$$(3.14)$$

The biological meaning of this choice can be understood as follows: An extreme solution of Eqs. (3.11)–(3.13) is when the entire source is consumed by a single neurite, a situation that closely resembles our expectations for the growth of a single axon. In that case, we have $S = G\alpha C_i$ (production equals consumption by neurite i), which we may write as $C_i = S/(\alpha G)$. With this as the choice for C_{scale}, we can expect the nondimensionalized concentration in any axon to be near 1. The time scale was chosen so that the nondimensionalized source term of Eq. (3.11) would be unity. The length scale was chosen to simplify the rate of growth in Eq. (3.13). The equations in nondimensional form are then

$$\frac{d\tilde{C}_0}{d\tilde{t}} = 1 - \sum_{i=1}^{N} \left(\chi_1 \frac{\tilde{C}_0 - \tilde{C}_i}{\tilde{L}_i} + \chi_2 \frac{d\tilde{L}_i}{d\tilde{t}} \tilde{C}_0 \right)$$

$$\frac{d\tilde{C}_i}{d\tilde{t}} = \chi_3 \left(\chi_1 \frac{\tilde{C}_0 - \tilde{C}_i}{\tilde{L}_i} + \chi_2 \frac{d\tilde{L}_i}{d\tilde{t}} \tilde{C}_0 - \frac{d\tilde{L}_i}{d\tilde{t}} \right) \quad (3.15)$$

$$\frac{d\tilde{L}_i}{d\tilde{t}} = \tilde{C}_i,$$

where a tilde marks a nondimensionalized quantity and the parameters are defined as

$$\chi_1 = \frac{DAG}{SV_{\text{soma}}}; \quad \chi_2 = \frac{FS}{\alpha G^2}; \quad \chi_3 = \frac{V_{\text{soma}}}{V_{\text{tip}}}. \quad (3.16)$$

The nine parameters of Eqs. (3.11)–(3.13) are now reduced to four (χ_1, χ_2, χ_3, and N). The parameters χ_3 and N are simple geometric parameters. All of the important biological parameters are subsumed into χ_1 and χ_2. The values of χ_1 and χ_2 for a real neuron cannot be determined until the identity of the determinant chemical is known. Nevertheless, we have observed that much of the behavior of the solutions to Eq. (3.15) is not strongly dependent on the exact values of χ_1 and χ_2, as long as the parameters are not near their critical values for axon development.

We used a Runge-Kutta-Fehlberg method to integrate Eq. (3.15). Typical initial conditions are that the initial concentrations are set to zero and the initial neurite lengths are set to small random values $[\tilde{L}_i(\tilde{t} = 0) < 10^{-2}]$, where "small" is defined relative to the typical length attained by the neurites that do not form the axon.

Axon Development

Within a certain parameter range, the numerical solutions of Eq. (3.15) show an instability leading to the formation of $N - 1$ very slowly growing neurites and a single quickly growing neurite, which we identify as the axon. Figure 3.7 shows the development of the determinant chemical concentrations and neurite lengths for a typical simulation. In figure 3.7a, we can see the rapid and continuous growth of the axon, while the other neurites remain short. The development of the chemical concentrations (figure 3.7b) shows more detail. Our model [Eq. (3.15)] specifies that the growth rate of each neurite is proportional to the concentrations of the determinant chemical at its tip; thus the development of the concentrations also describes the development of the neurite growth rates.

Initially, the concentrations in the soma (dashed line) and all the neurites (solid lines) rise together. At

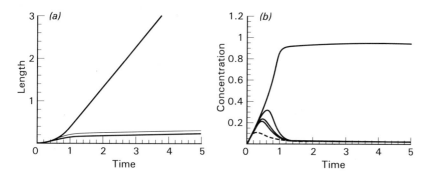

Figure 3.7
The development of one long neurite (the axon). Typical simulation results for a cell with four neurites. All values are dimensionless. The simulation parameters are $\chi_1 = 1$, $\chi_2 = 100$, $\chi_3 = 10$, and $N = 4$. (*a*) The development of neurite lengths. (*b*) The development of the concentrations of the determinant chemical in the soma (dashed line) and in the neurites (solid lines).

this stage, the neurites are very short and diffusion is sufficient to keep the concentrations equal. As the concentrations within the neurites increase, the soma concentration begins to fall, owing to the increase in the transfer term [Eq. (3.12)] from soma to neurites. During this stage, the concentrations within the neurites diverge, as one by one the neurite concentrations begin to fall. Eventually, only a single neurite is left with a high concentration. The concentration of this neurite then increases very rapidly and levels out at a value slightly less than 1, while the concentrations of the other neurites decrease and level out at a very low value. This progression is typical of axon development in these simulations and corresponds to neuronal developmental stages 2 and 3 as defined by Dotti et al. (1988).

Instability Leading to Axon Formation

Not all values of the parameters lead to axon formation. A stability map for Eq. (3.15) may be constructed as a function of the four parameters χ_1, χ_2, χ_3, and N. To construct a useful two-dimensional map, we held the geometric parameters constant at reason-able values ($\chi_3 = 10$ and $N = 4$) and concentrated on the stability as a function of the biological parameters χ_1 and χ_2 (see figure 3.8).

In the region labeled "no axon" in figure 3.8, the steady-state solution of Eq. (3.15) is that the concentrations (and therefore the growth rates) all have equal values of $\tilde{C}_i = 1/N$. In this region, all neurites continue to grow at equal rates with approximately equal lengths. In the "axon formation" region, a single neurite develops a concentration of near 1 while all other neurites have small concentrations. The position of the instability curve was defined to be there where the largest concentration reaches a value halfway between $\tilde{C}_i = 1/N$ and $\tilde{C}_i = 1$. The exact value defined for the transition is relatively unimportant because the transition between the two regions is rapid.

The shape of the instability curve contains information about the instability. There is a broad minimum extending across three orders of magnitude in χ_1; thus, aside from extreme low and high values, the instability is rather insensitive to the value of χ_1. In contrast, the dependence of the stability curve on χ_2 is

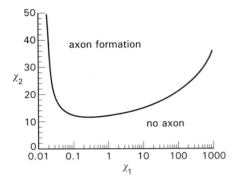

Figure 3.8
Stability map for the growth of axons. The map shown is for the values $N = 4$, $\chi_3 = 10$, and an initial length difference of 10^{-2}. The position of the curve depends on these parameters, but the shape of the curve does not.

quite sharp, with a transition to axon formation at approximately $\chi_2 = 10$ to 20. Recalling the definitions $\chi_1 = DAG/(SV_{\text{soma}})$ and $\chi_2 = FS/(\alpha G^2)$, it is evident that since $\chi_1 = 0$ is always in the stable (no axon) region, diffusion must be present for instability to occur. Similarly, since $\chi_2 = 0$ is always in the stable region, anterograde transport must also be present ($F > 0$) for the instability to arise.

We have modeled the source term S as a constant, but it is reasonable that in a real neuron this parameter could change during development. From the instability curve, we can understand how changes in S will affect the development of the instability. If S decreases, then χ_1 increases and χ_2 decreases. These changes in parameters could shift an axon-bearing cell across the instability line into the region of stability, perhaps corresponding to developmental stage 4, dendritic differentiation and growth (see later discussion).

Axotomy Experiments and Simulations
The underlying assumptions of our simulation—that all neurites are initially equally competent and that

a length-related instability determines the identity of the axon—are derived from the axotomy experiments of Dotti and Banker (1987) and Goslin and Banker (1989), and we should be able to reproduce the qualitative behavior seen in these experiments.

We simulate axotomy by reducing the length of the axon at an arbitrary time ($\tilde{t} = 2$) after the identity of the axon has been established (see figure 3.9). Since C_i represents the concentration at the tip of the neurite (where growth is occurring), we must also alter that value to simulate an axotomy. As a simplification, we assume that the concentration profile of the available determinant chemical is linear from the soma to the neurite tip; we calculate the new concentration from this profile and the new length of the cut neurite. Other models for the concentration profile may be used, but we have found that the development of the simulation after an axotomy is only dependent upon the chosen concentration profile if the axotomy leaves the original axon at a length close to that of the longest of the other neurites.

Figures 3.9a and 3.9b are two examples of axotomy simulations. In figure 3.9a, the axon is cut to a length that is still longer than (although comparable to) the length of the longest of the other neurites. The original axon quickly regains its rapid rate of growth and reforms the axon. In figure 3.9b, the original axon is cut to a length shorter than the other three neurites. In this case, the longest neurite begins to grow rapidly and forms a new axon, while the original axon regrows only slightly to a length comparable to the other short neurites. Immediately after the axotomy, all of the neurites experience a spurt of growth before a new axon emerges, which is in agreement with experiments (Goslin and Banker, 1989) in which the axon was cut to a length comparable to the other neurites.

Goslin and Banker (1989) observed that the latency for resumption of growth of an axon after axotomy

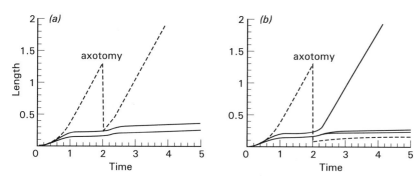

Figure 3.9
Axotomy of the growing axon. The dashed line represents the original axon. (*a*) The axon is cut to a length of 0.3, which is still longer than all the other neurites. The original axon reforms. (*b*) The axon is cut to a length of 0.1, which is shorter than that of the other neurites. The longest remaining neurite forms a new axon.

was a function of the difference in length between the two longest neurites, with a smaller difference requiring a longer latency. The latency in our simulations shows the same qualitative behavior (figure 3.10). At $\tilde{t} = 2$, when the simulation variables were close to a steady-state solution, we cut the axon to a random length and measured the time from axotomy to the beginning of rapid growth of a new axon, where rapid growth was arbitrarily defined as a growth rate halfway between $1/N$ and the maximum sustainable growth rate of 1.

3.4 Discussion

This chapter has been concerned with the modeling of the early stages (stages 1–3) of neuronal development and with the identification of possible physical instabilities underlying the evolution of form. These instabilities are generic and follow almost automatically from the known biology. However, the sheer variety of dendritic branching patterns implies that physical instabilities alone cannot account for the observed variety. For example, there is evidence that

some dendritic branching patterns, such as those of human cerebellar Purkinje arbors, are space filling, while many other branching patterns are fractal or have asymmetrical characteristics. In our simulations, we found that at intermediate times the neurons had fractal characteristics, but if we allowed our simulations to proceed for very long periods, the dendritic branching patterns tended to become space filling. To what extent such characteristics are generic and to what extent they depend on the details of the model (recall that the growth rule we used in our simulations, although plausible, is non-unique) is a difficult question to answer without more extensive simulations and more constraints from biology.

These observations show why modeling is so important for understanding the development of neuronal form. The degree of control that simulations give us over the great variety of biological parameters (genetic switches, pumping rates, channel conductivities and excitability, calcium-induced calcium release from stores, membrane rigidity, etc.) that might influence morphology means that at least in principle it may be possible to uncover their con-

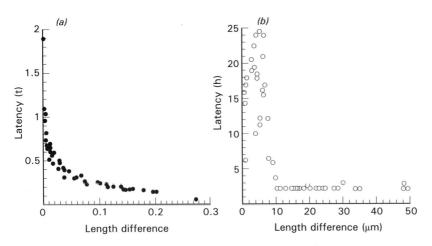

Figure 3.10
Latency of axon regeneration after axotomy versus the difference in length between the two longest remaining neurites. (*a*) Data from the simulation. (*b*) Latency data are from the axotomy experiments of Goslin and Banker (1989). Note that since the cells were not observed continuously, the minimum latency time measurable in those experiments was 2 hr.

tributions to morphogenesis by varying individual control parameters.

Future Modeling Studies

Perhaps the most important avenue for further theoretical work lies in separating genetic from epigenetic factors in morphogenesis. Biochemical reaction networks (see chapters 1 and 2) can be used to simulate genetic switching mechanisms occurring during development. The genetic component may express itself as a tendency for the dendritic arbor to branch at regular intervals, or as a specific time dependence for the production of a determinant chemical influencing axonal differentiation. Branching frequencies may also be influenced by extrinsic attractant and repellent substances. Also, the role of neuronal asymmetry needs to be investigated further, as well as synaptic activity, which can influence dendritic morphogenesis (Wong and Wong, 2000).

Future Experimental Studies

Both environmental and genetic factors can be expected to influence neuronal development, and as we uncover new biological mechanisms, these mechanisms can be incorporated into our simulations and their influence on growth and form investigated. The effect of different growth substrates (Esch et al., 1999), or of growth-stimulating signals, such as nerve growth factor (Ruit and Snider, 1991) or cyclic AMP (Zheng et al., 1994), can influence morphology; the more we can tease out the mechanisms involved, the more realistic our simulations will be. Environmental influences on axon differentiation could be important. For example, we found a very sensitive dependence of axon formation on a small (10 percent) variation in the parameter α controlling the growth rate of the neurites in Eq. (3.13) (Samuels et al., 1996). Clearly, a much deeper understanding is needed of the exact reaction kinetics, as well as the identity of the determinant chemicals. Although the choice of the neurite that forms the axon is random under homogeneous environmental conditions, this choice may easily become deterministic in an inhomogeneous environment. Axon formation by neurites in different locations may be inhibited or encouraged by their environment. More experimental work on these influences is needed.

Acknowledgments

This work was made possible by grants from the National Science Foundation (IBN-9221654), the North Atlantic Treaty Organization, and the Canadian Natural Sciences and Engineering Research Council.

References

Andersen, S. S., and Bi, G.-Q. (2000). Axon formation: A molecular model for the generation of neuronal polarity. *BioEssays* 22: 172–179.

Bass, P. W. (1999). Microtubules and neuronal polarity: Lessons from mitosis. *Neuron* 22: 23–31.

Bedlack, R. S., Jr., Wei, M.-D., Fox, S. H., Gross, E., and Loew, L. M. (1994). Distinct electric potentials in soma and neurite membranes. *Neuron* 13: 1187–1193.

Binder, L. I., Frankfurter, A., and Rebhun, L. I. (1984). The distribution of tau in the mammalian central nervous system. *J. Cell Biol.* 101: 1371–1378.

Blaustein, M. P. (1988). Calcium transport and buffering in neurons. *Trends Neurosci.* 11: 438–443.

Bradke, F., and Dotti, C. G. (1997). Neuronal polarity: Vectorial cytoplasmic flow precedes axon formation. *Neuron* 19: 1175–1186.

Bradke, F., and Dotti, C. G. (1999). The role of local actin instability in axon formation. *Science* 283: 1931–1934.

Burton, P. R. (1988). Dendrites of mitral cell neurons contain microtubules of opposite polarity. *Brain Res.* 473: 107–115.

Burton, P. R., and Paige, J. L. (1981). Polarity of axoplasmic microtubules in the olfactory nerve of the frog. *Proc. Natl. Acad. Sci. U.S.A.* 78: 3269–3273.

Caceres, A., and Kosik, K. S. (1990). Inhibition of neurite polarity by tau antisense oligonucleotides in primary cerebellar neurons. *Nature* 343: 461–463.

Campenot, R. B. (1977). Local control of neurite development by nerve growth factor. *Proc. Natl. Acad. Sci. U.S.A.* 74: 4516–4519.

Caserta, F., Stanley, H. E., Eldred, W. D., Daccord, G., Hausman, R. E., and Nittman, J. (1990). Physical mechanisms underlying neurite outgrowth: A quantitative analysis of neuronal shape. *Phys. Rev. Lett.* 64: 95–98.

Cohan, C., Connor, J. A., and Kater, S. B. (1987). Electrically and chemically mediated increases in intracellular calcium in neuronal growth cones. *J. Neurosci.* 7: 3588–3599.

Craig, A. M., and Banker, G. A. (1994). Neuronal polarity. *Annu. Rev. Neurosci.* 17: 267–310.

Craig, A. M., Wyborski, R. J., and Banker, G. A. (1995). Preferential addition of newly synthesized membrane protein at axonal growth cones. *Nature* 375: 592–594.

Davies, A. M. (2000). Neurotrophins: Neurotrophic modulation of neurite outgrowth. *Curr. Biol.* 10: R198–R200.

Dotti, C. G., and Banker, G. A. (1987). Experimentally induced alterations in the polarity of developing neurons. *Nature* 330: 254–256.

Dotti, C. G., and Banker, G. A. (1991). Intracellular organization of hippocampal neurons during the development of neuronal polarity. *J. Cell Sci.* Suppl. 15: 75–84.

Dotti, C. G., and Simons, K. (1990). Polarized sorting of viral glycoproteins to the axon and dendrites of hippocampal neurons in culture. *Cell* 62: 63–72.

Dotti, C. G., Sullivan C. A., and Banker, G. A. (1988). The establishment of polarity by hippocampal neurons in culture. *J. Neurosci.* 8: 1454–1468.

Esch, T., Lemmon, V., and Banker, G. (1999). Local presentation of substrate molecules directs axon specification by cultured hippocampal neurons. *J. Neurosci.* 19: 6417–6426.

Forscher, P. (1989). Calcium and polyphosphoinositide control of cytoskeletal dynamics. *TINS* 12: 469–478.

Forscher, P., and Smith, S. J. (1988). Actions of cytochalasins on the organization of actin filaments and microtubules in a neuronal growth cone. *J. Cell. Biol.* 107: 1505–1516.

Gierer, A., and Meinhardt, H. (1972). A theory of biological pattern formation. *Kybernetik* 12: 30–39.

Goodwin, B. C., and Trainor, L. E. H. (1988). Tip and whorl morphogenesis in *Acetabularia* by calcium-regulated strain fields. *J. Theor. Biol.* 130: 493–515.

Goslin, K., and Banker, G. (1989). Experimental observations on the development of polarity by hippocampal neurons in culture. *J. Cell Biol.* 108: 1507–1516.

Goslin, K., Schreyer, D. J., Skene, J. H. P., and Banker, G. A. (1988). Development of neuronal polarity: GAP-43 distinguishes axonal from dendritic cones. *Nature* 336: 672–674.

Grinvald, A., and Farber, I. A. (1981). Optical recording of calcium action potentials from growth cones of cultured neurons with a laser microbeam. *Science* 212: 1164–1167.

Harrison, L. G., Graham, K. T., and Lakowski, B. C. (1988). Calcium localization during *Acetabularia* whorl formation: Evidence supporting a two-stage hierarchical mechanism. *Development* 104: 255–262.

Heidemann, S. R., and Buxbaum, R. E. (1994). Mechanical tension as a regulator of axonal development. *Neurotoxicology* 15: 95–108.

Hentschel, H. G. E., and Fine, A. (1994). Instabilities in cellular dendritic morphogenesis. *Phys. Rev. Lett.* 73: 3592–3595.

Hentschel, H. G .E., and Fine, A. (1996). Diffusion-regulated control of cellular dendritic morphogenesis. *Proc. Roy. Soc. London B.* 263: 1–8.

Holliday, J., Adams, R. J., Sejnowski, T. J., and Spitzer, N. C. (1991). Calcium-induced release of calcium regulates differentiation of cultured spinal neurons. *Neuron* 7: 787–796.

Jaffe, L. F., and Poo, M.-M. (1979). Neurites grow faster towards the cathode than the anode in a steady field. *J. Exp. Zool.* 209: 115–128.

Kaibuchi, K., Kuroda, S., and Amano, M. (1999). Regulation of the cytoskeleton and cell adhesion by the Rho family GTPases in mammalian cells. *Annu. Rev. Biochem.* 68: 459–486.

Kater, S. B., Mattson, M. P., Cohan, C., and Connor, J. (1988). Calcium regulation of the neuronal growth cone. *Trends Neurosci.* 11: 315–321.

Lamoureux, P., Buxbaum, R. E., and Heidemann, S. R. (1989). Direct evidence that growth cones pull. *Nature* 340: 159–162.

Laux, T., Fukami, K., Thelen, M., Golub, T., Frey, D., and Caroni, P. (2000). GAP43, MARCKS and CAOP23 modulate PI(4,5)P2 at plasmalemmal rafts, and regulate cell cortex actin dynamics through a common mechanism. *J. Cell Biol.* 149: 1455–1471.

Lau, P.-M., Zucker, R.S., and Bentley, D. (1999). Induction of filopodia by direct local elevation of intracellular calcium ion concentration. *J. Cell Biol.* 145: 1265–1275.

Levi-Montalcini, R. (1976). The nerve growth factor: Its role in growth, differentation and function of the sympathetic adrenergic neuron. *Prog. Brain Res.* 45: 235–258.

Li, Z., Van Aelst, L., and Cline, H. T. (2000). Rho GTPases regulate distinct aspects of dendritic arbor growth in *Xenopus* central neurons *in vivo*. *Nature Neurosci.* 3: 217–225.

Lin, D., Edwards, A. S., Fawcett, J. P., Mbamalu, G., Scott, J. D., and Pawson, T. (2000). A mammalian PAR-3-PAR-6 complex implicated in Cdc42/Rac1 and aPKC signaling and cell polarity. *Nat. Cell Biol.* 2: 540–547.

Lohof, A. M., Quillan, M., Dan, Y., and Poo, M.-M. (1992). Asymmetric modulation of cytosolic cAMP activity induces growth cone turning. *J. Neurosci.* 12: 1253–1261.

Mattson, M. P. (1988). Neurotransmitters in the regulation of neuronal cytoarchitecture. *Brain Res. Rev.* 13: 179–212.

Mattson, M. P., Murain, M., and Guthrie, P. B. (1990). Localized calcium influx orients axon formation in embryonic hippocampal pyramidal neurons. *Dev. Brain Res.* 52: 201–209.

Matus, A. (1988). Microtubule-associated proteins: Their potential role in determining neuronal morphology. *Annu. Rev. Neurosci.* 11: 29–44.

McAdams, H. H., and Arkin, A. (1997). Stochastic mechanisms in gene expression. *Proc. Natl. Acad. Sci. U.S.A.* 94: 814–819.

Meberg, P. J., Kossel, A. H., Williams, C. V., and Kater, S. B. (1999). Calcium-dependent alterations in dendritic architecture of hippocampal pyramidal neurons. *Neuro-Report* 10: 639–644.

Meinhardt, H., and Gierer, A. (2000). Pattern formation by local self-activation and lateral inhibition. *BioEssays* 22: 753–760.

Montague, P. R., and Friedlander, M. J. (1989). Expression of an intrinsic growth strategy by mammalian retinal neurons. *Proc. Natl. Acad. Sci. U.S.A.* 86: 7223–7227.

Mullins, W. W., and Sekerka, R. F. (1963). Morphological stability of a particle growing by diffusion or heat flow. *J. Appl. Phys.* 34: 323–329.

Nobes, C. D., and Hall, A. (1999). Rho GTPases control polarity, protusion, and adhesion during cell movement. *J. Cell Biol.* 144: 1235–1244.

Odell, G. M., Oster, G., Alberch, P., and Burnside, B. (1981). The mechanical basis of morphogenesis. I Epithelial folding and invagination. *Dev. Biol.* 85: 446–462.

Pelce, P. (1993). Origin of cellular ionic currents. *Phys. Rev. Lett.* 71: 1107–1110.

Pelce, P., and Pocheau, A. (1992). Geometrical approach to the morphogenesis of unicellular algae. *J. Theor. Biol.* 156: 197–214.

Pellionisz, A. J. (1989). Fractal geometry of Purkinje neurons: Relationships among metrical and non-metrical geometries. *Soc. Neurosci. Abstr.* 15: 180.

Polleux, F., Morrow, T., and Ghosh, A. (2000). Semaphorin 3A is a chemoattractant for cortical apical dendrites. *Nature* 404: 567–573.

Ramon y Cajal, R. (1911). *Histologie du Systeme Nerveux de L'homme et des Vertebres.* Paris: Maloine.

Rebhan, M., Vacun, G., and Rosner, H. (1995). Complementary distribution of tau proteins in different phosphor-

ylation states within growing axons. *NeuroReport* 6: 429–432.

Robinson, K. R. (1985). The response of cells to electrical fields: A review. *J. Cell Biol.* 101: 2023–2027.

Ruit, K. G., and Snider, W. D. (1991). Administration or deprivation of nerve growth factor during development permanently alters neuronal geometry. *J. Comp. Neurol.* 314: 106–113.

Samuels, D. C., Hentschel, H. G. E., and Fine, A. (1996). The origin of neuronal polarization: A model of axon formation. *Phil. Trans. Roy. Soc. London. B.* 351: 1147–1156.

Schilling, K., Dickinson, M. H., Connor, J. A., and Morgan, J. I. (1991). Electrical activity in cerebellar culture determines Purkinje cell dendritic growth patterns. *Neuron* 7: 891–902.

Schoenfeld, T. A., and Obar, R. A. (1994). Diverse distribution and function of fibrous microtubule-associated proteins in the nervous system. *Int. Rev. Cyt.* 151: 67–137.

Sharp, D. J., Yu, W., Ferhat, L., Kuriyama, R., Rueger, D. C., and Baas, P. W. (1997). Identification of a motor protein essential for dendritic differentiation. *J. Cell Biol.* 138: 833–843.

Song, H.-J., and Poo, M.-M. (1999). Signal transduction underlying growth cone guidance by diffusible factors. *Curr. Opin. Neurobiol.* 9: 355–363.

Stebbings, H., and Hunt, C. (1983). Microtubule polarity in the nutritive tubes of insect ovarioles. *J. Cell Sci.* 6: 133–141.

Stenmark, H. (2000). Membrane traffic: Cycling lipids. *Curr. Biol.* 10: R57–R59.

Takeda, T., Ishikawa, A., Ohtomo, K., Kobayashi, Y., and Matsuoka, T. (1992). Fractal dimension of dendritic tree of cerebellar Purkinje cell during onto- and phylogenetic development. *Neurosci. Res.* 13: 19–31.

Tapon, N., and Hall, A. (1997). Rho, Rac and Cdc 42 GTPases regulate the organization of the actin cytoskeleton. *Curr. Opin. Cell Biol.* 9: 86–92.

Tessier-Lavigne, M., Placzek, M., Lumsden, A. G. S., Dodd, J., and Jessell, T. M. (1988). Chemotropic guidance of developing axons in the mammalian central nervous system. *Nature* 336: 775–778.

Turing, A. M. (1952). The chemical basis of morphogenesis. *Phil. Trans. Roy. Soc. London. B.* 237: 37–72.

Vallee, R. B., and Bloom, G. S. (1991). Mechanisms of fast and slow axonal transport. *Annu. Rev. Neurosci.* 14: 59–92.

Waterman-Storer, C. M., and Salmon, E. D. (1999). Positive feedback interactions between microtubule and actin dynamics during cell motility. *Curr. Opin. Cell Biol.* 11: 81–94.

Winckler, B., Forscher, P., and Mellman, I. (1999). A diffusion barrier maintains distribution of membrane proteins in polarized neurons. *Nature* 397: 698–701.

Witten, T. A., and Sander, L. M. (1981). Diffusion-limited aggregation, a kinetic critical phenomenon. *Phys. Rev. Lett.* 47: 1400–1403.

Wong, W. T., and Wong, R. O. L. (2000). Rapid dendritic movements during synapse formation and rearrangement. *Curr. Opin. Neurobiol.* 10: 118–124.

Yu, W., Cook, C., Sauter, C., Kuriyama, R., Kaplan, P. L., and Baas, P. W. (2000). Depletion of a microtubule-associated motor protein induces the loss of dendritic identity. *J. Neurosci.* 20: 5782–5791.

Zheng, J. Q., Felder, M., Connor, J. A., and Poo, M.-M. (1994). Turning of nerve growth cones induced by neurotransmitters. *Nature* 368: 140–144.

Formation of Dendritic Branching Patterns

Jaap van Pelt, Bruce P. Graham, and Harry B. M. Uylings

4

Understanding the enormous diversity of neuronal shapes and their impact on neuronal function is a major challenge in neuroscience. Much experimental effort has been dedicated to the reconstruction of three-dimensional neuronal morphologies, to the quantification of shape characteristics, and to the measurement of electrophysiological properties. Neurons attain their shapes as the result of a developmental process in which many cellular and molecular processes are involved. Understanding neuronal morphology therefore requires quantitative insight into these processes as well as a powerful framework for the description of morphology. This chapter reviews the different modeling approaches used in studying the morphology of dendritic branching patterns. Detailed examples are given of a stochastic dendritic growth model and models of intracellular mechanisms in neurite outgrowth.

4.1 Neurobiological Background

4.1.1 Neuronal Morphology

Neurons are characterized by the shape of their axonal and dendritic arborizations. Axons enable the neuron to deliver action potentials to local and remote target neurons (axons thus form the substrate of neuronal connectivity), while dendrites serve as target structures and receive and integrate incoming signals.

The contribution of an individual, active synapse located at a particular site on the dendritic tree, to the firing probability of the neuron depends on, among other factors, the amplitude of the postsynaptic potential, the characteristics of the path to the soma, the momentary state of the dendritic membrane and its ion channels, and the spatial and temporal relations with other active synapses. Dendritic morphology is therefore strongly involved in the electrical signal transduction properties of a neuron. For a detailed discussion on the functional role of dendritic morphology, see chapter 13.

Dendritic and axonal arborizations show an enormous diversity among and within neuron classes. It is tempting to assume that the morphology of a neuron contributes to its functional specialization, i.e., to its role in neuronal information processing. However, lack of a thorough understanding of how information is "encoded" or processed makes it very difficult to give a quantitative assessment of this role. The functional role of morphological specializations therefore remains largely unknown.

Neuronal morphology is the outcome of a developmental process, and understanding the characteristics of neuronal morphology requires an understanding of this process. To this end, one needs not only data and tools for efficiently and quantitatively describing neuronal morphologies, but also data and tools for describing neuronal morphogenesis.

4.1.2 Neurite Outgrowth and Neuronal Morphogenesis

In this section we give a brief overview of some of the major stages in neuronal development.

Cell Division and Migration

Rat cerebral cortex neurons are mainly generated from embryonic day 10 (E10) until birth. For the human brain, the period of major neocortical neuron formation is between approximately 6 and 18 weeks of gestation (Rakic, 1995; Uylings, 2000). According to the current view, pyramidal neurons migrate radially from the neocortical proliferative zone of the cerebral wall toward their final location, whereas the majority of the nonpyramidal neurons are derived from the medial ganglionic eminence via tangential migration (Parnavelas, 2000).

Neurite Outgrowth

After migration to their final location, neurons start to grow out neuritic processes. In the early phase of neuronal outgrowth, these neuritic processes differentiate; one of the neuron's neurites becomes an axon and the others become dendrites (see also chapter 3). Axons continue their advanced outgrowth rate, arborize, and migrate to their targets (axon guidance; see chapter 5), where they make synaptic connections. Pyramidal and nonpyramidal neurons accelerate their dendritic growth after the ingrowth of thalamic and other subcortical fibers. In rat cortex, the period of fastest dendritic outgrowth is between postnatal days 8 and 14, reaching mature dendritic extent around postnatal day 18 (Parnavelas and Uylings, 1980; Uylings et al., 1994; Koenderink and Uylings, 1995).

Role of Growth Cones in Neurite Elongation and Branching

Neurite elongation and branching are mediated by growth cones, i.e., specialized structures at the tip of growing neurites. Growth cones consist of a central core, lamellae sheets, and filopodia, all of which contain filamentous actin. Elongation of neurites proceeds by growth cone migration and requires the lengthening of the microtubule cytoskeleton (by polymerization of tubulin) in the trailing neurites. Dendritic branching is initiated by the splitting of a growth cone, a process that requires a reorganization of the growth cone's actin cytoskeleton (for a recent review of the molecular mechanisms involved in branching, see Acebes and Ferrus, 2000). The actin cytoskeleton is modified by the activity of a number of small GTPase enzymes, the Rho family, with Rac promoting lamellipodium and membrane ruffles, Cdc42 polymerizing actin into filopodia, and especially RhoA being involved in dendritic branching. Initial growth cone splitting and filamentous actin branch formation are further consolidated by the formation of a more rigid microtubular scaffold.

Neural branching is subject to fine modulation, with Rho proteins acting as molecular switches and integrating extracellular and intracellular signals to regulate rearrangement of the actin cytoskeleton. By altering neuronal network formation, mutations in the proteins involved in Rho-dependent signaling may possibly result in mental retardation (Ramakers, 2000). For the dependence of neurite outgrowth on electrical activity and intracellular calcium, see chapters 3 and 6. Growth cones have both a motor and a sensory function. They sense the local environment through receptor- and ion channel-mediated signaling, adhere to local cues, exert elastic tension, and react by internal reorganization. The actual behavior of a growth cone is the outcome of this multitude of processes and consists of an integration of environmental information (adhesivity, chemorepellants, chemoattractants) and internal processes [(de)polymerization of actin and microtubules, stabilization of microtubules by microtubule-associated proteins, transport of structural proteins, signaling pathways involving calcium and Rho proteins, electrical activity, and gene expression].

Overshoot and Regression

After a mature neuronal morphology has formed, further minor alterations may occur in different cell types, e.g., some further growth in layer II/III pyramidal neurons and some minor regression in layer IV multipolar nonpyramidal neurons (Uylings et al., 1994). Some neocortical neurons show pronounced dendritic regression (e.g., in the apical dendritic field in the callosal, small layer V pyramidal cell; Koester and O'Leary, 1992), but this appears to be the exception rather than the rule. A clear pattern of outgrowth, regression, and regrowth has been observed in rat cerebellar Purkinje cells (PC). Quackenbush et al. (1990) and Pentney (1986) have shown that PC networks sampled from 18-month-old rats had fewer terminal segments than those from 10-month-old rats, while PC networks from 28-month-old rats were intermediate in size. Woldenberg et al. (1993) analyzed the distributions of terminal segment numbers and argued for the simultaneous existence of growing and declining subpopulations of PC cells.

This brief overview of neuronal development shows that many mechanisms are involved in neurite outgrowth, including the intracellular machinery, the neuron's response to the extracellular environment, and neuronal electrical activity. Cline (1999) recently formulated a consensus view that dendritic structure and function develop as part of a continual dynamic process that balances the effects of neuronal activity, growth-promoting and growth-inhibiting proteins, and homeostatic mechanisms. In addition, the notion was emphasized that dendrites develop as part of a neural circuit, with development regulated by synaptic activity, activity-regulated proteins, and activity-induced genes.

All processes and mechanisms involved in dendritic development exert their effects on neurite elongation and branching through the influence they have on the actin and microtubule cytoskeleton, which makes understanding the cytoskeletal mechanisms in dendritic morphogenesis of crucial importance.

4.2 Questions and Approaches in Modeling Neuronal Morphology

4.2.1 Questions

The preceding section made it clear that understanding morphological diversity in neurons is an extremely challenging goal. Experimental approaches for longitudinal, in vivo studies of neurite outgrowth and neuronal morphogenesis are still limited, although high-resolution time-lapse confocal imaging techniques offer promising prospects. But even if all experimental data are available, computational approaches remain essential for describing the processes in their quantitative interrelationships and their consequences for neuronal development.

Important questions that can be addressed by computational approaches include (1) how can neuronal morphology and developmental changes be described, (2) how does neuronal morphology emerge from the dynamic behavior of growth cones, and (3) which mechanisms are involved and how do they contribute to growth cone behavior and neurite outgrowth? There is not a single computational approach to all of these questions; each question requires its own particular strategy, which will also depend on the level of detail of the description.

Addressing the first question requires a set of shape factors that capture the characteristic shape properties of neuronal morphology. Neuronal morphogenesis may then be described by the way in which these morphological characteristics change over time. On the basis of these quantitative descriptions, we may search for the most efficient algorithms that can produce arborizations that have morphological characteristics similar to the neuronal ones.

The second question can be approached by quantifying the dynamic actions of growth cones and studying what variety of morphologies will emerge from these actions. As a first step, we may describe growth cone elongation and branching as the outcomes of a stochastic process. Such an approach will provide insight into how the growth cones' elongation and branching probabilities translate into typical shape properties of the model dendritic trees produced. These model trees may then be compared with neuronal dendritic trees, in an attempt to find agreement by optimizing the stochastic growth rules. In further refinements, we may approximate growth cone dynamics at a finer level of detail by including state- and time-dependent conditions in the stochastic growth rules.

The third question is the most complex one because it concerns a multitude of mechanisms and their interactions. As a first step, we may focus on a particular mechanism and explore its role in neurite outgrowth. For instance, we may focus on the cytoskeleton and build computational tools for quantitatively studying cytoskeletal dynamics and their contribution to the behavior of growth cones. Alternatively, we may focus on regulatory mechanisms; this requires computational tools for quantitatively describing the biochemical and signaling pathways that target the cytoskeletal dynamics. From a more general point of view, we may ask whether neurite outgrowth is subject to basic (biophysical) constraints, as imposed by conservation of matter and energy or by limited resources (possibly leading to competitive phenomena). Addressing this question requires biophysical models of production, transport, use, and decay of key proteins involved in neurite outgrowth.

An interesting question is also whether the mechanisms involved in neurite outgrowth are operating in an orchestrated way or more or less independently. This question is far from trivial, because even when two mechanisms lack direct interaction and thus may seem independent, the complexity of the full system may include indirect links, resulting in an effective dependence (homeostasis of, e.g., electrical activity or the intracellular calcium concentration may produce such effective correlative behavior; see also chapter 6). Related questions are whether growth cones may be considered as operating independently from one another, only under control of local mechanisms, and to what extent axonal and dendritic outgrowth is correlated.

4.2.2   *Parameterization of Neuronal Morphology*

Neurons may be readily distinguished by the global shape of their dendritic field. Recently, Fiala and Harris (1999) reviewed different schemes for characterizing the global shape of dendritic fields: (1) in terms of extent, density, and polarity (e.g., bipolar, multipolar); (2) on the basis of a selective, sampling, or space-filling appearance; and (3) in terms of regular 3-D geometric bodies [spherical, laminar, cylindrical, (bi)conical, or fan shaped].

For a detailed description of neuronal arborizations, it is necessary to include both topological and metrical characteristics. The topological structure is defined by the number of segments and the connectivity pattern of the segments as a rooted tree. It determines, for example, the distribution of segments versus centrifugal order (figure 4.1), the division of segments in subtrees at branch points (indicated by the tree asymmetry index; Van Pelt et al., 1992), and the frequencies of different types of branch points (e.g., Horsfield et al., 1987; Sadler and Berry, 1983). Metrical properties include the lengths and diameters of segments, their curvature, the three-dimensional embedding of the arborization, and further details, such as the number

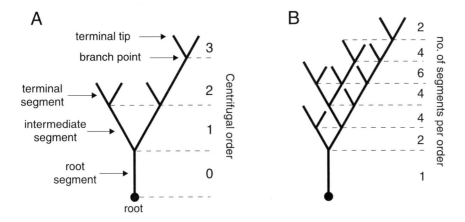

Figure 4.1
(*A*) Tree elements. Intermediate and terminal segments, labeled by a centrifugal ordering scheme. (*B*) Distribution of number of segments versus centrifugal order.

and shape of dendritic spines. Metrical characterization can be given in terms of length and diameter distributions and spatial densities of segments or branch points. Measures that are a combination of topological and metrical properties include radial distribution functions of dendritic length or the number of branch points (for a review, see Uylings et al., 1986). Morphological characterization thus requires many measures, the choice of which depends on the required detail of the analysis. For example, Ipiña et al. (1987) applied multivariate analysis techniques to a set of ten variables to analyze developmental effects and aging, and Cannon et al. (1999) used a large set of dendritic measures to differentiate among different populations of hippocampal neurons.

4.2.3 Stochastic Models of Neuronal Arborizations

Different algorithms have been developed for generating neuronal arborizations with shape properties (statistically) similar to their empirically observed counterparts. One class of models focuses on the radial

distribution of dendrites, in terms of intersections with spheres (e.g., Ten Hoopen and Reuver, 1971), or of segment number per centrifugal order (e.g., Kliemann, 1987). Focusing on the topological structure, Devaud et al. (2000) proposed a parsimonious description for the variation in segment order distributions among different dendritic trees by assuming a simple two-parameter model for the centrifugal order-dependent splitting probabilities. Using this model, they found good matching with cultured honeybee (olfactory) antennal lobe neurons.

Another class of models, which focuses on both topological and metrical aspects, uses the empirical distribution functions for segment lengths and diameters, as well as their correlations, to generate random dendritic morphologies by a repeated process of random sampling of these distributions (e.g., Hillman, 1979, 1988; Burke et al., 1992; Tamori 1993; Ascoli and Krichmar, 2000).

A third class of models, stochastic growth models, uses hypothetical, stochastic growth rules for branching and elongation in the generation of random dendritic trees. Topological growth models aim at explaining topological variation by assuming that the branching probability depends on the type of segment (intermediate or terminal) (e.g., Berry et al., 1975) as well as on the centrifugal order (see figure 4.1) of the segment (Van Pelt and Verwer, 1986). These studies indicated how the mode of branching determines the eventual variation in dendritic topological structures and showed that the empirically observed topological variation in dendritic trees was consistent with branching of predominantly terminal segments, as was also found in an extensive study of dendrites of motor neurons (Dityatev et al., 1995).

Segment length distribution, a metrical property, depends on both the branching and the elongation process. Van Pelt et al. (2001a) extended the topological growth model by including neurite elongation in the already optimized branching process. By this separation of elongation and branching, they were able to match many topological and metrical shape properties of the dendrites of a variety of cell types, including basal dendrites of Wistar rat cortical layer 5 large pyramidal neurons (Van Pelt and Uylings, 1999a), basal dendrites of Wistar rat cortical layer 5 small pyramidal neurons (Van Pelt and Uylings, 1999b), basal dendrites of S1 rat cortical layer 2/3 pyramidal neurons (Van Pelt et al., 2001a), guinea pig cerebellar Purkinje cell dendritic trees (Van Pelt et al., 2001a), and cat deep-layer superior colliculus neurons (Van Pelt et al., 2001b). These studies demonstrated that the empirically observed morphological variability in dendrites can emerge from a growth process in which branching and elongation events show variation as well, being described by the probability functions in the model. For these results, it was necessary to assume (1) that branching depends on the total number of terminal segments and the centrifugal order of each terminal segment, and (2) that after branching, newly formed (stabilized) daughter segments have

initial lengths, in order to account for the small number of short intermediate segments observed in all segment length distributions. Empirical growth curves for the number of terminal segments provide the possibility of gauging the time scale for the branching process and predicting absolute elongation rates.

Other metrical studies include those of Nowakowski et al. (1992), who studied segment length distributions by assuming an increasing but saturating branching probability with segment length. They found that it was necessary to include an inhibition of branching for some distance beyond a branch point. Ireland et al. (1985) studied the growth of apical dendritic trees in rat entorhinal cortex by means of analytical, time-dependent functions for elongation and branching and found that terminal growth velocities decrease with time. Li et al. (1992, 1995) developed a model for neurite elongation and branching that includes interactions with a morphogen gradient; they found evidence for the existence of lateral inhibition as well as a role for filopodial tension in branching. Studying the lengths of intermediate and terminal segments during phases of growth, decline, and regrowth of rat Purkinje cell dendritic trees, Woldenberg et al. (1993) found evidence for Fibonacci scaling in segment lengths.

4.2.4 *Mechanistic Models of Neurite Outgrowth and Neuronal Morphogenesis*

Stochastic growth models do not consider the underlying mechanisms involved in neurite elongation and branching. A different class of models can be distinguished that is concerned with these mechanisms. Because of the multitude of mechanisms involved, these models generally focus on particular mechanisms and explore their implications for neurite outgrowth.

Van Veen and Van Pelt (1994) studied elongation and branching under the control of production, transport, and polymerization of tubulin. They predicted that small differences in polymerization rates would result in competitive phenomena in the elongation rates of daughter segments at branch points. This model has recently been extended by Van Ooyen et al. (2001) (see section 4.3.4).

Graham et al. (1998) and Graham and Van Ooyen (2001) have investigated the possible intracellular origins of the dependence of branching on the number of terminal segments and their centrifugal order (see sections 4.3.2 and 4.3.3).

Hely et al. (2001) have modeled both the rate of terminal branching and the rate of elongation as functions of the stability of microtubule bundles in the growth cone. The stability depends on the phosphorylation state of microtubule-associated protein MAP2, which is in turn determined indirectly by calcium influx. Dephosphorylated MAP2 favors elongation by promoting microtubule polymerization and bundling. Phosphorylation of MAP2 disrupts its cross-linking of microtubules and thus destabilizes the microtubule bundles and promotes branching. A wide variety of tree characteristics are produced by the model, depending on the relative rates of phosphorylation and dephosphorylation of MAP2, the production and transport of MAP2, and calcium influx.

In her doctoral thesis, Aeschlimann (2000) introduced a biophysical model of the sensory function and the motor behavior of growth cones. On the basis of this model, she made quantitative predictions for elastic and inelastic elongation and for the shear stress and bending forces within growth cones.

4.2.5 Modeling Arborizations in Other Biological and (Geo)physical Areas

Branching patterns are common structures in nature, and in many fields of research stochastic and mecha-

nistic models have been used. There has been methodological cross-fertilization between these fields of research. For instance, topological studies of river systems in the 1960s and 1970s (e.g., Shreve, 1966; Dacey and Krumbein, 1976) stimulated topological growth studies of neuronal branching patterns (e.g., Berry et al., 1975; Hollingworth and Berry, 1975), lung branching patterns (Horsfield and Woldenberg, 1986; Horsfield et al., 1987), microvessel networks (Ley et al., 1985), and plant root systems (Fitter et al., 1991). In turn, further generalizations of the topological models for neuronal arborizations have produced a new view on random river topology and headward growth (Van Pelt et al., 1989).

4.3 Detailed Examples of Models of Neuronal Morphology

4.3.1 Dendritic Growth Model

The dendritic growth model aims at describing the morphology and variability of dendritic trees for a wide variety of neuron types (for reviews, see Van Pelt and Uylings, 1999a; Van Pelt et al., 2001a). Briefly, the model describes dendritic growth by elongation and branching of segments. Because in reality many intracellular and extracellular mechanisms are involved in the behavior of growth cones (which mediate elongation and branching), it is assumed that elongation and branching can be described as stochastic processes. The branching process is defined by a branching probability for each terminal segment:

$$p_i(\gamma) = (B/N)C_i 2^{-S\gamma} n_i^{-E} \tag{4.1}$$

evaluated at each time bin i $(i = 1, \ldots, N)$ during the developmental period T. In Eq. (4.1), the branching probability is assumed to depend on the growing number of terminal segments, n_i, according to parameter E, and on the centrifugal order γ of the segment, according to parameter S. The basic branching parameter B denotes the expected number of branching events at an isolated segment in the full period. The ratio B/N denotes the basic branching probability per time bin. Parameter $C_i = n_i / \sum_{j=1}^{n_i} 2^{-S\gamma_j}$ is a normalization constant and must be evaluated at each time bin with a summation over all n_i terminal segments. The number of time bins, N, can be chosen arbitrarily but so that the branching probability per time bin remains much smaller than 1, making the probability of more than one branching event per time bin negligibly small. To describe the branching process in continuous time, the time bin scale needs to be mapped onto an absolute time scale. Time bins will have equal durations in a linear mapping, but different durations in a nonlinear mapping. The equation for the branching probability per time bin transforms into a branching probability per unit of time:

$$p_t(\gamma) = D(t)C_t 2^{-S\gamma} n_t^{-E}, \tag{4.2}$$

where $D(t)$ denotes the basic branching rate parameter per unit of time [$D(t)$ is not constant for a nonlinear mapping].

After a branching event, newly formed daughter segments are given a gamma-distributed, randomly chosen initial length with mean $\bar{l}_{in}$ and standard deviation $\sigma_{l_{in}}$, and a gamma-distributed, randomly chosen elongation rate. The developmental period may consist of a first phase of elongation and branching and a subsequent phase of elongation only, with elongation rates v_{be} and v_e, respectively, both with a coefficient of variation cv_v. A summary of the model parameters is given in table 4.1.

During outgrowth, the total number of terminal segments increases at each branching event. If the branching probability is independent of the total number of terminal segments, branching is prolific,

leading to very large trees. This is shown in figure 4.2A for parameter value $E = 0$, resulting in an exponentially declining degree distribution. For positive values of E, the branching probability decreases with an increasing number of terminal segments, leading to degree distributions with a modal shape that becomes narrower at larger E values. By its control of the increase in the total number of terminal segments, parameter E may be interpreted as representing competition during dendritic branching.

Application of the Dendritic Growth Model to Wistar Rat Cortical Layer IV Multipolar Nonpyramidal Neurons

The morphological data on Wistar rat visual cortex layer IV nonpyramidal neurons are obtained from a developmental study by Parnavelas and Uylings (1980), with detailed reconstructions for different age groups. Growth starts in the first postnatal week and shows a continuing increase in the number of segments and total dendritic length up to postnatal (PN) day 16. After this growth phase, terminal segments show further elongation up to at least PN90.

In the present study, the growth model is applied to the PN16 data set and studied for the following shape parameters: total dendritic length, number and lengths of intermediate and terminal segments, path lengths, centrifugal order of the segments, and the tree asymmetry index (as a measure of the topological structure, or connectivity pattern, of the segments). The observed mean and standard deviation (SD) values are listed in table 4.2; the frequency distributions are shown in figure 4.3 as dashed histograms. First, the branching process was studied by optimizing the branching parameters B, E, and S. The results are illustrated in figure 4.4A, showing the predicted growth curve of the number of terminal segments. The panel shows an excellent matching with the empirically observed data at PN16, but a mismatch with

Table 4.1
Summary of parameters used in the dendritic growth model

	Aspect of growth	Related to
Optimizing parameters		
B	Basic branching parameter	Segment number
E	Size dependence in branching	Segment number
S	Order dependence in branching	Topology
$\alpha_{l_{in}}$ (μm)	Initial length—offset in gamma distribution	Segment length
$\bar{l}_{in}$ (μm)	Initial length—mean	Segment length
$\sigma_{l_{in}}$ (μm)	Initial length—SD	Segment length
$\bar{v}_{be}$ (μm/hr)	Mean elongation rate in branching and elongation phase	Segment length
$\bar{v}_{e}$ (μm/hr)	Mean elongation rate in elongation phase	Segment length
cv_v	Coefficient of variation in elongation rates	Segment length
Experimental parameters		
T_0 (hr)	Start of growth	
T_{be} (hr)	End of branching and elongation phase	
T_e (hr)	End of elongation phase	
$\bar{d}_t$ (μm)	Terminal segment diameter—mean	Segment diameter
σ_{d_t} (μm)	Terminal segment diameter—SD	Segment diameter
$\bar{e}$	Branch power—mean	Segment diameter
σ_e	Branch power—SD	Segment diameter

Notes: A distinction is made between optimizing parameters, whose values are subjected to optimization, and experimental parameters, whose values are taken (in)directly from experimental observations. Note that the segment diameter parameters are not part of the growth model, but are used afterward to assign diameter values to the skeleton trees produced by the model. It is assumed that the gamma distributions for the elongation rates have zero offset ($\alpha_v = 0$). SD, standard deviation.

the data in earlier age groups at 4, 6, 8, 10, 12, and 14 days PN, when time bins of equal duration are assumed. Applying a nonlinear mapping of time bins to absolute time scale (figure 4.4B illustrates an exponential mapping with exponent 3) results in a growth curve for the model that matches in both mean and SD the different age groups very closely (figure 4.4C). By a transformation to absolute time, the time course of the basic branching rate $D(t)$ can be predicted; it shows a rapid decline in the first week of development (figure 4.4D) (see also Van Pelt and Uylings, 2002).

The elongation process was studied by optimizing the parameters α_{in}, $\bar{l}_{in}$, and $\sigma_{l_{in}}$ for the initial lengths at

the time of branching, and the parameters v_{be} and cv_v for the sustained elongation rate. The shape properties of the random trees generated with these optimized parameters (table 4.3) are listed in table 4.2 for their mean and SD, and plotted as dashed histograms in figure 4.3. These outcomes show that the model trees conform closely to the observed dendrites in most of their statistical shape properties.

The decline in the branching probability with an increasing number of terminal segments via parameter E may be interpreted as a competition effect when growth cones compete for some limited resource. The positive value of $E = 0.106$ thus suggests that such

A

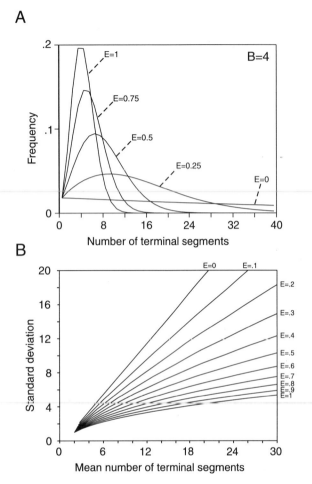

Figure 4.2

(A) Distributions of the number of terminals per dendritic tree. The model trees are randomly generated by the dendritic growth model. The distributions, calculated for several values of the parameter E, demonstrate the control of the competition parameter E over the proliferation of the number of terminal segments during dendritic branching. (B) Relation between the mean and standard deviation of the distribution of the number of terminal segments of trees randomly generated by the dendritic growth model for different values of E.

competitive behavior also occurs during outgrowth of these nonpyramidal dendrites.

4.3.2   Intracellular Signal Model

The dendritic growth model (see section 4.3.1) reveals that the branching probability may change with increasing numbers of terminals in the growing tree and with the centrifugal order of each terminal. Although this model characterizes these dependencies, it does not specify any particular biophysical mechanisms that may be their underlying cause. The intracellular signal model is an attempt at a growth model in which branching is modulated by the number of terminals and their centrifugal order, but which also may have a physical instantiation in terms of intracellular processes (Graham et al., 1998).

The intracellular signal model describes the growth process as one in which the probability that a terminal branches is proportional to the amount of some substance, v, in the terminal. Branching is still described as a stochastic process, but this process is now modulated according to the distribution of v throughout the growing tree. The substance is produced in the cell body and is transported along the tree to the terminal tips. The branching probability of a terminal segment i is $p_i = (B/N)v_i$. The production and distribution of v is given by

$$v_0 = n^{(1-E)} \tag{4.3}$$

$$v_1 = \frac{n_1^{(1-S)}}{n_1^{(1-S)} + n_r^{(1-S)}} v_p \tag{4.4}$$

$$v_r = \frac{n_r^{(1-S)}}{n_1^{(1-S)} + n_r^{(1-S)}} v_p, \tag{4.5}$$

where B, N, E, S, and n are as for the dendritic growth model; v_0 is the amount of substance in the cell body; v_1 (v_r) is the amount in the left (right)

Table 4.2

Mean and SD of dendritic shape parameters of observed and modeled visual cortex layer IV nonpyramidal neurons in the PN16 Wistar rat

Shape variables	Observations			Model outcomes	
	No.	Mean	SD	Mean	SD
Degree	238	3.29	2.5	3.17	2.4
Tree asymmetry	84	0.47	0.21	0.45	0.23
Centrifugal order	1330	1.92	1.45	1.85	1.43
Total dendritic length (μm)	238	169	164	159	127
Terminal segment length (μm)	782	38.6	35.1	38.6	39.1
Intermediate segment length (μm)	537	17.7	19.2	16.8	20.4
Path length (μm)	782	75.2	48.4	70.3	40.8

daughter segment at a branch point; ν_p is the amount in the parent segment to a branch point; and n_l (n_r) is the number of terminals in the subtree emanating from the left (right) segment. The probability that a particular terminal i will branch is calculated by applying the above equations iteratively at each time step, starting with the root segment, 0. The specification of time bins and the calculation of segment elongation rates are as for the dendritic growth model.

The production of ν in the cell body is a function of the number of terminals, according to the parameter E. This modulates the terminal branching probabilities identically to the effect of E in the dendritic growth model. The distribution of ν at branch points according to the value of S results in a modulation of branching probabilities that is a close approximation to the dendritic growth model's dependence on centrifugal order. The difference lies in the fact that the dendritic growth model requires global knowledge at each terminal tip of the number of terminals in the tree and their centrifugal orders. The intracellular signal model relies only on local knowledge at each branch point of the size of the subtrees emanating from it. Such knowledge may be gained from a retro-

grade signal in the form of a molecule produced at each terminal and transported to the cell body. Another possible signal arises from changes in segment diameters as the tree grows. Many actual dendritic trees exhibit an approximate power law relationship between the diameters of parent and daughter segments at branch points. If during growth parent segments increase their diameters according to such a power law as new terminal branches are created, then the diameter of any segment is an indication of the size of the subtree below it. Thus a transport mechanism that splits the amount of substance available at a branch point according to the relative diameters of the daughter branches would instantiate this model.

Examples of the tree characteristics produced by the intracellular signal model are given in figures 4.5 and 4.6. These figures show the degree and centrifugal order of trees grown using the dendritic growth model parameters optimized to data drawn from either rat multipolar nonpyramidal cells (figure 4.5; see the dendritic growth model example) or the basal dendrites of large layer V rat cortical pyramidal neurons (figure 4.6; Van Pelt and Uylings, 1999a). In both examples, the degree and centrifugal order dis-

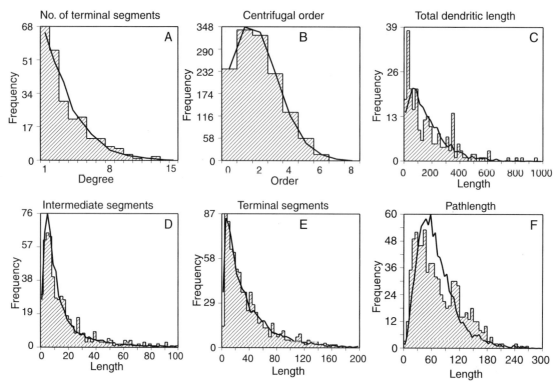

Figure 4.3

Frequency distributions of dendritic tree shape parameters of visual cortex layer IV multipolar nonpyramidal neurons in the Wistar rat at postnatal (PN) day 16 (shaded histograms) and in model-generated trees (continuous thick lines), using the optimized parameter values given in table 4.1. The panels show (*A*) the frequency distributions for the number of terminal segments, (*B*) centrifugal order of the segments, (*C*) total dendritic tree length, (*D*) intermediate segment length, (*E*) terminal segment length, and (*F*) path length.

tributions closely match those produced by the dendritic growth model.

4.3.3 Diffusional Model

The intracellular signal model does not explicitly specify a transport mechanism for the branch-determining substance. The diffusional model explores how the transport of a branch-determining substance by diffusion affects tree growth (Graham

and Van Ooyen, 2001). The spatial production, consumption, decay, and diffusion of the substance results in a branching process that also shows dependence upon both the number of terminals in the growing tree and their centrifugal order.

In this model, the branch-determining substance has concentration C_i at terminal i in the growing tree. Terminal segments elongate at fixed rates as determined for the dendritic growth model. The branching probability of terminal i is $p_i = (B/N)C_i$. The sub-

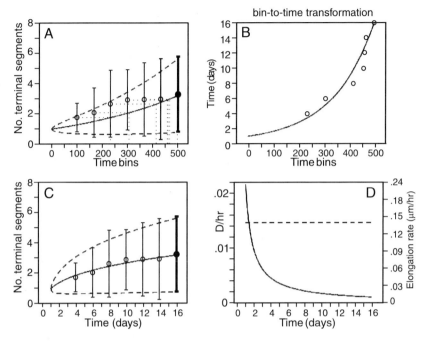

Figure 4.4

(*A*) Comparison of the observed mean (open circles) and standard deviation (error bars) of the number of terminal dendritic segments at different time points [i.e., 4, 6, 8, 10, 12, 14, and 16 days postnatal (PN)] during development of Wistar rat cortical layer IV multipolar nonpyramidal neurons, with the growth curve predicted by the dendritic growth model for the parameters optimized for the 16-day PN group. The time bins represent a linear time scale with time bin 500 corresponding to 16 days PN. The dotted lines indicate the time bins at which the model's growth curves attain values similar to the observed data points. Parameters: $B = 1.26$; $E = 0.106$. (*B*) A nonlinear exponential mapping of time bins onto absolute time in order to match the model's predicted growth curve through the observed data points. (*C*) Comparison of the model's predicted and observed growth curve for the number of terminal segments, plotted against an absolute time scale. Note the good matching of the standard deviations. (*D*) Time course of the basic branching rate per hour during dendritic development, as predicted by the growth model for the exponential bin-to-time transformation. A constant mean elongation rate has been assumed.

Table 4.3

Optimized values for growth parameters (see table 4.1) to match the statistical shape properties of cortical layer IV nonpyramidal cell basal dendrites in the PN16 Wistar rat, given in table 4.2

Growth model parameters								Experimental	
B	E	S	α_{in}	$\bar{l}_{in}$ (µm)	$\sigma_{l_{in}}$	$\bar{v}_{be}$ (µm/hr)	cv_v	T_0 (hr)	T_{be} (hr)
1.26	0.106	0	0	4	3	0.16	0.9	24	384

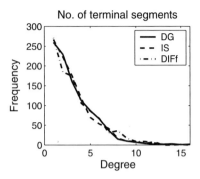

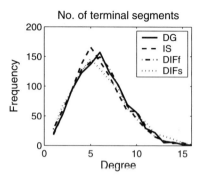

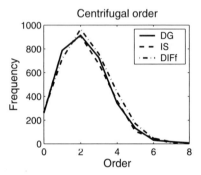

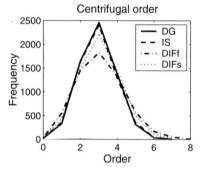

Figure 4.5

Distributions of the number of terminals in the fully grown trees (degree) and the segment centrifugal order for 1000 trees grown using either the dendritic growth (DG), intracellular signal (IS), or diffusional (DIFf) models. The dendritic growth model used parameters optimized to the branching phase of rat multipolar nonpyramidal cell dendrites [$B = 1.26$, $E = 0.106$, $S = 0$, $N = 500$; mean elongation rate, 0.16 μm/hr (CV = 0.9)]. These identical parameter values were used with the intracellular signal model. For the diffusional model, the decay parameters were set to match the required E value, giving $I = 1.0$, $\gamma_0 = 0.96$, and $\gamma_i = 0.04$; segment diameters were 1 μm throughout and diffusion was $D = 600$ μm^2/hr.

Figure 4.6

Distributions of the number of terminals in the fully grown trees (degree) and the segment centrifugal order for 1000 trees grown using either the dendritic growth (DG), intracellular signal (IS), or diffusional (DIFf and DIFs) models. The dendritic growth model used parameters optimized to the branching phase of rat large layer V pyramidal cell basal dendrites [$B = 3.85$, $E = 0.74$, $S = 0.87$, $N = 264$; mean elongation rate, 0.22 μm/hr (CV = 0.28); Van Pelt and Uylings, 1999a]. These identical parameter values were used with the intracellular signal model. For the diffusional model, the decay parameters were set to match the required E value, giving $I = 1.0$, $\gamma_0 = 0.45$, and $\gamma_i = 0.55$; segment diameters were 1 μm throughout and diffusion was either fast ($D = 600$ μm^2/hr; DIFf) or slow ($D = 2$ μm^2/hr; DIFs).

stance is produced at rate I in the cell body (location 0) and decays there at rate γ_0. The substance also decays (or is consumed by the branching process) at rate γ_i in terminal i. The substance diffuses between its site of production and the terminal tips at rate D. The changes in concentration over time at these locations when there are n terminals in the growing tree are given by

$$\frac{dC_0}{dt} = I - \gamma_0 C_0 + \sum_{i=1}^{n} \frac{DA_i}{L_i V_0}(C_i - C_0) \qquad (4.6)$$

$$\frac{dC_i}{dt} = -\gamma_i C_i + \frac{DA_i}{L_i V_i}(C_0 - C_i), \qquad (4.7)$$

where L_i is the intracellular longitudinal distance between terminal i and the cell body, A_i is the available cross-sectional area (assumed to be uniform along the length of the branch), and V_i is the volume into which diffusion takes place at the terminal.

In the steady state, the cell body and terminal concentrations can be calculated explicitly. If diffusion is rapid relative to segment elongation, then

$$C_0 \approx C_i \approx \frac{I}{\gamma_0 + n\gamma_i}. \qquad (4.8)$$

In this situation, the entire dendritic tree acts as one large compartment in which the concentration of the branch-determining substance decreases as the number of terminals, n, increases. Thus the branching probability also decreases with the number of terminals. Although it is not identical to the dendritic growth model dependence, n^{-E}, the effect of any given value of E can be approximated by selecting appropriate values for the production rate I and the decay rates γ_0 and γ_i. If diffusion is slow, branch lengths and cross-sectional areas will affect terminal concentrations. More distant terminals may have a lower concentration than those closer to the cell body. Terminals clustered closely on a subtree may

interact so that they have lower concentrations than isolated terminals, owing to their combined decay rates. Branch diameters will influence the amount of substance transported into subtrees, as described for the intracellular signal model. This results in centrifugal order effects similar to that determined by parameter S of the dendritic growth model.

The model is implemented in a computer simulation by calculating the concentrations C_i at the distal ends of every segment (terminal and intermediate) in the tree and C_0 in the cell body. All concentrations are measured in the small volume of the last 1 μm of the segment, and concentration differences are measured over the length of each segment. For the results shown here, the segment diameter is a uniform 1 μm throughout and diffusion is proportional to the total cross-sectional area of a segment.

Comparisons with the dendritic growth and intracellular signal models are shown in figures 4.5 and 4.6. The dendrites of the rat multipolar nonpyramidal cells show no branching dependence on centrifugal order (table 4.3, $S = 0$). Consequently, a fast diffusion rate produces a close match in the degree and centrifugal order distribution with the dendritic growth model (figure 4.5). Rat layer V pyramidal neuron basal dendrites do show a branching dependence on centrifugal order ($S = 0.87$; Van Pelt and Uylings, 1999a). In this case, fast diffusion does not produce a good match to the centrifugal order distribution for the dendritic growth model (figure 4.6) because it shows no effect of centrifugal order on branching probability. This results in the centrifugal order distribution being skewed toward higher values than found in the actual trees. A better distribution results when diffusion is much slower, resulting in a reduction in branching probability with centrifugal order (figure 4.6). If segment diameters are set according to a power law, i.e., to approximate the real dendrites, the centrifugal order effect is lessened because of the increased transport of substance into larger subtrees (results not shown). However, it is possible that the transport area does not increase proportionately with the anatomical area.

A candidate for the branch-determining substance in the intracellular signal and diffusional models is tubulin. This molecule is produced in the cell body and transported along dendrites by a combination of diffusion and active transport (for references, see Van Veen and Van Pelt, 1994). In the growth cones, it is assembled into microtubules, leading to elongation and possibly branching, depending on the stability of the microtubules. Trees that result from the relative rates of elongation and branching determined by microtubule stability, as influenced by the phosphorylation state of MAP2, have been explored in the model of Hely et al. (2001) (see section 4.2.4). In the diffusional model, for realistic elongation rates, centrifugal order effects appear only at unreasonably slow diffusion rates. It remains to be explored whether an active transport component could lead to an inhomogeneous distribution of tubulin in a growing tree.

4.3.4 Elongation Model

The elongation model explores how intracellular molecular transport may affect the growth of particular dendritic segments (Van Veen and Van Pelt, 1994; Van Ooyen et al., 2001). The model is essentially the same as the diffusional model, except that now the concentration of the substance, C_i, in terminal i determines the elongation rate of the segment, rather than its branching probability. If the substance is identified as tubulin, then the rate of change in segment length L_i is a function of the relative rates of assembly, a_i, and disassembly, b_i, of microtubules. The model is described by the following equations (Van Ooyen et al. 2001):

$$\frac{dL_i}{dt} = a_i C_i - b_i \tag{4.9}$$

$$\frac{dC_0}{dt} = I - \gamma_0 C_0 + \sum_{i=1}^{n} \frac{D}{L_i}(C_i - C_0) - \sum_{i=1}^{n} f C_0 \tag{4.10}$$

$$\frac{dC_i}{dt} = b_i - a_i C_i - \gamma_i C_i + \frac{D}{L_i}(C_0 - C_i) + f C_0 \tag{4.11}$$

for uniform cross-sectional area and diffusional volume, and with an active transport rate f. In Van Veen and Van Pelt (1994) there is no degradation of tubulin—which is biologically not plausible and which makes the mathematical analysis more difficult—and no active transport of tubulin.

The model reveals that small differences in (dis)-assembly rates between two branches emanating from the same cell body, and thus competing for the same pool of tubulin, can result in retardation in the growth of one branch while the other grows. If the elongating branch stops growing (say, it has reached its synaptic target), the dormant branch then starts growing after some delay. Such apparent competitive growth is seen in cells grown in culture (see Costa et al., 2002). It remains to be determined whether this is actually due to competition for tubulin.

4.3.5 Compartmental Models

Moving closer to the underlying biophysics of growth and branching requires the specification of ever more detailed models. A first step in this direction is to extend the compartmental modeling framework commonly used to investigate the intracellular (particularly electrical) properties of morphologically static neurons. This framework subdivides the structure of a neuron into small compartments so that quantities of interest (such as membrane potential or molecular concentration) are assumed to have a constant value throughout a compartment.

A compartmental model of the diffusional model (section 4.3.3) requires dividing each segment of the dendritic tree into short compartments. The model must calculate the concentration of the branch-determining substance in each compartment and its diffusion between compartments. The novelty of the situation is that the morphology of the tree is not fixed, so that new compartments must be added or subtracted as the tree grows. Appropriate algorithms for the addition and deletion of compartments are under investigation (Graham and Van Ooyen, 2001). For stable calculation of diffusion and accurate determination of concentrations, finite compartment sizes must be maintained and care must be taken to ensure conservation of material during growth. Such compartmental models allow the investigation of more complex situations in which inhomogeneities along segment lengths can be incorporated. These include the degradation and interaction of molecules as they are transported, or the formation and maintenance of synaptic connections.

4.4 Discussion

This chapter has focused on the question of how to understand the emergence of neuronal morphology from a developmental point of view. The modeling approaches have clearly contributed to this understanding. The dendritic growth model has shown how morphological variability arises from stochasticity in growth cone elongation and branching. The intracellular origins of the global assumptions in the dendritic growth model have been explored, and this has provided further insight into the possible biophysical mechanisms underlying the dependence of the branching probability on the number of terminal segments and their centrifugal order.

Our models may also be seen as an attempt to link different levels of biological organization, by integrating (in a quantitative way) phenomena and mechanisms at the level of molecules, growth cones, and dendritic morphology. Clearly, these are only first steps toward a full, quantitative understanding of how neuronal morphology arises from the cellular machinery in interaction with, and in response to, the many intracellular and local environmental factors.

Important considerations in all modeling studies are the spatial and temporal scales of abstraction. In the dendritic growth model, neuronal development has been approximated by a sustained process of elongation and branching, thereby implicitly assuming a time scale at which "rapid" alterations are averaged out. The models that implement intracellular processes look at a more detailed level and may therefore be more suitable for describing dendritic remodeling, dendritic regression, and activity-dependent plasticity.

Future Modeling Studies

Further modeling work will increasingly be confronted with the multitude and complexity of the processes involved in dendritic growth. Several strategies may be followed in future work. First, we may search for general principles of organization, assuming that the concerted actions of the many mechanisms involved serve simple and robust functional goals. Homeostasis (of, e.g., the level of electrical activity or the intracellular calcium concentration; see chapter 6) is one such functional goal. Second, we may focus on particular biophysical processes and explore their effects on dendritic growth. Examples of such processes are the (de)polymerization of cytoskeletal elements and the production and transport of structural proteins (e.g., tubulin, MAPs). Third, we may follow a brute force approach by including in a computational model all actors and structures known at present to be involved in growth cone behavior and neurite outgrowth. Such an approach will undoubtedly lead to a highly complex model, but it may allow the computational study of neurite outgrowth in relation to any parameter involved.

Future Experimental Studies

In order to validate the models, predictions from the modeling studies must be complemented by experimental investigations. Our models have made a number of predictions that can be tested experimentally.

Given a population of dendritic trees of neurons at some stage of development, the dendritic growth model can calculate, from the observed standard deviation and mean of the number of terminal segments in the population, the value of the competition parameter E and from this predict (see figure 4.2B) what the standard deviation should be at another stage of development (e.g., at a later stage of development, when the mean number of terminal segments has increased). This thus provides a relatively straightforward way of testing the dendritic growth model. Another prediction of the dendritic growth model is that after splitting of a growth cone, the daughter branches should already have a small initial length. The model also predicts quantitative values for the elongation rate of the growth cone (see table 4.3). The intracellular signal and diffusional models predict the likely effects of the differential transport of a growth-determining substance, such as tubulin, on the branching patterns formed by the growing neurite.

A significant outcome of the elongation model is the prediction of competition between elongating neurites (of the same neuron). Such apparent competitive neurite growth has indeed been observed in neurons grown in culture (Costa et al., 2002). To test whether this is due to competition for tubulin, as our model suggests, the concentration of tubulin in growth cones should be monitored. The model predicts that the concentration of tubulin in growth cones that are not growing out should be below the critical value [i.e., the concentration of tubulin at which assembly ($a_i C_i$) just equals disassembly (b_i)].

The model by Hely et al. (2001) predicts what the form of the dose–response functions relating calcium with phosphorylation and dephosphorylation should be to obtain trees in which the terminal segments are longer than the proximal segments or to obtain trees in which the terminal segments are shorter than the proximal segments. Thus, to test the model, these dose–response functions can be measured in neurons with different branching patterns.

References

Acebes, A., and Ferrus, A. (2000). Cellular and molecular features of axon collaterals and dendrites. *Trends Neurosci.* 23: 557–565.

Aeschlimann, M. (2000). "Biophysical Models of Axonal Pathfinding." PhD thesis, University of Lausanne, Faculty of Science, Switzerland.

Ascoli, G. A., and Krichmar, J. L. (2000). L-Neuron: A modeling tool for the efficient generation and parsimonious description of dendritic morphology. *Neurocomputing* 32–33: 1003–1011.

Berry, M., Hollingworth, T., Anderson, E. M., and Flinn, R. M. (1975). Application of network analysis to the study of the branching patterns of dendritic fields. *Ad. Neurol.* 12: 217–245.

Burke, R. E., Marks, W. B., and Ulfhake, B. (1992). A parsomonious description of motoneuron dendritic morphology using computer simulation. *J. Neurosci.* 12: 2403–2416.

Cannon, R. C., Wheal, H. V., and Turner, D. A. (1999). Dendrites of classes of hippocampal neurons differ in structural complexity and branching patterns. *J. Comp. Neurol.* 413: 619–633.

Cline, H. T. (1999). Development of dendrites. In *Dendrites*, G. Stuart, N. Spurston, and M. Häusser, eds. pp. 35–67. Oxford: Oxford University Press.

Costa, L. da F., M. Manoel, E. T. M., Faucereau, F., Chelly, J., Van Pelt, J., and Ramakers, G. J. A. (2002). A shape analysis framework for neuromorphometry. *Network: Comput. Neural Syst.* 13: 283–310.

Dacey, M. F., and Krumbein, W. C. (1976). Three growth models for stream channel networks. *J. Geol.* 84: 153–163.

Devaud, J. M., Quenet, B., Gascuel, J., and Masson, C. (2000). Statistical analysis and parsimonious modelling of dendrograms of *in vitro* neurones. *Bull. Math. Biol.* 62: 657–674.

Dityatev, A. E., Chmykhova, N. M., Studer, L., Karamian, O. A., Kozhanov, V. M., and Clamann, H. P. (1995). Comparison of the topology and growth rules of motoneuronal dendrites. *J. Comp. Neurol.* 363: 505–516.

Fiala, J. C., and Harris, K. M. (1999). Dendritic structure. In *Dendrites*, G. Stuart, N. Spurston, and M. Häusser, eds. pp. 1–28. Oxford: Oxford University Press.

Fitter, A. H., Stickland, T. R., Harvey, M. L., and Wilson, G. W. (1991). Architectural analysis of plant root systems 1. Architectural correlates of exploitation efficiency. *New Phytol.* 118: 375–382.

Graham, B., and Van Ooyen, A. (2001). Compartmental models of growing neurites. *Neurocomputing* 38–40: 31–36.

Graham, B., Hely, T., and Van Ooyen, A. (1998). An internal signalling model of the dendritic branching process. *Eur. J. Neurosci.* 10. Suppl.: 274.

Hely, T. A., Graham, B. P., and Van Ooyen, A. (2001). A computational model of dendrite elongation and branching based on MAP2 phosphorylation. *J. Theor. Biol.* 210: 375–384.

Hillman, D. E. (1979). Neuronal shape parameters and substructures as a basis of neuronal form. In *The Neurosciences*, F. O. Schmitt and F. G. Worden, eds. pp. 477–498. Cambridge, Mass.: MIT Press.

Hillman, D. E. (1988). Parameters of dendritic shape and substructure: Intrinsic and extrinsic determination? In *Intrin-sic Determinants of Neuronal Form and Function*, R. J. Lasek and M. M. Black, eds. pp. 83–113. New York: A. R. Liss.

Hollingworth, E. M., and Berry, M. (1975). Network analysis of dendritic fields of pyramidal cells in neocortex and Purkinje cells in the cerebellum of the rat. *Phil. Trans. Roy. Soc. London B.* 270: 227–264.

Horsfield, K., and Woldenberg, M. J. (1986). Branching ratio and growth of tree-like structures. *Resp. Physiol.* 63: 97–107.

Horsfield, K., Woldenberg, M. J., and Bowes, C. L. (1987). Sequential and synchronous growth models related to vertex analysis and branching ratios. *Bull. Math. Biol.* 49: 413–429.

Ipiña, S. L., Ruiz-Marcos, A., Del Rey, F. E., and De Escobar, G. M. (1987). Pyramidal cortical cell morphology studied by multivariate analysis: Effects of neonatal thyroidectomy, ageing and thyroxine-substitution therapy. *Dev. Brain Res.* 37: 219–229.

Ireland, W., Heidel, J., and Uemura, E. (1985). A mathematical model for the growth of dendritic trees. *Neurosci. Lett.* 54: 243–249.

Kliemann, W. A. (1987). Stochastic dynamic model for the characterization of the geometrical structure of dendritic processes. *Bull. Math. Biol.* 49: 135–152.

Koenderink, M. J. Th., and Uylings, H. B. M. (1995). Postnatal maturation of the layer V pyramidal neurons in the human prefrontal cortex. A quantitatitve Golgi study. *Brain Res.* 678: 233–243.

Koester, S. E., and O'Leary, D. D. M. (1992). Functional classes of cortical projection neurons develop dendritic distinctions by class-specific sculpting of an early common pattern. *J. Neurosci.* 12: 1382–1393.

Ley, K., Pries, A. R., and Gaehtgens, P. (1985). Topological structure of rat mesenteric microvessel networks. *Microvascular Res.* 32: 315–332.

Li, G.-H., Qin, C.-D., and Wang, Z.-S. (1992). Neurite branching pattern formation: Modeling and computer simulation. *J. Theor. Biol.* 157: 463–486.

Li, G.-H., Qin, C.-D., and Wang, L.-W. (1995). Computer model of growth cone behavior and neuronal morphogenesis. *J. Theor. Biol.* 174: 381–389.

Nowakowski, R. S., Hayes, N. L., and Egger, M. D. (1992). Competitive interactions during dendritic growth: A simple stochastic growth algorithm. *Brain Res.* 576: 152–156.

Parnavelas, J. G. (2000). The origin and migration of cortical neurones: New vistas. *Trends Neurosci.* 23: 126–131.

Parnavelas, J. G., and Uylings, H. B. M. (1980). The growth of non-pyramidal neurons in the visual cortex of the rat: A morphometric study. *Brain Res.* 193: 373–382.

Pentney, R. J. (1986). Quantitative analysis of dendritic networks of Purkinje neurons during aging. *Neurobiol. Aging.* 7: 241–248.

Quackenbush, L. J., Ngo, H., and Pentney, R. J. (1990). Evidence for nonrandom regression of dendrites of Purkinje neurons during aging. *Neurobiol. Aging.* 11: 111–115.

Rakic, P. (1995). A small step for the cell, a giant leap for mankind: A hypothesis of neocortical expansion during evolution. *Trends Neurosci.* 18: 383–388.

Ramakers, G. J. A. (2000). Rho proteins and the cellular mechanisms of mental retardation. *Am. J. Med. Genet.* 94: 367–371.

Sadler, M., and Berry, M. (1983). Morphometric study of the development of Purkinje cell dendritic trees in the mouse using vertex analysis. *J. Micros.* 131: 341–354.

Shreve, R. L. (1966). Statistical law of stream numbers. *J. Geol.* 74: 17–37.

Tamori, Y. (1993). Theory of dendritic morphology. *Phys. Rev.* E48: 3124–3129.

Ten Hoopen, M., and Reuver, H. A. (1971). Growth patterns of neuronal dendrites—an attempted probabilistic description. *Kybernetik* 8: 234–239.

Uylings, H. B. M. (2000). Development of the cerebral cortex in rodents and man. *Eur. J. Morphology.* 38: 309–312.

Uylings, H. B. M., Ruiz-Marcos, A., and Van Pelt, J. (1986). The metric analysis of three-dimensional dendritic tree patterns: A methodological review. *J. Neurosci. Meth.* 18: 127–151.

Uylings, H. B. M., Van Pelt, J., Parnavelas, J. G., and Ruiz-Marcos, A. (1994). Geometrical and topological characteristics in the dendritic development of cortical pyramidal and nonpyramidal neurons. *Progr. Brain Res.* 102: 109–123.

Van Ooyen, A., Graham, B. P., and Ramakers, G. J. A. (2001). Competition for tubulin between growing neurites during development. *Neurocomputing* 38–40: 73–78.

Van Pelt, J., and Verwer, R. W. H. (1986). Topological properties of binary trees grown with order-dependent branching probabilities. *Bull. Math. Biol.* 48: 197–211.

Van Pelt, J., and Uylings, H. B. M. (1999a). Natural variability in the geometry of dendritic branching patterns. In *Modeling in the Neurosciences: From Ionic Channels to Neural Networks*, R. R. Poznanski, ed. pp. 79–108. Amsterdam: Harwood Academic.

Van Pelt, J., and Uylings, H. B. M. (1999b). Modeling the natural variability in the shape of dendritic trees: Application to basal dendrites of small rat cortical layer 5 pyramidal neurons. *Neurocomputing* 26–27: 305–311.

Van Pelt, J., and Uylings, H. B. M. (2002). Branching rates and growth functions in the outgrowth of dendritic branching patterns. *Network: Comput. Neural Syst.* 13: 261–281.

Van Pelt, J., Woldenberg, M. J., and Verwer, R. W. H. (1989). Two generalized topological models of stream network growth. *J. Geol.* 97: 281–299.

Van Pelt, J., Uylings, H. B. M., Verwer, R. W. H., Pentney, R. J., and Woldenberg, M. J. (1992). Tree asymmetry—a sensitive and practical measure for binary topological trees. *Bull. Math. Biol.* 54: 759–784.

Van Pelt, J., Van Ooyen, A., and Uylings, H. B. M. (2001a). Modeling dendritic geometry and the development of nerve connections. In *Computational Neuroscience: Realistic Modeling for Experimentalist*, E. DeSchutter and R. C. Cannon, eds. (CD-ROM). pp. 179–208. Boca Raton, Fla.: CRC Press.

Van Pelt, J., Schierwagen, A., and Uylings, H. B. M. (2001b). Modeling dendritic morphological complexity of deep layer cat superior colliculus neurons. *Neurocomputing* 38–40: 403–408.

Van Veen, M. P., and Van Pelt, J. (1994). Neuritic growth rate described by modeling microtubule dynamics. *Bull. Math. Biol.* 56: 249–273.

Woldenberg, M. J., O'Neill, M. P., Quackenbush, L. J., and Pentney, R. J. (1993). Models for growth, decline and regrowth of the dendrites of rat Purkinje cells induced from magnitude and link-length analysis. *J. Theor. Biol.* 162: 403–429.

Axon Guidance and Gradient Detection by Growth Cones

5

Geoffrey J. Goodhill and Jeffrey S. Urbach

Growing axons can find appropriate targets in the developing nervous system with remarkable precision. They do this using a variety of molecular cues, including concentration gradients. Axon guidance is a very active area of experimental research, and several of the molecules involved have recently been identified. This chapter reviews these recent data and some of the theoretical models proposed to account for these data. It particularly focuses on constraints on guidance by a target-derived diffusible factor, and on signal-to-noise constraints on gradient detection by growth cones.

5.1 Neurobiological Background

A crucial stage in the construction of the nervous system is the appropriate wiring up of its components. As the brain develops, axons often have to grow over long distances to find their appropriate targets. One can distinguish between initial pathfinding, by which axons grow out to the right general target region, and more specific processes, such as topographic map formation (see chapter 11), by which axons find the right location within the target region. Axons perform both these tasks by integrating a number of different molecular cues in their environment. These cues are primarily detected by the growth cone, a complex and sensitive structure at the tip of the developing axon. Several families of molecules have been identified as playing important roles in guidance in many different parts of the nervous system, and just a few general mechanisms can now account, at least qualitatively,

for many guidance phenomena both in vivo and in vitro (Tessier-Lavigne and Goodman, 1996). A major challenge is now to understand these mechanisms and molecules from a more quantitative perspective.

5.1.1 Types of Mechanisms in Axon Guidance

Most of the molecular mechanisms now known to be involved in axon guidance can be characterized by three traits, although many molecules, depending on the circumstances, can take on a number of different roles.

1. *Permissive or inhibitory.* Conceptually one of the simplest ways of directing axons along specific paths is to express molecules in three-dimensional patterns that label some parts of the environment as permissive for growth and others as inhibitory for growth. For instance, a "railroad track" of a molecule that encourages growth can join a population of axons and their target, or axons can be restricted to a narrow path by surrounding inhibitory signals. Molecules that simply encourage the growth of cells and/or axons are sometimes called trophic factors.

2. *Attractive or repulsive.* Here the molecule is effective for guidance when it is present in a concentration gradient, so that it provides a vector signal for the axon to move in one direction or another in a process known as chemotaxis. This can be a positive (attractive) or a negative (repulsive) signal. For instance, a molecule released by a target structure can set up a gradient by diffusion, which provides information for distant axons as to the location of the target.

Molecules that provide such directional signals are sometimes called tropic factors.

3. *Contact-mediated or diffusible.* Another characteristic is whether the molecule is relatively fixed, e.g., expressed on cell membranes or bound to the extracellular matrix (ECM), or whether it diffuses more freely through the ECM. In the former case, the growth cone must remain in direct contact with the relevant tissue to be guided, whereas in the latter this need not be the case. As previously mentioned, an important class of freely diffusing molecules consists of target-derived diffusible factors.

Two additional mechanisms help to simplify the wiring problem. Although the final target for an axon may be quite distant, its pathway is often broken up into several shorter segments, with possibly different guidance mechanisms operating on each segment. In addition, axons generated after the first pioneering population need only follow the track laid down already, e.g., by fasciculating (bundling) with the pioneering axons.

5.1.2   Families of Molecules Involved in Axon Guidance

A much larger number of molecules play a role in axon guidance than it is possible to review here. Instead, we focus on just a few families discovered in the past few years that provide paradigm examples of how tropic factors are relevant in neural development.

Netrin-1
One of the most important systems for the study of axon guidance mechanisms in recent years has been the growth of axons to and across the midline in the developing spinal cord. For instance, commissural axons initially generated in the dorsal spinal cord grow ventrally toward the floor plate and then cross to

the contralateral side. Tessier-Lavigne and colleagues (Kennedy et al., 1994; Serafini et al., 1994) purified a 75-kD protein, Netrin-1, based on its ability to attract commissural axons. They also showed that it was diffusible and that it was expressed by the floor plate at appropriate times during development. Analysis of the Netrin-1 knockout mouse provided further evidence that Netrin-1 indeed plays a crucial role in attracting commissural axons to the midline in normal in vivo development (Serafini et al., 1996).

A common theme among different families of axon guidance molecules is that they are reused in many different places and times during development. This is abundantly true of Netrin-1 , which has also been shown to guide cerebellofugal axons in the rostral hindbrain toward the floor plate (Shirasaki et al., 1995), alar plate axons in the myelencephalon toward the mesencephalon (Shirasaki et al., 1996), retinal ganglion cell axons to the optic disk (Deiner et al., 1997), and cortical axons to subcortical targets (Richards et al., 1997). Netrin-1 has also been shown to provide important repellent guidance signals in some circumstances; for instance, ventrally generated trochlear motor neurons are repelled by Netrin-1 and floor plate (Colamarino and Tessier-Lavigne, 1995), as are dorsally projecting cranial motor axons (Varela-Echavarria et al., 1997).

An initial hypothesis, based on work on Netrin-1 homologs in *Caenorhabditis elegans*, was that Netrin-1 may be signaling through a different receptor in these cases. In the examples of attractive guidance cited above, Netrin-1 binds to the receptor DCC (deleted in colorectal cancer) with an affinity of about 1 nM (Keino-Masu et al., 1996), whereas in *C. elegans* it exerts a repulsive effect via binding to the receptor Unc5. However, more recent work has shown that the determination of an attractive versus repulsive response is much more complex than this; DCC and Unc5 must come together to produce a response, and it is the intracellular domain alone of the receptor that determines the sign of the response (Hong et al., 1999). In addition, as discussed in section 5.1.3, manipulations in the levels of cyclic nucleotides in the growth cone can convert attraction to repulsion and vice versa (Song et al., 1997).

Ephrins and Eph Receptors

The ephrin family of ligands, acting repulsively through receptors of the Eph family, plays an important role in the initial activity-independent stage of topographic map formation in the retinotectal system (see chapter 11). Unlike netrin-1, these ligands are substrate bound. Increasing nasal-to-temporal gradients of Eph receptors are present in the retina, while increasing anterior-to-posterior gradients of ephrin are present in the tectum (Cheng et al., 1995; Drescher et al., 1995; Feldheim et al., 1998). These patterns provide a substrate for the "chemoaffinity" hypothesis of Sperry (1963), and there is evidence from misexpression and knockout experiments that the resulting topographic map is strongly affected by the shape of Eph and ephrin gradients (Friedman and O'Leary, 1996; Feldheim et al., 2000; Goodhill, 2000). Ephrin gradients also seem to be important in map formation in the thalamus and cortex (Feldheim et al., 1998; Vanderhaeghen et al., 2000). In addition, ephrins have been suggested to play a role in regulating branching in the cortex (Castellani et al., 1998), although this is controversial (Yabuta et al., 2000). For reviews, see Flanagan and Vanderhaeghen (1998) and Goodhill and Richards (1999).

Semaphorins and Slits

The semaphorins (also called collapsins) are a very large family of axon guidance molecules (Raper, 2000). They bind to the neuropilin and plexin families of receptors and have been implicated in the correct targeting of sensory projections to the spinal cord (Messersmith et al., 1995). Although so far their in vivo function mostly appears to be repulsive, under some circumstances they can also act as attractive cues (Song et al., 1998). They can be both diffusible and substrate bound. Members of the Slit family of guidance molecules are expressed at the midline in both the spinal cord and the brain (Brose and Tessier-Lavigne, 2000). They are usually repulsive and, acting through receptors of the Robo family, play an important role in regulating whether axons can cross the midline (e.g., Shu and Richards, 2001), and where they project to beyond the midline (Rajagopalan et al., 2000; Simpson et al., 2000; Goodhill, 2003). Surprisingly, members of the Slit family have also been implicated in controlling the branching of dorsal root ganglion axons (Wang et al., 1999).

5.1.3 Growth Cones

The growth cone integrates guidance cues from the extracellular environment and guides the developing axon to its target. The morphology of the growth cone varies somewhat but typically consists of a central zone at the terminus of the axon containing organelles and microtubule bundles and surrounded by highly dynamic spikelike protrusions, the filopodia, and weblike veils, the lamellipodia. Both filopodia and lamellipodia consist primarily of actin filaments. The filopodia are narrow, 0.2–0.5 μm, and can be up to 40 μm in length. When axon growth follows a well-defined path, e.g., during fasciculation, the growth cones tend to be small and compact, while at decision points they are widely spread out (reviewed in Rehder and Kater, 1996). The contribution of filopodia to axonal navigation has been suggested through observation of the interactions of filopodia with substrate-bound guidance cues both in vitro and in vivo (O'Connor et al., 1990; Myers and Bastiani, 1993; Steketee and Tosney, 1999). When the filopodia are eliminated, growth cones cannot navigate their environment and do not respond to either substrate-bound or diffusible guidance cues (Bentley and Toroian-Raymond, 1986; Chien et al., 1993).

While some of the biochemical mechanisms underlying the dynamics of filopodia and lamellipodia have been identified (Suter and Forscher, 1998), the pathways responsible for chemotaxis are mostly unknown. However, an intriguing recent finding is that attraction can be converted to repulsion and vice versa by altering levels of cyclic nucleotides within the growth cone (Song et al., 1997, 1998; Ming et al., 1997; Song and Poo, 1999). For instance, the normally attractive response of *Xenopus* spinal axons to a Netrin-1 gradient can be converted to repulsion by lowering levels of cAMP, and the normally repulsive response of the same axons to a gradient of Sema III can be converted to attraction by raising levels of cyclic guanosine monophosphate (cGMP). Calcium, both from outside the growth cone and from intracellular stores, seems to play a crucial role in the signaling pathways (Hong et al., 2000; Zheng, 2000), but much work remains to be done to trace the complete paths from receptor binding to directed movement.

5.1.4 Relation of Axon Guidance to Other Fields of Developmental Biology

Mechanisms of axon guidance may be closely related to mechanisms regulating other aspects of neural development. Two specific examples are cell migration and activity-dependent plasticity. During development, neurons often make long journeys from their initial birthplace to their final location. In some cases they seem to be following gradients of exactly the same families of molecules that guide axons to their targets (e.g., Wu et al., 1999). In addition, the signal transduction pathways in such cellular chemotaxis

may be similar to those involved in growth cone chemotaxis (Parent and Devreotes, 1999; Dekker and Segal, 2000).

Once the basic architecture of the nervous system has formed, synapses are modified by patterns of neural activity. Besides changing their strength within a fixed architecture, fine-scale structural changes can occur by the making of new contacts and the breaking of old ones. This local searching of space for the most appropriate nearby contacts may involve both the generation of filopodia (Jontes and Smith, 2000) and the tropic guidance of axons and dendrites by factors released in an activity-dependent manner. For instance, the release of brain-derived neurotrophic factor (BDNF) is activity dependent (McAllister et al., 1999), and BDNF can also act as a tropic factor (Song et al., 1997).

5.2 Review of Models

Several types of models have been developed that address axon guidance phenomena at a variety of levels. As with all modeling, the goals are to construct mathematical frameworks that impose order on complex phenomena, to propose quantitatively rigorous hypotheses, and to produce testable predictions. Such goals are becoming increasingly relevant as the number of molecules implicated in axon guidance multiplies and the qualitative mechanisms invoked to explain the data become increasingly subtle.

5.2.1 The Growth Cone

Van Veen and Van Pelt (1992) modeled the dynamics of filopodia on the growth cone and showed how the morphological characteristics of neurites, such as the amount of branching, could emerge from the interaction of the growth cone with its environment. Buettner and colleagues have also developed models of filopodial and microtubular dynamics (e.g. Buettner, 1995; Odde and Buettner, 1998) based on experimentally determined distributions for such parameters as rates of filopodial initiation, extension, and retraction, filopodial length, and angular orientation. These models suggest that the dynamics of the growth cone and axon–target encounters are determined by the dynamics of the filopodia and microtubules.

Hely and Willshaw (1998) modeled microtubule dynamics and the interactions between microtubules and F-actin, and found that these interactions may play an important role in determining axonal growth rate and in microtubule invasion into the growth cone once the target cell is reached. Meinhardt (1999) proposed a model for how amplification of a weak gradient signal occurs within the growth cone. This is based on a reaction-diffusion system in which a small inhomogeneity in an initially uniform system is amplified by the interaction of a short-range activator with a longer-range inhibitor. A second type of reaction with a longer time constant is invoked to return the system to a uniform state so that the directional preference of the growth cone can change with time.

5.2.2 Diffusible Factors

One branch of modeling has attempted to understand certain axon guidance phenomena in terms of the physics of diffusible factors. Goodhill (1997, 1998) modeled guidance by a target-derived diffusible factor by imagining a small target that releases factors at a constant rate into a large uniform volume. By calculating the places and times at which certain sensing constraints are satisfied (i.e., both the absolute concentration at, and the fractional change across, the growth cone exceed minimum thresholds), he derived the maximum distance over which an axon could be guided in such a scenario. This is discussed further in section 5.3.1.

Hentschel and Van Ooyen (1999) investigated a possible role for diffusion in controlling axon fasciculation. They considered a population of axons being guided by a target-derived diffusible factor, and hypothesized that in addition each axon releases a diffusible attractant that pulls it toward the other axons, leading to fasciculation as they grow together toward the target. In order to account for defasciculation at the target, they hypothesized that each axon also releases a repulsive factor for other axons at a rate dependent on the concentration of the target-derived factor. As the axons approach the target, this repulsive force overcomes the attractive force, leading to defasciculation.

5.2.3 Retinotectal Maps

Perhaps the richest set of models of axon guidance are those addressing map formation in the retinotectal system (see chapter 11). Sperry (1963) qualitatively proposed the "chemospecificity" hypothesis: that graded distributions of molecules are somehow matched to graded distributions of complementary molecules in the tectum so as to form a topographic map. This inspired a great deal of experimental work to test the hypothesis, based on surgical manipulation experiments in which parts of the retina or tectum were removed, rotated, or translocated (reviewed in Udin and Fawcett, 1988; Goodhill and Richards, 1999). In parallel, several theoretical models were developed that implemented different versions of Sperry's hypothesis, and for a period there was a relatively close association between models and data. Prestige and Willshaw (1975) proposed a competitive model in which all axons from one end of the retina compete to innervate space at one end of the tectum. They showed that some kind of normalization process is essential for this to produce smooth maps (see also Goodhill, 2000). Willshaw and Von der Malsburg

(1979) proposed the "tea trade model," in which distributions of molecules in the retina were hypothesized to be transported up retinal axons and then diffused within the tectum. A molecular version of Hebb's rule is then sufficient to produce a smooth map. Other models also considered various types of molecular matching mechanisms (e.g., Fraser, 1980; Whitelaw and Cowan, 1981; Gierer, 1987). However, with a few exceptions (e.g., Fraser and Perkel, 1990; Weber et al., 1997; Honda, 1998), most models since then have focused on the activity-dependent aspects of retinotectal map formation. No models have yet attempted to address the complexity of the rapidly evolving literature on the role of Eph and ephrin gradients.

5.2.4 Sensing at the Growth Cone

In order to move reliably in response to a molecular gradient of a diffusible factor, the growth cone of the developing axon must perform some relatively sophisticated signal analysis to overcome the noise inherent in a measurement of molecules that move about randomly through Brownian motion. The fundamental statistical limitations on gradient detection by a small sensing device were originally described by Berg and Purcell (1977) in the context of understanding chemotaxis in leukocytes and bacteria. The presence of a ligand gradient will produce a variation in the average occupancy of receptors across the sensing device. Regardless of the specific signal transduction mechanisms involved, quantitative limits on the guidance process can be obtained from the fact that at any instant in time, the actual occupancy of the receptors will differ from the average. If these fluctuations are large compared with the difference in the average that arises from the concentration gradient, the sensing device will not be able to obtain a clean guidance signal from the gradient in receptor occupancy at any

one instant. This random noise can be overcome by making a sufficient number of statistically independent measurements. Berg and Purcell (1977) performed an analysis of the size of these fluctuations, which makes predictions about the minimum steepness of gradient the sensing device can detect as a function of various parameters. This approach has been applied specifically to growth cones (Goodhill and Urbach, 1999; Urbach and Goodhill, 1999) and is described further in section 5.3.2.

5.3 Some Models in More Detail

In this section, we focus on constraints on guidance by a target-derived diffusible factor, and on signal-to-noise constraints on gradient detection by growth cones.

5.3.1 Guidance by a Target-Derived Diffusible Factor

As a simple first step toward understanding axon guidance by target-derived diffusible factors both in vivo and in vitro, Goodhill (1997, 1998) considered a point source continuously releasing a factor with a diffusion constant D cm^2/sec, at a rate of q mol/sec, into an infinite, spatially uniform three-dimensional volume. Initially, no decay of the factor was assumed. For radially symmetrical Fickian diffusion in three dimensions, the diffusion equation has the form

$$\frac{\partial C(r, t)}{\partial t} = D\left[\frac{\partial^2 C(r, t)}{\partial r^2} + \frac{2}{r}\frac{\partial C(r, t)}{\partial t}\right], \quad (5.1)$$

where $C(r, t)$ is the concentration at distance r from the source at time t, and the source is at $r = 0$. This has the solution for the above boundary conditions of

$$C(r, t) = \frac{q}{4\pi Dr}\, \text{erfc}\, \frac{r}{\sqrt{4Dt}} \quad (5.2)$$

(e.g., Crank, 1975), where $\mathrm{erfc}(x) = 1 - \mathrm{erf}(x)$, and $\mathrm{erf}(x)$ is the error function: $\mathrm{erf}(x) = (2/\sqrt{\pi}) \cdot \int_0^x e^{-\xi^2}\, d\xi$. As $t \to \infty$, $C(r,t) \to q/4\pi Dr$.

If the factor is reversibly bound to the substrate through which the growth cone is growing, so that the amount S bound at any time equals a constant R times the local free concentration C, then it is straightforward to show that the relevant solution of the diffusion equation is the same as for the case without binding, but with an effective diffusion constant of $D/(R+1)$; the speed of diffusion is simply reduced (Crank, 1975). Motivated by the signal-to-noise arguments discussed in section 5.3.2, we assume that the important constraint on gradient detection is the fractional change $\Delta C/C$ across the growth cone. As long as the gradient is not too steep, this change across a growth cone width a is given by $(\Delta C/C) = [(\partial C/\partial r)(a/C)]$.

For Eq. (5.2), this can be straightforwardly calculated:

$$\frac{\Delta C}{C} = \frac{a}{r}\left[1 + \frac{r}{\sqrt{\pi Dt}}\frac{e^{-r^2/4Dt}}{\mathrm{erfc}(r/\sqrt{4Dt})}\right]. \qquad (5.3)$$

(This expression is actually negative, meaning that in $\partial C/\partial r$, C decreases as r increases; the minus sign has been omitted for clarity.) $\Delta C/C$ has an identical form for the two cases of reversible binding and irreversible decay considered above. This function has the perhaps surprising characteristic that for fixed r, $\Delta C/C$ decreases with t. That is, the largest gradient at any distance occurs immediately after the source starts releasing factor (see figure 5.1a,b). For large t, $\Delta C/C$ asymptotes at a/r. Thus: (1) At small times after the start of production, the factor is unevenly distributed. The concentration C falls quickly to almost zero moving away from the source, the gradient is steep, and the percentage change $\Delta C/C$ across the growth cone is large everywhere. (2) As time proceeds, the

factor becomes more evenly distributed, C increases everywhere, but $\Delta C/C$ decreases everywhere. (3) For large times, C tends to an inverse variation with the distance from the source r, while $\Delta C/C$ tends to a/r independent of all other parameters.

Equation (5.3) gives the size of the true gradient, while Eq. (5.6) (see later) gives the size of the smallest gradient the growth cone can detect. These two can be compared to find the regions of parameter space in which guidance is possible. Goodhill (1997, 1998) considered the simpler case where Eq. (5.6) is approximated by a step function. Based on data for leukocyte chemotaxis such as that of Zigmond (1977, 1981), he assumed that gradient detection occurs when $\Delta C/C \geq p$ and $C \geq C_{\min}$, where p is a threshold assumed to be independent of C. [The high concentration limit, where all receptors are saturated, does not significantly constrain guidance in this case (Goodhill, 1997).] Given appropriate estimates for the parameters D, q, a, p, and $C_{\min}$, the positions and times for which the gradient calculated above satisfies these criteria were examined.

The constraints arising from Eqs. (5.2) and (5.3) are plotted in figure 5.1. The cases of $D = 10^{-7}$ cm^2/sec and $D = 5 \times 10^{-7}$ cm^2/sec are shown in figures 5.1C and 5.1D, respectively. For large times (several days) after the start of factor production, the maximum distance from the target for which axon guidance is possible is independent of the diffusion constant and is about 1 mm (figure 5.1). This value fits well with what has been observed in 3-D collagen gel cultures and with the fact that target and growth cone are not separated by more than a few hundred microns in vivo for the guidance of axons from the trigeminal ganglion to the maxillary process in the mouse (Lumsden and Davies, 1983, 1986), or of commissural axons in the spinal cord to the floor plate (Tessier-Lavigne et al., 1988). This limit on maximum guidance distance is due to the requirement that there be a

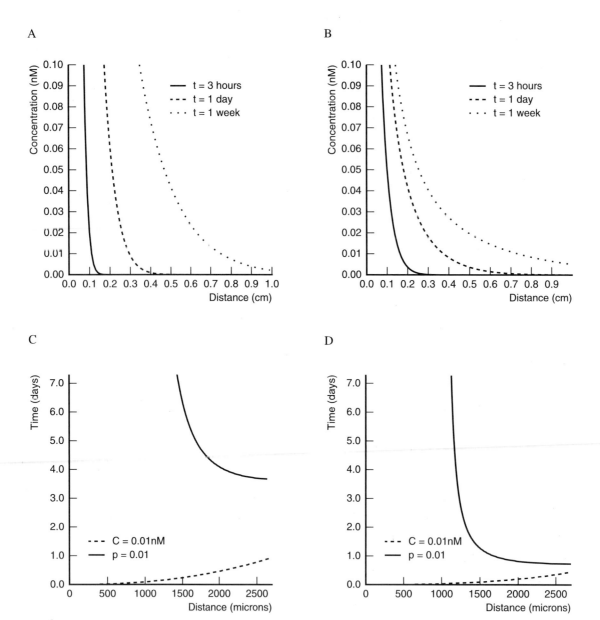

Figure 5.1

Gradients for guidance by a target-derived diffusible factor. (A), (B) Concentration profiles. (C), (D) Interaction of constraints. (A) and (C) are for a larger molecule with a diffusion constant $D = 10^{-7}$ cm^2/sec. (B) and (D) are for a smaller molecule with $D = 5 \times 10^{-7}$ cm^2/sec. In (C) and (D), each graph shows, at each distance, the time at which two constraints are satisfied: the low concentration limit, where not enough receptors are bound for a gradient signal to be detected (assumed to be $K_D/100$, with $K_D = 1$ nM, and the fractional change constraint (assumed to be $p \geq \Delta C/C = 1$ percent). The region between the two curves in each graph is where guidance is possible. In both cases, the guidance limit imposed by the fractional change constraint once the gradient has stabilized is 1 mm. However, the guidance range is extended at earlier times, when the fractional change constraint has yet to take full effect. This is particularly apparent for the slowly diffusing molecule ($D = 10^{-7}$ cm^2/sec).

minimum change in concentration across the growth cone. The minimum concentration constraint is easily satisfied at large times after the start of factor production. At earlier times, however, the factor is more unevenly distributed, being more concentrated around the source. This makes the fractional change larger than at later times, increasing the range over which guidance can occur (figure 5.1). Depending on the parameters, the model predicts that guidance may be possible at distances of several millimeters before the distribution of factor equilibrates. This is particularly true for a large molecule that diffuses slowly; the change across the growth cone remains larger for a longer time. It is conceivable that such a mechanism might be utilized in vivo to extend the guidance range beyond the 1-mm limit imposed once the gradient has stabilized.

5.3.2   Statistical Constraints on Gradient Detection

As described in section 5.2.4, Berg and Purcell (1977) performed a classic analysis of gradient detection by a small sensing device (e.g., a growth cone). Reliable guidance by a chemical gradient requires that the statistical fluctuations in the measured concentration be less than the difference in average concentration on opposite sides of the sensing device that is due to the gradient. In particular, Berg and Purcell did several increasingly sophisticated calculations for the fractional error $\Delta C_{noise}/C$, the ratio of the root-mean-square fluctuation in the concentration estimate to the mean concentration.

The simplest analysis is as follows: Random fluctuations in molecular positions will produce fluctuations in the concentration in the vicinity of the growth cone, given roughly by

$$\frac{\Delta C_{noise}}{C} = \frac{1}{\sqrt{N}}, \tag{5.4}$$

where N is the number of molecules around the cell. This is approximately equal to $1/\sqrt{VC}$, where $V = a^3$ is the volume of the cell and C is the average ligand concentration. However, this is for an instantaneous measurement. If the concentration is instead averaged over M measurements suitably spaced over time, the fractional error is reduced by roughly $1/\sqrt{M}$. "Suitably spaced" means allowing sufficient time for a molecule that is counted in one measurement to diffuse away before the next measurement, about a^2/D. Thus if the total averaging time available is T, then the number of independent measurements is $T/(a^2/D)$, so

$$\frac{\Delta C_{noise}}{C} = \frac{1}{\sqrt{aDTC}}. \tag{5.5}$$

Using more sophisticated methods, Berg and Purcell also calculated $\Delta C_{noise}/C$ for a measuring instrument that relies on the statistics of binding to a finite number of receptors to assess concentration. In the three-dimensional case, they find

$$\frac{\Delta C_{noise}}{C} = \sqrt{\frac{1}{2\pi T D a \frac{N_r s}{N_r s + \pi a} \frac{CC_{1/2}}{C + C_{1/2}}}}, \tag{5.6}$$

where N_r is the number of receptors on the growth cone, s is the effective receptor area, and $C_{1/2}$ is the concentration of ligand molecules at which half the receptors are bound, i.e., the dissociation constant K_D (measured in units of molecules per volume). Note that the fractional uncertainty given by Eq. (5.6) tends to a constant as C increases above $C_{1/2}$. On the one hand, there are more ligand molecules available for making a comparison, and the uncertainty decreases. On the other hand, an increasingly high proportion of receptors will be bound, and the uncertainty increases. In Eq. (5.6), these two tendencies exactly cancel, and the uncertainty tends to a constant for large C.

The application of this model to growth cone sensing produces results that are in good agreement with the limited experimental data available on the chemotactic sensitivity of growth cones (Goodhill and Urbach, 1999; Urbach and Goodhill, 1999). Consider a growth cone sampling a spherical volume of radius 10 μm, in a ligand concentration of 1 nM [roughly the value of the dissociation constant for many receptor-ligand pairs implicated in axon guidance (see the references cited in Goodhill and Baier, 1998)]. This volume contains about 2500 molecules, and the fractional error in an instantaneous measurement of the gradient in this case is therefore about $1/\sqrt{2500}$, or 2 percent. However, growth cones take a time on the order of a minute to show a response to a gradient signal (Zheng et al., 1996), which suggests they may be averaging concentration measurements over a time T of about 100 sec. For a freely diffusing molecule the size of netrin in a liquid, the diffusion constant $D \approx 10^{-6}$ cm^2/sec, whereas in vivo $D \approx 10^{-7}$ cm^2/sec (reviewed in Goodhill, 1997). For proteins bound to cell membranes, experimental values have been obtained in the range $D = 10^{-8}$ cm^2/sec to $D = 10^{-11}$ cm^2/sec (Wiegel, 1983); for instance, Poo (1982) found a value of $D = 2.6 \times 10^{-9}$ cm^2/sec for ACh receptors diffusing over the surface of embryonic muscle cells. Taking therefore $D \approx 10^{-9}$ cm^2/sec for membrane-bound ligands yields values from Eq. (5.5) in the (3-D) liquid, in vivo, and in membrane-bound cases of 0.5 percent, 1 percent, and 10 percent, respectively. The model therefore predicts that much shallower gradients of a diffusible factor are detectable than of a bound factor.

Receptor numbers have been measured in the case of, for example, neurotrophin receptors on embryonic sensory neurons (not growth cones themselves), yielding numbers between a few hundred and about 50,000 (reviewed in Meakin and Shooter, 1992). If $N_r = 10,000$, $s = 50$ Å, and $K_D = 1$ nM, this yields values for the fractional gradient error given by Eq. (5.6) that are very similar to those for the rough calculation using Eq. (5.5). (Note that the gradient error varies only as the square root of most of the parameters, so that a variation that is an order of magnitude in a parameter yields a change of only a factor of 3 in the gradient error.) The number of receptors for which the sphere collects half the available flux is $N_r = 10,000$, even though only a tiny fraction of the surface area is covered by receptors (Berg and Purcell, 1977). If $N_r = 100$, the minimum detectable gradient would be ten times greater.

What is the optimal gradient shape in the Berg and Purcell model for guiding an axon over the maximum possible distance, and what is this maximum distance? The optimal gradient has a percent concentration change across each growth cone diameter that is equal to the minimum required for gradient detection. Setting $(a/C)(dC/dx)$ equal to Eq. (5.6) and integrating, Goodhill and Urbach (1999) calculated the maximum guidance distance, assuming there exists a maximum concentration $C - C_{max}$ above which the gradients cannot be detected. This assumption was motivated by experimental data both from leukocytes and from the experiments of Ming et al. (1997), which indicated that owing to factors not considered in the Berg and Purcell model, gradient detection is not possible above a certain high concentration limit of about ten to one hundred times $C_{1/2}$. The resulting formula for the maximum distance x_{max} is

$$x_{max} = a\sqrt{2\alpha}\,\log\!\left[2\,\sqrt{\bar{C}_{max}^2 + \bar{C}_{max}} + 2\bar{C}_{max} + 1/2\right],$$

$$(5.7)$$

where $\bar{C} = C/C_{1/2}$, $\bar{C}_{max} = C_{max}/C_{1/2}$, $\alpha = 2\pi(T/T_D)N_{1/2}f$, $T_D = a^2/D$, $N_{1/2} = C_{1/2}a^3$, and $f = N_r s/(N_r s + \pi a)$. Assuming $\bar{C}_{max} \gg 1$, this reduces to $x_{max} = a\sqrt{2\alpha} \log(4\bar{C}_{max})$. Using the estimated parameter values given earlier and setting $\bar{C}_{max} = 100$ yields $x_{max} \approx 1$ cm. This is not far from the estimate based on a much cruder model of the gradient sensing process discussed in section 5.3.1 (Goodhill and Baier, 1998; Goodhill, 1998), but it makes specific predictions for how this varies with parameters such as T, D, and N_r. The natural length scale for this problem is $a\sqrt{\alpha}$, which gives the distance between $\bar{C} = 1$ and $\bar{C} = 0$ for a linear gradient with a slope equal to the minimum detectable gradient at $\bar{C} = 1$.

5.4 Discussion

Axon guidance is an extremely active area of experimental work. New and interesting results appear almost monthly, and the overall story continually becomes more complicated. This provides great opportunities for theoretical modeling to be on the cutting edge and to feed back directly into experiments, but it also brings great challenges. Modelers of axon guidance must keep up with data on a broad front in order to extract key insights and constraints from many diverse areas. These areas include the biochemistry of many different families of guidance molecules, the in vivo mode of action of guidance molecules (such as knockout and misexpression studies), the molecular mechanisms of growth cone movement, signal transduction within the growth cone, and analogous results from related systems such as leukocytes, *Dictyostelium*, and bacteria.

The area of axon guidance within which models have so far had the greatest impact is retinotectal map formation. In the 1970s and 1980s, many experimentalists in this area were keenly aware of the relevant theoretical literature, and theoretical results had some influence on the experiments performed. With the discovery of Eph and ephrin gradients in the 1990s,

the attention of a new generation of experimentalists has focused on the "dual gradient" model of Gierer (1987). Although this model does not attempt to explain most of the older surgical manipulation data, it does provide an appealing picture for how measured Eph and ephrin gradients could produce a map (see also chapter 11). An important piece of the model that is currently missing experimentally is an attractive gradient to counterbalance the repulsive one, and this prediction is actively being pursued by some experimentalists. Other areas where models have successfully accounted for experimental data are growth cone morphology and how this changes at, for instance, decision points, and the approximately 1-mm limit on the maximum guidance range of a target-derived diffusible factor.

Future Modeling Studies

Some attractive areas for future theoretical work include the following:

1. Models of retinotectal map formation that take into account the full range of available data, from surgical manipulations to gene knockout and misexpression studies (see chapter 11)

2. Models of growth cone movement in response to patterned molecular cues that can explain the paths taken by axons in specific in vivo scenarios, such as in the tectum or toward and across the midline

3. Models of the statistical limitations on gradient sensing that go beyond the relatively simple picture of Berg and Purcell (1977) (cf. DeLisi et al., 1982; Lauffenburger, 1982; DeLisi and Marchetti, 1983)

4. Models of the signal transduction mechanisms underlying growth cone chemotaxis that take into account recent discoveries not included in generic reaction-diffusion proposals.

Future Experimental Studies

One of the strongest proposals for new experimental work suggested by these modeling studies is the need for a more quantitative assay for measuring axonal response to gradients. This is needed to determine the minimum steepness of gradient that can be detected by growth cones, how this varies with concentration, and how growth cone morphology varies as a function of gradient parameters. Some studies (e.g., Baier and Bonhoeffer, 1992; Rosentreter et al., 1998, Bagnard et al., 2000) have attempted to set up controlled gradients of molecules such as ephrins, which are expressed on cell surfaces. However, these are created by depositing a gradient of ground-up cell membranes rather than the actual guidance molecules themselves, and work only for substrate-bound factors. Poo and colleagues (see papers cited in Song and Poo, 1999) have developed a technique for slow ejection of diffusible factors into a culture dish so as to establish a gradient near a growth cone. Although this has been very useful for examining the acute response of growth cones to gradients, it does not address the chronic response and it is not possible to arbitrarily vary the gradient shape. In order to address these problems, we are currently developing a new technique for establishing precisely controlled gradients in collagen gels, which should provide a variety of novel quantitative results with which to constrain theories of gradient detection (Rosoff et al., 2001).

References

Bagnard, D., Thomasset, N., Lohrum, M., Püschel, A. W., and Bolz, J. (2000). Spatial distributions of guidance molecules regulate chemoattraction and chemorepulsion of growth cones. *J. Neurosci.* 20: 1030–1035.

Baier, H., and Bonhoeffer, F. (1992). Axon guidance by gradients of a target-derived component. *Science* 255: 472–475.

Bentley, D., and Toroian-Raymond, A. (1986). Disoriented pathfinding by pioneer neuron growth cones deprived of filopodia by cytochalasin treatement. *Nature* 323: 712–715.

Berg, H. C., and Purcell, E. M. (1977). Physics of chemoreception. *Biophys. J.* 20: 193–219.

Brose, K., and Tessier-Lavigne, M. (2000). Slit proteins: Key regulators of axon guidance, axonal branching, and cell migration. *Curr. Opin. Neurobiol.* 10: 95–102.

Buettner, H. M. (1995). Computer simulation of nerve growth cone filopodial dynamics for visualization and analysis. *Cell Motil. Cytoskel.* 32: 187–204.

Castellani, V., Yue, Y., Gao, P. P., Zhou, R., and Bolz, J. (1998). Dual action of a ligand for Eph receptor tyrosine kinases on specific populations of axons during the development of cortical circuits. *J. Neurosci.* 18: 4663–4672.

Cheng, H. J., Nakamoto, M., Bergemann, A. D., and Flanagan, J. G. (1995). Complementary gradients in expression and binding of Elf-1 and Mek4 in development of the topographic retinotectal projection map. *Cell* 82: 371–381.

Chien, C. B., Rosenthal, D. E., Harris, W. A., and Holt, C. E. (1993). Navigational errors made by growth cones without filopodia in the embryonic *Xenopus* brain. *Neuron* 11: 237–251.

Colamarino, S. A., and Tessier-Lavigne, M. (1995). The axonal chemoattractant netrin-1 is also a chemorepellent for trochlear motor axons. *Cell* 81: 621–629.

Crank, J. (1975). *The Mathematics of Diffusion*, 2nd ed. Oxford: Clarendon Press.

Deiner, M. S., Kennedy, T. E., Fazeli, A., Serafini, T., Tessier-Lavigne, M., and Sretavan, D. W. (1997). Netrin-1 and DCC mediate axon guidance locally at the optic disc: Loss of function leads to optic nerve hypoplasia. *Neuron* 19: 575–589.

Dekker, L. V., and Segal, A. W. (2000). Signals to move cells. *Science* 287: 982–985.

DeLisi, C., and Marchetti, F. (1983). A theory of measurement error and its implications for spatial and temporal gradient sensing during chemotaxis II. The effects of nonequilibriated receptor binding. *Cell Biophys.* 5: 237–253.

DeLisi, C., Marchetti, F., and Del Grosso, G. (1982). A theory of measurement error and its implications for spatial and temporal gradient sensing during chemotaxis. *Cell Biophys.* 4: 211–229.

Drescher, U., Kremoser, C., Handwerker, C., Loschinger, J., Noda, M., and Bonhoeffer, F. (1995). In-vitro guidance of retinal ganglion-cell axons by RAGS, a 25 KDa tectal protein related to ligands for Eph receptor tyrosine kinases. *Cell* 82: 359–370.

Feldheim, D. A., Vanderhaeghen, P., Hansen, M. J., Frisén, J., Lu, Q., Barbacid, M., and Flanagan, J. G. (1998). Topographic guidance labels in a sensory projection to the forebrain. *Neuron* 21: 1303–1313.

Feldheim, D. A., Kim, Y. I., Bergemann, A. D., Frisen, J., Barbacid, M., and Flanagan, J. G. (2000). Genetic analysis of ephrin-A2 and ephrin-A5 shows their requirement in multiple aspects of retinocollicular mapping. *Neuron* 25: 563–574.

Flanagan, J. G., and Vanderhaeghen, P. (1998). The ephrins and Eph receptors in neural development. *Annu. Rev. Neurosci.* 21: 309–345.

Fraser, S. E. (1980). A differential adhesion approach to the patterning of neural connections. *Dev. Biol.* 79: 453–464.

Fraser, S. E., and Perkel, D. H. (1990). Competitive and positional cues in the patterning of nerve connections. *J. Neurobiol.* 21: 51–72.

Friedman, G. C., and O'Leary, D. D. M. (1996). Retroviral misexpression of engrailed genes in the chick optic tectum perturbs the topographic targeting of retinal axons. *J. Neurosci.* 16: 5498–5509.

Gierer, A. (1987). Directional cues for growing axons forming the retinotectal projection. *Development* 101: 479–489.

Goodhill, G. J. (1997). Diffusion in axon guidance. *Eur. J. Neurosci.* 9: 1414–1421.

Goodhill, G. J. (1998). Mathematical guidance for axons. *Trends Neurosci.* 21: 226–231.

Goodhill, G. J. (2000). Dating behavior of the retinal ganglion cell. *Neuron* 25: 501–503.

Goodhill, G. J. (2003). A theoretical model of axon guidance by the robo code. *Neur. Comput.*, in press.

Goodhill, G. J., and Baier, H. (1998). Axon guidance: Stretching gradients to the limit. *Neur. Comput.* 10: 521–527.

Goodhill, G. J., and Richards, L. J. (1999). Retinotectal maps: Molecules, models, and misplaced data. *Trends Neurosci.* 22: 529–534.

Goodhill, G. J., and Urbach, J. S. (1999). Theoretical analysis of gradient detection by growth cones. *J. Neurobiol.* 41: 230–241.

Hely, T. A., and Willshaw, D. J. (1998). Short term interactions between microtubules and actin filaments underlie long term behaviour in neuronal growth cones. *Proc. Roy. Soc. London B* 265: 1801–1807.

Hentschel, H. G. E., and Van Ooyen, A. (1999). Models of axon guidance and bundling during development. *Proc. Roy. Soc. London B.* 266: 2231–2238.

Honda, H. (1998). Topographic mapping in the retinotectal projection by means of complementary ligand and receptor gradients: A computer simulation study. *J. Theor. Biol.* 192: 235–246.

Hong, K., Hinck, L., Nishiyama, M., Poo, M.-M., and Tessier-Lavigne, M. (1999). A ligand-gated association between cytoplasmic domains of UNC5 and DCC family receptors converts netrin-induced growth cone attraction to repulsion. *Cell* 92: 205–215.

Hong, K., Nishiyama, M., Henley, J., Tessier-Lavigne, M., and Poo, M.-M. (2000). Calcium signalling in the guidance of nerve growth by netrin-1. *Nature* 403: 93–98.

Jontes, J. D., and Smith, S. J. (2000). Filopodia, spines, and the generation of synaptic diversity. *Neuron* 27: 11–14.

Keino-Masu, K., Masu, M., Hinck, L., Leonardo, E. D., Chan, S. S.-Y., Culotti, J. G., and Tessier-Lavigne, M. (1996). *Deleted in Colorectal Cancer* (DCC) encodes a netrin receptor. *Cell* 87: 175–185.

Kennedy, T. E., Serafini, T., de al Torre, J. R., and Tessier-Lavigne, M. (1994). Netrins are diffusible chemotropic factors for commissural axons in the embryonic spinal cord. *Cell* 78: 425–435.

Lauffenburger, D. A. (1982). Influence of external concentration fluctuations on leukocyte chemotactic orientation. *Cell Biophys.* 4: 177–209.

Lumsden, A. G. S., and Davies, A. M. (1983). Earliest sensory nerve fibres are guided to peripheral targets by attractants other than nerve growth factor. *Nature* 306: 786–788.

Lumsden, A. G. S., and Davies, A. M. (1986). Chemotropic effect of specific target epithelium in the developing mammalian nervous system. *Nature* 323: 538–539.

McAllister, A. K., Katz, L. C., and Lo, D. C. (1999). Neurotrophins and synaptic plasticity. *Annu. Rev. Neurosci.* 22: 295–318.

Meakin, S. O., and Shooter, E. M. (1992). The nerve growth family of receptors. *Trends. Neurosci.* 15: 323–331.

Meinhardt, H. (1999). Orientation of chemotactic cells and growth cones: Models and mechanisms. *J. Cell Sci.* 112: 2867–2874.

Messersmith, E. K., Leonardo, E. D., Shatz, C. J., Tessier-Lavigne, M., Goodman, C. S., and Kolodkin, A. L. (1995). Semaphorin III can function as a selective chemorepellent to pattern sensory projections in the spinal cord. *Neuron* 14: 949–959.

Ming, G.-L., Song, H.-J., Berninger, B., Holt, C. E., Tessier-Lavigne, M., and Poo, M.-M. (1997). cAMP-Dependent growth cone guidance by netrin-1. *Neuron* 19: 1225–1235.

Myers, P. Z., and Bastiani, M. J. (1993). Growth cone dynamics during the migration of an identified commissural growth cone. *J. Neurosci.* 13: 127–143.

O'Connor, T. P., Duerr, J. S., and Bentley, D. (1990). Pioneer growth cone steering decisions mediated by single filopodial contacts in situ. *J. Neurosci.* 10: 3935–3946.

Odde, D. J., and Buettner, H. M. (1998). Autocorrelation function and power spectrum of two-state random processes used in neurite guidance. *Biophys. J.* 75: 1189–1196.

Parent, C. A., and Devreotes, P. N. (1999). A cell's sense of direction. *Science* 284: 765–770.

Rajagopalan, S., Vivancos, V., Nicolas, E., and Dickson, B. J. (2000). Selecting a longitudinal pathway: Robo receptors specify the lateral position of axons in the *Drosophila* CNS. *Cell* 103: 1033–1045.

Poo, M.-M. (1982). Rapid lateral diffusion of functional ACh receptors in embryonic muscle cell membrane. *Nature* 295: 332–334.

Prestige, M. C., and Willshaw, D. J. (1975). On a role for competition in the formation of patterned neural connexions. *Proc. Roy. Soc. London B.* 190: 77–98.

Raper, J. A. (2000). Semaphorins and their receptors in vertebrates and invertebrates. *Curr. Opin. Neurobiol.* 10: 88–94.

Rehder, V., and Kater, S. B. (1996). Filopodia on neuronal growth cones: Multi-functional structures with sensory and motor capabilities. *Sem. Neurosci.* 8: 81–88.

Richards, L. J., Koester, S. E., Tuttle, R., and O'Leary, D. D. M. (1997). Directed growth of early cortical axons is influenced by a chemoattractant released from an intermediate target. *J. Neurosci.* 17: 2445–2458.

Rosentreter, S. M., Davenport, R. W., Löschinger, J., Huf, J., Jung, J., and Bonhoeffer, F. (1998). Response of retinal ganglion cell axons to striped linear gradients of repellent guidance molecules. *J. Neurobiol.* 37: 541–562.

Rosoff, W. J., Esrick, M. A., Gardner, D., Savich, J., Richards, L. J., Urbach, J. S., and Goodhill, G. J. (2001). A novel method for establishing defined gradients in gels. *Soc. Neurosci. Abstr.* 27. Program. No. 795.1.

Serafini, T., Kennedy, T. E., Galko, M. J., Mirzayan, C., Jessell, T. M., and Tessier-Lavigne, M. (1994). The netrins define a family of axon outgrowth-promoting proteins homologous to *C. elegans* UNC-6. *Cell* 78: 409–424.

Serafini, T., Colamarino, S. A., Leonardo, E. D., Wang, H., Beddington, R., Skarnes, W. C., and Tessier-Lavigne, M. (1996). Netrin-1 is required for commissural axon guidance in the developing vertebrate nervous system. *Cell* 87: 1001–1014.

Shirasaki, R., Tamada, A., Katsumata, R., and Murakami, F. (1995). Guidance of cerebellofugal axons in the rat embryo: Directed growth toward the floor plate and subsequent elongation along the longitudinal axis. *Neuron* 14: 961–972.

Shirasaki, R., Mirzayan, C., Tessier-Lavigne, M., and Murakami, F. (1996). Guidance of circumferentially growing axons by netrin-dependent and -independent floor plate chemotropism in the vertebrate brain. *Neuron* 17: 1079–1088.

Shu, T., and Richards, L. J. (2001). Cortical axon guidance by the glial wedge during the development of the corpus callosum. *J. Neurosci.* 21: 2749–2758.

Simpson, J. H., Bland, K. S., Fetter, R. D., and Goodman, C. S. (2000). Short range and long range guidance by Slit and its Robo receptors: A combinatorial code of Robo receptors controls lateral position. *Cell* 103: 1019–1032.

Song, H., and Poo, M.-M. (1999). Signal transduction underlying growth cone guidance by diffusible factors. *Curr. Opin. Neurobiol.* 9: 955–969.

Song, H. J., Ming, G. L., and Poo, M.-M. (1997). cAMP-induced switching in turning direction of nerve growth cones. *Nature* 388: 275–279.

Song, H., Ming, G., He, Z., Lehmann, M., Tessier-Lavigne, M., and Poo, M.-M. (1998). Conversion of neuronal growth cone responses from repulsion to attraction by cyclic nucleotides. *Science* 281: 1515–1518.

Sperry, R. W. (1963). Chemoaffinity in the orderly growth of nerve fiber patterns and connections. *Proc. Natl. Acad. Sci., U.S.A.* 50: 703–710.

Steketee, M. B., and Tosney, K. W. (1999). Contact with isolated sclerotome cells steers sensory growth cones by altering distinct elements of extension. *J. Neurosci.* 19: 3495–3506.

Suter, D. M., and Forscher, P. (1998). An emerging link between cytoskeletal dynamics and cell adhesion molecules in growth cone dynamics. *Curr. Opin. Neurobiol.* 8: 106–116.

Tessier-Lavigne, M., and Goodman, C. S. (1996). The molecular biology of axon guidance. *Science* 274: 1123–1133.

Tessier-Lavigne, M., Placzek, M., Lumsden, A. G. S., Dodd, J., and Jessell, T. M. (1988). Chemotropic guidance of developing axons in the mammalian central nervous system. *Nature* 336: 775–778.

Udin, S. B., and Fawcett, J. W. (1988). Formation of topographic maps. *Annu. Rev. Neurosci.* 11: 289–327.

Urbach, J. S., and Goodhill, G. J. (1999). Limitations on detection of gradients of diffusible chemicals by axons. *Neurocomputing* 26–27: 39–43.

Vanderhaeghen, P., Lu, Q., Prakash, N., Frisen, J., Walsh, C. A., Frostig, R. D., Flanagan, J. G. (2000). A mapping label required for normal scale of body representation in the cortex. *Nat. Neurosci.* 3: 358–365.

Van Veen, M., and Van Pelt, J. (1992). A model for outgrowth of branching neurites. *J. Theor. Biol.* 159: 1–23.

Varela-Echavarria, A., Tucker, A., Püschel, A. W., and Guthrie, S. (1997). Motor axon subpopulations respond differentially to the chemorepellents netrin-1 and semaphorin D. *Neuron* 18: 193–207.

Wang, K. H., Brose, K., Arnott, D., Kidd, T., Goodman, C. S., Henzel, W., and Tessier-Lavigne, M. (1999). Biochemical purification of a mammalian slit protein as a positive regulator of sensory axon elongation and branching. *Cell* 96: 771–784.

Weber, C., Ritter, H., Cowan, J., and Obermayer, K. (1997). Development and regeneration of the retinotectal map in goldfish: A computational study. *Phil. Trans. Roy. Soc. London B.* 352: 1603–1623.

Whitelaw, V. A., and Cowan, J. D. (1981). Specificity and plasticity of retinotectal connections: A computational model. *J. Neurosci.* 1: 1369–1387.

Wiegel, F. W. (1983). Diffusion and the physics of chemoreception. *Phys. Rep.* 95: 283–319.

Willshaw, D. J., and Von der Malsburg, C. (1979). A marker induction mechanism for the establishment of ordered neural mappings: Its application to the retinotectal problem. *Phil. Trans. Roy. Soc. London B.* 287: 203–243.

Wu, W., Wong, K., Chen, J.-H., Jiang, Z.-H., Dupuis, S., Wu, J. Y., and Rao, Y. (1999). Directional guidance of neuronal migration in the olfactory system by the protein Slit. *Nature* 400: 331–336.

Yabuta, N. H., Butler, A. Y., and Callaway, E. M. (2000). Laminar specificity of local circuits in barrel cortex of ephrin-A5 knockout mice. *J. Neurosci.* 20: RC88.

Zheng, J. Q. (2000). Turning of nerve growth cones induced by localized increases in intracellular calcium ions. *Nature* 403: 89–93.

Zheng, J. G., Wan, J.-J., and Poo, M.-M. (1996). Essential role of filopodia in chemotropic turning of nerve growth cone induced by a glutamate gradient. *J. Neurosci.* 16: 1140–1149.

Zigmond, S. H. (1977). Ability of polymorphonuclear leukocytes to orient in gradients of chemotactic factors. *J. Cell. Biol.* 75: 606–616.

Zigmond, S. H. (1981). Consequences of chemotactic peptide receptor modulation for leukocyte orientation. *J. Cell. Biol.* 88: 644–647.

Activity-Dependent Neurite Outgrowth: Implications for Network Development and Neuronal Morphology

6

Arjen van Ooyen, Jaap van Pelt, Michael A. Corner, and
Stanley B. Kater

Empirical studies have shown that high levels of neuronal activity can cause neurites to retract, whereas lower levels allow further outgrowth. Using simulation studies, we have explored the possible implications of such activity-dependent neurite outgrowth for network development and neuronal morphology. These implications include a transient phase of high connectivity during development, the presence of multiple stable end states of development at different connectivity levels, and the emergence of size differences between the neuritic fields of excitatory and inhibitory cells. These phenomena, which are also observed in developing cultures of cerebral cortex cells, emerge in the model without assuming predetermined, time-scheduled mechanisms.

6.1 Introduction

Electrical activity, in the form of action potentials and synaptically driven fluctuations in membrane potential, plays an important role in the development of neurons into functional neural networks. This activity-dependent maturation begins even before the onset of sensory responses, and is driven by intrinsically generated patterns of electrical discharges. Most studies, theoretical as well as empirical, have largely focused on activity-dependent changes in synaptic strength, but many other processes that determine network connectivity and neuronal function are, on a variety of time scales, also modulated by electrical activity. These include naturally occurring cell death (see chapter 9), neurite outgrowth and branching,

synaptogenesis, elimination of synapses, changes in the number and effectiveness of ion channels and neurotransmitter receptors (see chapter 8), and even gene expression (for reviews, see Van Ooyen, 1994; Corner et al., 2002).

As a result of these activity-dependent processes, a reciprocal influence exists between the development of neuronal form, function, and connectivity on the one hand ("slow dynamics"; different time scales are involved, but they are all slow relative to the time scale of the dynamics of electrical activity) and neuronal and network activity on the other hand ("fast dynamics"). Thus, the activity patterns generated by a developing network can modify the organization of the network and the functional characteristics of the neurons, leading to altered activity patterns, which in turn can further modify structural and functional characteristics.

Electrical activity exerts its effects on multiple time scales (hours, e.g., number and effectiveness of neurotransmitter receptors; days or weeks, e.g., neurite outgrowth and synapse formation) as well as on multiple levels of organization [individual synapses, e.g., long-term potentiation (LTP) and depression (LTD); whole cell, e.g., neuronal excitability and neurite outgrowth; population of cells, e.g., balance of excitation and inhibition]. In many cases, the way in which activity modifies network connectivity and neuronal function contributes to the homeostasis of neuronal activity (for reviews, see Van Ooyen, 1994; Turrigiano, 1999; Abbott and Nelson, 2000; Corner et al., 2002). When the activity of a neuron is high, neuronal connectivity and excitability are modified by activity-dependent processes such as neurite outgrowth (see section 6.2), changes in ionic conductances (Turrigiano et al., 1994, 1995) and neurotransmitter receptors (Turrigiano et al., 1998; see also chapter 8), and changes in the balance of excitation and inhibition (Corner and Ramakers, 1992; Turrigiano, 1999) so as to decrease activity. When the activity of a neuron is low, on the other hand, neuronal connectivity and excitability will be modified so as to increase activity.

In this chapter we focus on activity-dependent neurite outgrowth and explore its implications for network development and neuronal form. Section 6.2 describes the empirical studies, which have established that high levels of neuronal activity often cause neurites to retract (mediated by changes in intracellular calcium levels), whereas lower levels allow further outgrowth. Section 6.3 briefly reviews modeling studies of activity-dependent neurite outgrowth and some other activity-dependent processes. In section 6.4 we describe in detail a model of activity-dependent neurite outgrowth.

6.2 Activity-Dependent Neurite Outgrowth— Neurobiological Background

Neurite elongation, branching, and steering are under the control of the growth cone, a specialized structure at the tip of a growing neurite (Letourneau et al., 1991). Growth cones consist of a central zone containing organelles and microtubules (long polymers of tubulin that form a continuous core within the neurite) and surrounded by spikelike protrusions (filopodia) and fan-shaped sheets (lamellipodia), both of which are filled with actin filaments. Polymerization of tubulin into microtubules—which for the most part takes place at the growth cone—provides the

driving force for neurite elongation. The actin cytoskeleton mainly serves to control branching and steering (Acebes and Ferrús, 2000), but also participates in elongation. The tension generated by the rearward flow of fibrillar actin (F-actin) in the growth cone slows down elongation, probably by affecting the rate of microtubule polymerization (Buxbaum and Heidemann, 1992; Lin and Forscher, 1995).

Neurite outgrowth is sensitive to the calcium concentration within the growth cone, $[Ca^{2+}]_{in}$. Calcium affects the polymerization and depolymerization of microtubules, both directly (e.g., Schilstra et al., 1991) and via its influence on microtubule-associated proteins (MAPs). By interacting with microtubules, MAPs regulate many aspects of microtubule dynamics, not only polymerization and depolymerization, but also bundling, spacing, and interactions with actin filaments (Maccioni and Cambiazo, 1995). Calcium modulates MAP function by regulating the phosphorylation state of MAPs (Sánchez et al., 2000).

The dynamics of the actin cytoskeleton are also influenced by calcium. The (de)polymerization of F-actin and the formation of cross-linked meshworks and bundles of actin filaments are controlled by actin-binding proteins, many of which are modulated by calcium (Forscher, 1989). The Rho proteins are particularly important for growth cone morphology and neurite branching (Aspenstrom, 1999; Cline, 1999; Li et al., 2000). These are involved in the organization of the actin cytoskeleton and are essential for the formation of lamellipodia and filopodia (Tapon and Hall, 1997; Aspenstrom, 1999; see also chapters 3 and 4). Rho proteins, too, may be modulated by calcium (Ramakers et al., 1998; Chen et al., 1998).

Because of the strong dependence of neurite outgrowth on calcium, any factor that can change $[Ca^{2+}]_{in}$—such as depolarization, neurotransmitters (Cohan et al., 1987; Berridge, 1998), neurotrophins

(Stoop and Poo, 1996), and cell adhesion molecules (Bixby et al., 1994), which affect calcium influx through voltage- and ligand-gated calcium channels —will be able to affect neurite outgrowth (for reviews, see Goldberg and Grabham, 1999; McAllister, 2000).

Large increases in $[Ca^{2+}]_{in}$—caused, for example, by high levels of neuronal electrical activity (action potentials) or by depolarization induced by neurotransmitters or depolarizing media—arrest neurite outgrowth and can even cause retraction (e.g., Cohan and Kater, 1986; Fields et al., 1990; Mattson and Kater, 1989; Mattson et al., 1988; Torreano and Cohan, 1997). Growth cones can generate transient elevations of $[Ca^{2+}]_{in}$ as they migrate (Gomez and Spitzer, 1999), and, consistent with the results mentioned earlier, the rate of axon elongation is inversely proportional to the frequency of these spontaneous calcium transients (Gu and Spitzer, 1995; Gomez and Spitzer, 1999). However, decreases in $[Ca^{2+}]_{in}$ (e.g., as a result of lowered neuronal electrical activity) can also suppress neurite elongation (e.g., Mattson and Kater, 1987; Mattson, 1988; Lankford and Letourneau, 1991; Al-Mohanna et al., 1992; Ramakers et al., 2001), while in some cases neurite elongation can be promoted by elevations of $[Ca^{2+}]_{in}$ above the resting level (e.g., Kuhn et al., 1998).

These seemingly contradictory results are accommodated by Kater's calcium hypothesis for neurite outgrowth, which states that deviations in either direction from an optimal $[Ca^{2+}]_{in}$ slow down neurite elongation (Kater et al., 1988; Kater and Mills, 1991). Deviations within the physiological range are already effective, so that calcium signaling in the growth cone is indeed physiologically relevant (reviewed in Goldberg and Grabham, 1999).

The bell-shaped dependence of neurite elongation on $[Ca^{2+}]_{in}$ implies that one cannot predict a priori whether a given change in $[Ca^{2+}]_{in}$ has a growth-promoting or growth-inhibiting effect (Kater and Mills, 1991). This will depend on the magnitude of the change as well as on the existing resting calcium levels. Moreover, the optimal $[Ca^{2+}]_{in}$ for outgrowth may be different for different neurons (Kater et al., 1988; Kater and Mills, 1991). Furthermore, it is worthwhile to distinguish between the effects of calcium on the morphology of the growth cone (number and extent of its protrusions—filopodia and lamellipodia) and the effects of calcium on neurite elongation. For example, focal increases in $[Ca^{2+}]_{in}$ can induce the formation of protrusions (Davenport and Kater, 1992), which may slow the forward movement of the growth cone, and thus neurite elongation, through lateral interactions with the environment (Goldberg and Grabham, 1999).

In addition to affecting neurite elongation, calcium levels and depolarization influence neurite branching (Schilling et al., 1991; Sanes and Takács, 1993; Ramakers et al., 1998, 2001). Both increased and reduced branching have been observed following depolarization (for reviews, see Cline, 1999; Acebes and Ferrús, 2000; McAllister, 2000). So, as for elongation, there may be an optimal calcium level for branching, which could be different for different cell types.

The stimuli that change $[Ca^{2+}]_{in}$ in the growth cone can act on the level of a single neurite or on the level of the whole cell. For example, local application of glutamate to a single dendrite results in regression of that dendrite (Mattson et al., 1988), whereas somatic action potentials may simultaneously regulate the behavior of all the growth cones and neurites of a given neuron (Cohan and Kater, 1986; Kater and Guthrie, 1990), for example, by propagation of action potentials into the dendrites (Stuart and Sakmann, 1994) or by other mechanisms (electrotonic spread, calcium dynamics).

6.3   Review of Models

Hentschel and Fine (1996) used Kater's calcium hypothesis for neurite outgrowth to study the emergence of dendritic forms from initially spherical cells (see chapter 3). In their model, local outgrowth of the cell membrane is taken to depend on the local concentration of calcium close to the internal surface of the membrane.

Whereas Hentschel and Fine (1996) used Kater's hypothesis to study the development of dendritic forms in single, isolated cells, we used Kater's hypothesis to study the development of synaptically connected networks from initially unconnected cells (see section 6.4).

Abbott and Jensen (1997) studied calcium-dependent neurite outgrowth in combination with calcium-regulated conductances (see chapter 8). As in our model, they used the concept of a circular neuritic field to model outgrowth and connectivity, but instead of modeling firing frequency (see section 6.4.1), they used spiking neurons and also explicitly modeled the dynamics of intracellular calcium. Starting from random initial conditions, the model cells develop spontaneously into a coupled network displaying a complex pattern of activity. Although the cells are governed by identical equations, they differentiate within these networks and show a wide variety of intrinsic characteristics.

Raijmakers and Molenaar (1999) applied activity-dependent neurite outgrowth to the maturation of connections in an ART (adaptive resonance theory) neural network model of memory, and Eglen et al. (2000) studied the role of neurite outgrowth in the formation of retinal mosaics (see chapter 7).

Activity-dependent neurite outgrowth contributes to the homeostasis of neuronal activity (see section

6.1). Other forms of homeostatic plasticity include mechanisms for regulating the intrinsic excitability of neurons (ionic conductances, neurotransmitter receptors; for a review, see Corner et al., 2002) and mechanisms for stabilizing total synaptic strength (Turrigiano, 1999). Modeling studies have begun to explore the implications of these forms of plasticity for network development and function (e.g., LeMasson et al., 1993; Marder et al., 1996; Horn et al., 1998; Golowasch et al., 1999; see also chapter 8). In contrast to Hebbian mechanisms, where changes in synaptic strength occur in a synapse-specific manner, homeostatic plasticity acts on the neuronal level. Neurite outgrowth, too, operates on a higher level than that of a single synapse. A change in a cell's axonal or dendritic extent means that the connectivity with many other cells is changed simultaneously. During development, homeostatic plasticity ensures that neurons remain responsive to their inputs and allows Hebbian plasticity to modify synaptic strengths selectively (Turrigiano, 1999). During aging, homeostatic plasticity can account for maintenance of memory systems that undergo synaptic turnover and degradation (Horn et al., 1998).

6.4   Detailed Description of a Model of Activity-Dependent Neurite Outgrowth

The goal of the model that we present in this section (see Van Ooyen and Van Pelt, 1994; Van Ooyen et al., 1995) is not to reproduce any particular system in detail, but rather to explore the range of phenomena that could result from activity-dependent neurite outgrowth. In the model, the growth of both excitatory and inhibitory neurons is activity dependent, and all the cells are initially unconnected. Development into a connected network (slow dynamics) takes place only under the influence of the activity (fast dynamics) that is generated by the network itself (no external input). Growing neurons are modeled as expanding neuritic fields, and neurons become connected when their neuritic fields overlap. The outgrowth of each neuron depends upon its own level of electrical activity, according to Kater's theory for the control of neurite outgrowth (thus we do not model calcium explicitly, assuming that $[Ca^{2+}]_{in}$ is proportional to the level of neuronal electrical activity). The dependence of outgrowth on activity in an individual neuron is such that when activity is higher than a critical value, its neuritic fields retract (reducing its connectivity with other cells and thus, in general, its activity), and when it is lower, its neuritic fields expand (increasing its connectivity and activity). Thus, by adapting the size of its neuritic field, a neuron attempts to maintain a certain level of activity (homeostasis), the aforementioned critical value, at which the neuron's neuritic field remains stationary.

The model is inspired in part by cultures of dissociated cerebral cortex cells (e.g., Van Huizen et al., 1985; for a review, see Marom and Shahaf, 2002), whose development into a network (without external input) by neurite outgrowth and synaptogenesis shows many similarities with that in vivo. The phases through which the cultured networks pass during development—with respect to electrical activity and connectivity—as well the effects of various treatments, such as chronically blocking activity, have been described extensively (see section 6.4.4). In section 6.4.4, the phenomena observed in the model are compared with those observed in culture.

6.4.1 Neuron Model

The shunting model (Grossberg, 1988), transformed into dimensionless equations, is used to describe neuronal activity. In this model, excitatory inputs drive the membrane potential toward a maximum (the excitatory saturation potential), while inhibitory inputs drive the membrane potential toward a minimum (the inhibitory saturation potential):

$$\frac{dX_i}{dT} = -X_i + (1 - X_i) \sum_{k=1}^{N} W_{ik} F(X_k)$$

$$- (H + X_i) \sum_{l=1}^{M} W_{il} F(Y_l) \qquad (6.1)$$

$$\frac{dY_j}{dT} = -Y_j + (1 - Y_j) \sum_{k=1}^{N} W_{jk} F(X_k)$$

$$- (H + Y_j) \sum_{l=1}^{M} W_{jl} F(Y_l), \qquad (6.2)$$

where X_i and Y_j are the membrane potentials of, respectively, the excitatory cell i and the inhibitory cell j, expressed in units of excitatory saturation potential; N and M are the total number of excitatory and inhibitory cells, respectively; H is the ratio of inhibitory to excitatory saturation potential; T is time expressed in units of membrane time constant; the Ws denote the connection strengths (all $W \geq 0$; k and l are the indices of the excitatory and inhibitory driver cells, respectively; i and j are the indices of the excitatory and inhibitory target cells, respectively); and $F(X)$ is the mean firing rate, which is taken to be a sigmoidal function of the membrane potential:

$$F(X) = \frac{1}{1 + e^{(\theta - X)/\alpha}}, \qquad (6.3)$$

where α determines the steepness of the function and θ represents the firing threshold. The low firing rate when the membrane potential is subthreshold represents spontaneous activity.

6.4.2 Outgrowth and Connectivity

Neurons are randomly placed on a two-dimensional surface. Each neuron is given a circular "neuritic field," the radius of which is variable. When two such fields overlap, both neurons become connected with a strength proportional to the area of overlap:

$$W_{ij} = A_{ij}S, \tag{6.4}$$

where $A_{ij} = A_{ji}$ is the amount of overlap, representing the total number of synapses formed reciprocally between neurons i and j ($A_{ii} = 0$); and S is a constant of proportionality, representing the average synaptic strength. Strength may depend on the type of connection. We distinguish S^{ee}, S^{ei}, S^{ie}, and S^{ii}, where, for example, S^{ei} is the inhibitory-to-excitatory synaptic strength.

In this abstraction, no distinction has been made between axons and dendrites, so the connections among excitatory cells and among inhibitory cells are symmetrical. The main findings of the model do not change, however, if separate axonal and dendritic fields are implemented (see the section on differences among cells and section 6.5).

In the model, the outgrowth of each neuron, whether excitatory or inhibitory, depends in an identical way upon electrical activity:

$$\frac{dR_i}{dT} = \rho G[F(X_i)], \tag{6.5}$$

where R_i is the radius of the neuritic field of neuron i, $F(X_i)$ is the firing frequency of neuron i, and ρ determines the rate of outgrowth. The outgrowth function G is defined as

$$G[F(X_i)] = 1 - \frac{2}{1 + e^{[\varepsilon - F(X_i)]/\beta}}, \tag{6.6}$$

where ε is the value of $F(X_i)$ for which $G = 0$, and β determines the steepness of the function. Depending

on $F(X_i)$, a neuritic field will grow out [$G > 0$ when $F(X_i) < \varepsilon$], retract [$G < 0$ when $F(X_i) > \varepsilon$], or remain constant [$G = 0$ when $F(X_i) = \varepsilon$] (see figure 6.1A). All cells will thus attempt to get a neuritic field size for which the input from overlapping cells is such that $F(X_i) = \varepsilon$; in other words, ε is a "homeostatic setpoint." Equation (6.6) is a phenomenological description of the theory of Kater et al. (see section 6.2). It is not a complete bell-shaped curve, but the precise rates of outgrowth are not essential for the results as long as with high activity, neuritic fields retract and with low activity, they expand. What could produce new results is if a growth function is used for which cells also retract when neuronal activity is below a certain level (figure 6.1B). However, using such a function yields similar results, provided that the initial activity is high enough (see the section on an alternative growth function).

6.4.3 Parameters

The fraction of inhibitory cells, $M/(N + M)$, is taken in the range of 0.1–0.2 (e.g., Meinecke and Peters, 1987). For the rest, all the parameter values are the same for excitatory and inhibitory cells. The outgrowth of neurons is on a time scale of days or weeks (Van Huizen et al., 1985; Schilling et al., 1991), so connectivity is quasi-stationary on the time scale of membrane potential dynamics. To avoid unnecessarily slowing down the simulations, ρ is chosen as large as possible under the quasi-stationary approximation. In most simulations, we use $\rho = 0.0001$. As nominal values for the other parameters, we chose $H = 0.1$, $\theta = 0.5$, $\alpha = 0.1$, $\beta = 0.1$, and $\varepsilon = 0.6$.

6.4.4 Results of the Model

In most of the following subsections, the results of the model are given first, followed by the empirical

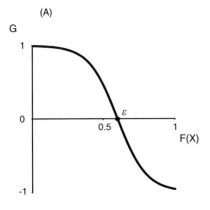

(A)

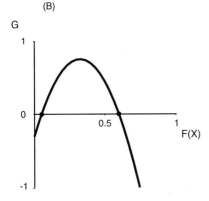

(B)

Figure 6.1
Growth functions relating the rate of neurite outgrowth (G) with firing rate $[F(X)]$. (A) The growth function [see Eq. (6.6)] as used in most of the simulations. The neuritic field of a neuron retracts when the neuronal firing rate is above ε. (B) A growth function for which the neuritic field of a neuron also retracts when the firing rate falls below some critical value.

data from cultures of dissociated cells that can be accounted for by the model.

Overshoot in Excitatory Networks

Model

Simulation shows that one of the implications of activity-dependent outgrowth, together with a hysteresis relationship between network connectivity and activity (see the following discussion), is that a developing network goes through a phase in which the connectivity, or number of synapses, is considerably higher than in the final, stable situation; i.e., the network exhibits overshoot in connectivity (figure 6.2A). For a purely excitatory network $(M = 0)$, this result can be predicted directly from Eq. (6.1). For a given connectivity matrix $\mathbf{W}$, the equilibrium points are solutions of

$$0 = -X_i + (1 - X_i) \sum_{k=1}^{N} W_{ik} F(X_k) \qquad \forall i. \qquad (6.7)$$

If all cells have the same ε and the variations in X_i are small relative to $\bar{X}$, the average membrane potential of the network, we find (Van Ooyen and Van Pelt, 1994):

$$0 \cong -\bar{X} + (1 - \bar{X}) \bar{W} F(\bar{X}), \qquad (6.8)$$

where $\bar{W}$ is the average connection strength. Based on this approximation,

$$\bar{W} = \frac{\bar{X}}{(1 - \bar{X}) F(\bar{X})} \qquad 0 \le \bar{X} < 1. \qquad (6.9)$$

Thus, this equation, which defines a manifold (plotted in figure 6.2B), gives the equilibrium value(s) of $\bar{X}$ for a given value of $\bar{W}$. Equilibrium states on branch CD of the manifold are unstable with respect to $\bar{X}$; equilibrium states on the other branches are stable. Because changes in $\bar{W}$—arising from

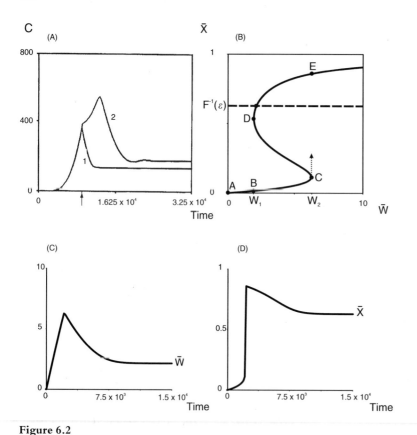

Figure 6.2

Overshoot in connectivity during development. (*A*) Total connectivity $C = \sum_{p=1, q=1}^{N+M} A_{pq}$ in (1) a network without inhibition and (2) a network with inhibition. The arrow indicates the onset of network activity in the network with inhibition. (From Van Ooyen et al., 1995.) (*B*) Hysteresis relationship between average membrane potential $\bar{X} = (1/N) \sum_{i=1}^{N} X_i$ at steady state for a given value of $\bar{W}$, and average connection strength $\bar{W} = (1/N) \sum_{i=1, k=1}^{N} W_{ik}$ in a purely excitatory network [see Eq. (6.9)]. The intersection point with the line $\bar{X} = F^{-1}(\varepsilon)$ is the equilibrium point of the system, at which $\bar{W}$ remains constant. (From Van Ooyen and Val Pelt, 1994.) See section 6.4.4. (*C*) To arrive at the equilibrium point in (*B*), a developing network, starting at point *A* in (*B*), has to go through a phase in which $\bar{W}$ is higher than in equilibrium. (*D*) The average membrane potential during development.

outgrowth and retraction of neuritic fields—are very much slower than changes in $\bar{X}$, $\bar{W}$ can be considered as quasi-stationary on the time scale of the membrane potential dynamics. In other words, in the time that $\bar{X}$ relaxes to its equilibrium value (for a given $\bar{W}$), $\bar{W}$ hardly changes. The slow evolution of $\bar{X}$, determined by changes in $\bar{W}$, therefore takes place along the equilibrium manifold defined by Eq. (6.9); this manifold is sometimes referred to as the slow manifold.

At the intersection point with the line $\bar{X} = F^{-1}(\varepsilon)$ (F^{-1} is the inverse of F), $\bar{W}$ remains constant; above and below that line, it decreases and increases, respectively [see Eq. (6.6)]. Consider the case in which the intersection point is on the branch DE (figure 6.2B). In a developing network, connectivity and activity are initially low, and $\bar{W}$ increases; $\bar{X}$ follows the branch ABC until it reaches W_2, at which point $\bar{X}$ jumps to the upper branch, thus exhibiting a transition from a quiescent to an activated state. But the activity in the network is then so high that the neuritic fields begin to retract and $\bar{W}$ to decrease, and so $\bar{X}$ moves along the upper branch from E to the intersection point. Thus, in order to arrive at an equilibrium point on the branch DE, a developing network has to go through a phase in which the average connectivity is higher than in the final situation. In other words, a higher connectivity is needed to trigger activity in a quiescent network than to sustain it once the network has been activated (hysteresis). The existence of such hysteresis (i.e., the S-shaped curve of figure 6.2B) hinges upon the firing rate function F having a firing threshold and low but nonzero values for subthreshold membrane potentials.

Empirical

A general feature of nervous system development, in vivo and in vitro, is that many structural elements show an initial overproduction followed by an elimination during further development. These so-called

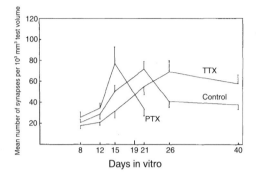

Figure 6.3
Cultures of dissociated cerebral cortex cells show a transient overproduction of synapse numbers (Control). Chronic blockade of activity (by tetrodotoxin, TTX) largely prevents synapse elimination, whereas intensification of activity (by picrotoxin, PTX, which blocks inhibition) accelerates the process. (After Van Huizen et al., 1985, 1987a.)

overshoot phenomena occur, for example, with respect to total dendritic length (Uylings et al., 1990), number of dendrites and axons (e.g., Gorgels et al., 1989), and synapse numbers (e.g., O'Kusky, 1985). The mechanism underlying the generation of overshoot in the model may provide part of the explanation for overshoot phenomena, at least for those that have been observed in vitro. For example, cultures of dissociated cerebral cortex cells show a transient overproduction of synapses during development (Van Huizen et al., 1985, 1987a), with a phase of neurite outgrowth and synapse formation during the first 3 weeks in vitro being followed by a substantial elimination of synapses during the week thereafter (figure 6.3). The development of electrical activity in these cultures also shows a good correspondence with the model. With increasing synaptic density, single-neuron firing and network activity abruptly appear within a window of a few days (Habets et al., 1987). Electrical activity appears to control both neurite outgrowth and synapse elimination. Chronic blockade of electrical activity enhances neurite outgrowth (Van Huizen and Romijn, 1987) and prevents subsequent synapse elimination (Van Huizen et al., 1985). Developing cerebellar cultures have also been shown to exhibit a sequence of events similar to that in the model (Schilling et al., 1991).

Periodic Behavior in Excitatory Networks

Model

The level of electrical activity (or $[Ca^{2+}]_{in}$) above which neurites retract may be different for different classes of neurons (Guthrie et al., 1988; Kater et al., 1988). In terms of the model, this means that there can be variation among cells in ε, the level of electrical activity above which the neuritic field retracts. Under these conditions, complex periodic behavior can occur, with individual cells displaying oscillations that differ in frequency and amplitude (figure 6.4C,D) (Van Ooyen and Van Pelt, 1996). The precise behavior depends on the spatial distribution of the cells and the distribution of ε values over the cells. Note that the oscillations in activity are caused by retraction and outgrowth of the cells' neuritic fields and thus occur on a time scale of days. The network as a whole still shows overshoot in connectivity, but instead of going to a stable connectivity level, the network connectivity (and activity) keeps oscillating to some degree.

Empirical

Using a multielectrode setup to record activity patterns in developing cultures of dissociated cortex cells, Van Pelt et al. (2003) have observed fluctuations in the level of electrical activity of individual cells, with periods of increased activity sometimes lasting as long as several days. Whether these are correlated with periodic changes in neurite outgrowth and connectivity, as the model suggests, is now being investigated.

Overshoot in Mixed Networks

Model

We now consider networks that contain both excitatory and inhibitory cells. Simulation shows that overshoot still takes place in the presence of inhibition, and can even be enhanced (figure 6.2A). To counterbalance inhibition, a higher excitatory connectivity is necessary to reach the level of electrical activity at which the average connectivity starts declining.

Whereas in purely excitatory networks the decline in connectivity begins shortly after the onset of network activity, in mixed networks the decline in overall connectivity can be considerably delayed relative to the onset of network activity (figure 6.2A). In parts of the network with many inhibitory cells, excitatory cells can still be growing out, while in parts with fewer inhibitory cells they are already retracting. For the overshoot curve this implies that average connectivity can still increase markedly after the onset of network activity. Blocking inhibition will thus advance the process of overshoot.

Empirical observations show that the development of inhibition tends to lag behind that of excitation (reviewed in Corner et al., 2002). Modeling the delayed development of inhibition by giving the inhibitory cells a lower outgrowth rate results in a less pronounced excitatory overshoot and a growth curve of the number of inhibitory connections that no longer exhibits any overshoot. The inhibitory cells develop into a network that is already approaching a more or less stable level of electrical activity, and will therefore simply grow out until their overlap is such that $F(X_i) = \varepsilon$.

Empirical

The observation in the model that the decline in connectivity can occur earlier in purely excitatory networks than in mixed networks (figure 6.2A) is

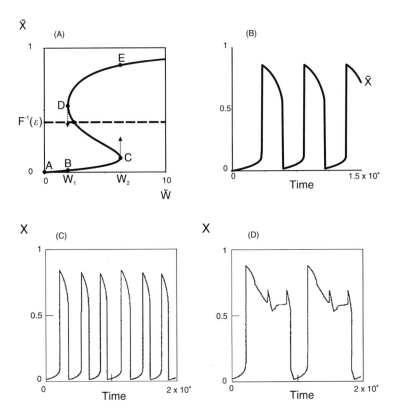

Figure 6.4

Oscillations in electrical activity as a result of periodic changes in neurite outgrowth. (*A*) See also figure 6.2B. If the value of ε is such that the intersection point of the line $\bar{X} = F^{-1}(\varepsilon)$ and the hysteresis curve is on the branch *CD*, regular oscillations occur, in both average connectivity and average membrane potential, that follow the path *ABCEDBCED*.... The period of these oscillations is determined by the value of ρ [see Eq. (6.5)]. (*B*) The oscillations in the average membrane potential. (*C, D*) If the value of ε is not approximately equal for all cells in the network [so that the approximation on which (*A*) is based cannot be made], complex periodic behavior can occur. Here are shown the oscillations in membrane potential for two different cells in a network in which there is variation in ε values. The cell in (*C*) has $\varepsilon = 0.15$; the cell in (*D*) has $\varepsilon = 0.68$. (From Van Ooyen and Van Pelt, 1996.)

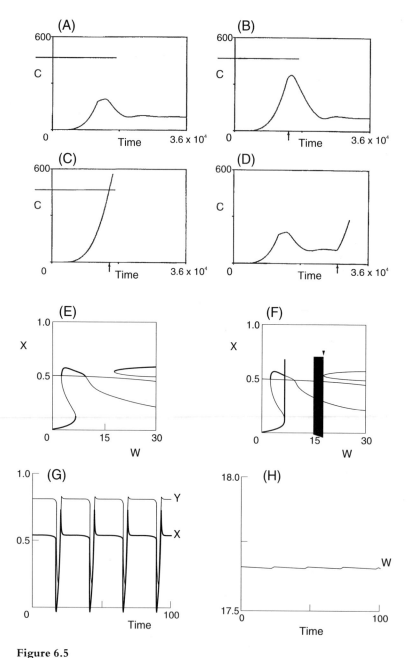

Figure 6.5
If the connection strength in a network with excitatory and inhibitory cells is larger than a critical value, connectivity will not be reduced to the normal equilibrium value. (*A*) Normal development in a mixed network. C = total connectivity =

in agreement with the observation in culture that chronic blockade of GABAergic (i.e., γ-aminobutyric acid) transmission advances the process of synapse elimination (Van Huizen et al., 1987a) (figure 6.3). The presence of inhibition can account for the observation in culture that the decline in connectivity is delayed relative to the onset of network activity (Van Huizen et al., 1985).

Delayed development of inhibition, which in the model causes the number of inhibitory connections to fail to exhibit overshoot, offers a putative explanation for the observation in culture that the synapses on dendritic shafts (presumably inhibitory; Shepherd, 1990) show no overshoot during development, whereas the synapses on dendritic spines, which are mostly excitatory, account for the overshoot phenomenon (Van Huizen et al., 1985).

Multistability, Critical Period, and Periodic Behavior in Mixed Networks

Model

Under all initial conditions, purely excitatory networks go to the same global end state with respect to electrical activity and average connection strength.

Mixed networks, however, do not necessarily do so. In a network with a moderate level of inhibition and an initial average connection strength (of both excitatory and inhibitory connections) that is larger than a critical value, connectivity will not be reduced to the normal equilibrium value but instead will continue to increase (figure 6.5C; Van Ooyen et al., 1995). Basically, this is because the increased inhibitory connection strength stimulates further outgrowth. Van Oss and Van Ooyen (1997) studied this effect in a simplified model consisting of one excitatory and one inhibitory unit. These units can be interpreted as single cells or as representing populations of cells. The excitatory unit (with membrane potential X) is connected to itself and to the inhibitory unit, while the inhibitory unit (with membrane potential Y) is connected only to the excitatory unit and not to itself. Furthermore, it is assumed that the connection between the excitatory and the inhibitory unit is symmetrical (this assumption is not essential for the results) and that the connection strength between the excitatory and the inhibitory unit is a fraction p of the connection strength W of the excitatory unit to itself. The latter assumption is reasonable because when a cell's neuritic field increases, its connection strength with both

$\sum_{p=1,\,q=1}^{N+M} A_{pq}$. (B, C) Electrical activity is blocked until the time indicated by the arrow. In this period, connectivity can only increase. If the level of connectivity thus reached is higher than the level indicated by the horizontal line, connectivity is not reduced, but continues to increase when activity is allowed to return. (D) Activity is blocked in a normally developed network at the time indicated by the arrow. (E) The slow manifold of X, where both $dX/dT = 0$ and $dY/dT = 0$, and the W-nullcline (the thin, nearly horizontal line), where $dW/dT = 0$, in the simplified model (see the section on multistability in mixed networks). The bold lines indicate stable equilibrium points with respect to X and Y (when W is regarded as a parameter), and the thin lines indicate the unstable ones. The intersections of the manifold and the W-nullcline are the equilibrium points of the system. (F) The slow manifold (without showing stability) and the W-nullcline. The bold lines are trajectories: one, starting at $W = 0$ and $X = 0$, approaches the normal point attractor at low W; the other one, starting at $W = 15$ and $X = 0$, approaches the limit cycle attractor at high W (see arrow). Starting at $W = 15$, X and Y oscillate while W slowly increases until the oscillations "touch" the fold of the slow manifold, at which point there is no further net increase in W. Note that since the changes in W are much slower than in X and Y, no separate oscillations are visible. (G) Time plot of the limit cycle attractor in (F) showing X (bold line) and Y. (H) Time plot of the limit cycle attractor in (F) showing W. (From Van Ooyen et al., 1995, and Van Oss and Van Ooyen, 1997.)

excitatory and inhibitory cells increases. The simplified model in differential equations is

$$\frac{dX}{dT} = -X + (1 - X)WF(X) - (H + X)pWF(Y) \tag{6.10}$$

$$\frac{dY}{dT} = -Y + (1 - Y)pWF(X) \tag{6.11}$$

$$\frac{dW}{dT} = q(\varepsilon - bW^2 - X), \tag{6.12}$$

where H is the inhibitory saturation potential and q determines the rate at which the connection strength increases. Because the rate of neurite outgrowth [ρ in Eq. (6.5)]—and thus the rate at which the connection strength changes—is low compared with the dynamics of the membrane potential, q is small. Compared with Eq. (6.6), the growth function [Eq. (6.12)] is simplified to $(\varepsilon - X)$, which has the same qualitative behavior as Eq. (6.6). To prevent connectivity from increasing indefinitely, a saturation term $-bW^2$ is added, where b is small. Without this term, however, the model gives essentially the same results. The model is simple enough to be studied by bifurcation analysis and elaborate enough to show the same phenomena as in the full network model [Eqs. (6.1), (6.2), and (6.5)].

Bifurcation analysis shows that in most cases there is a point attractor (attractor A) at a low connectivity level and a limit cycle attractor at a high connectivity level (attractor B) (figure 6.5E,F). Attractor B is interpretable as a "pathological" state; the limit cycle has fast, epileptiform oscillations in electrical activity (figure 6.5G). Under the normal initial conditions for a developing network (namely, a low level of connectivity), the system develops normally and ends up, via an overshoot in connectivity, in attractor A, whereas a high initial connectivity will cause the system to end up in attractor B (a high initial con-

nectivity can be brought about by, for example, blocking electrical activity for a certain time during network formation). Furthermore, the model shows that the higher the level of inhibition during development (e.g., number of inhibitory cells, strength of inhibitory synapses), the more likely the system is to end up in attractor B.

Empirical

The presence of two stable attractors at different connectivity levels can explain the observation in culture that following chronic blockade of electrical activity (thus causing enhanced neurite outgrowth and a high density of synapses, i.e., a high level of connectivity), there is no subsequent elimination of the excess synapses when the block is finally removed and activity returns to control levels (Van Huizen et al., 1987b, Van Huizen and Romijn, 1987).

In the model, the higher the level of inhibition during development, the more likely the system is to end up in pathological attractor B with strong oscillatory activity (epileptiform activity). This is in line with the following observation: Hypoxic-ischemic encephalopathy (HIE, i.e., brain damage as result of lack of oxygen) induced in rat pups can lead to permanent epileptiform activity later on in adulthood (Romijn et al., 1994), but this epileptiform activity is not the result of a preferential loss of inhibitory elements such as GABAergic nerve endings; indeed, there is a preferential survival of inhibitory elements in the damaged areas (Romijn et al., 1993).

Differences among Cells and Differentiation between Excitatory and Inhibitory Cells

Model

The neuritic field size adapts to the local cell density, resulting in small fields in dense areas and larger ones in sparse areas. Cell death in a mature network will

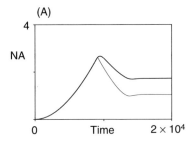

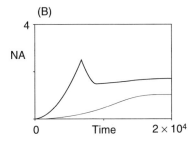

Figure 6.6
The average neuritic field area (NA) of excitatory cells (thick lines) becomes smaller than that of inhibitory cells (thin lines). The network has torus boundary conditions. In (*A*), the inhibitory cells have the same outgrowth rate as the excitatory cells; in (*B*), the inhibitory cells have a lower outgrowth rate than the excitatory cells. (From Van Ooyen et al., 1995.)

result in a compensatory increase in the neuritic fields of the surviving neurons. After excitatory cell loss, electrical activity decreases, and cells will begin to grow out until they all have the same activity level as before [$F(X_i) = \varepsilon$]. To compensate for the lost cells, a larger neuritic field is necessary.

Although in the model there are no intrinsic differences in growth properties between excitatory and inhibitory cells, their neuritic fields will nevertheless become different. The neuritic fields of inhibitory cells tend to become smaller than those of excitatory cells (figures 6.6 and 6.7), and the mechanism accounting for this is as follows: Each cell will attain a

neuritic field size for which the input from overlapping cells is such that $F(X_i) = \varepsilon$. An excitatory cell that receives inhibition therefore needs more excitatory input than a cell that is not inhibited, and thus grows a larger neuritic field. As a consequence, each inhibitory cell will become surrounded by large, strongly connected excitatory cells, whereas, since the same growth rules apply to inhibitory cells, the inhibitory cell itself can remain small because a small neuritic field yields sufficient overlap with its large surrounding cells. In other words, an inhibitory cell becomes small by increasing the size of its direct neighbors. These neighbors, in turn, will become surrounded by relatively small cells, and so on (figure 6.7).

When separate axonal and dendritic fields are used in the model, differentiation between excitatory and inhibitory cells still occurs (the other results are also robust to modeling axonal and dendritic fields separately; see section 6.5). Let R_i^{d} be the radius of the dendritic field of cell i, and R_i^{a} that of its axonal field. As a dendritic field receives input from axonal fields, Eq. (6.4) becomes

$$W_{ij} = O(R_i^{\mathrm{d}}, R_j^{\mathrm{a}})S$$
$$W_{ji} = O(R_j^{\mathrm{d}}, R_i^{\mathrm{a}})S,$$
(6.13)

where $O(\ )$ gives the area of overlap and S represents synaptic strength. The growth of both types of fields is governed by Eq. (6.6) in which, in order to have axonal fields larger than dendritic fields, the growth rate of the latter is given a smaller value ($\rho^{\mathrm{d}} < \rho^{\mathrm{a}}$). With this procedure, inhibitory cells still become smaller than excitatory cells. Even if the axonal field of an inhibitory cell is kept at a constant size, its dendritic field becomes smaller than the fields of the excitatory cells. Excitatory cells receiving input via the (constant) axonal field of a neighboring inhibitory cell will get large dendritic and axonal fields, so that the dendritic

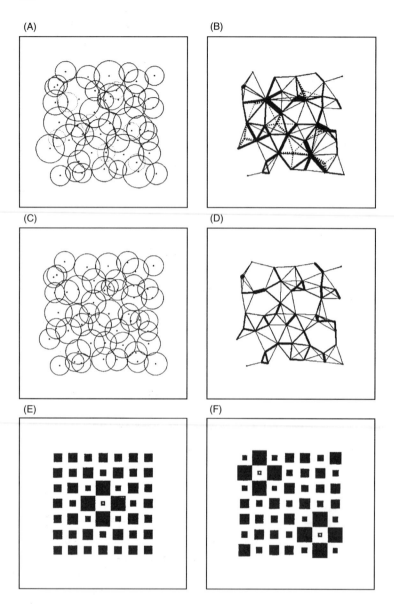

Figure 6.7
Patterns of neuritic field sizes imposed by inhibitory cells. All networks have torus boundary conditions. (*A*) Mature network with excitatory and inhibitory cells. A dotted line indicates an inhibitory cell. (*B*) Graph showing the connections in the network of (*A*). The line width is proportional to the connection strength. A dashed line indicates a connection between an inhibitory and an excitatory cell. Connections that cross boundaries are not shown. (*C*) Same placing of cells as in (*A*), but all former inhibitory cells are now excitatory. (*D*) Graph showing connections in network of (*C*). (*E*) Cells on grid positions. The diameter of a square is proportional to the area of the neuritic field. Scaled to maximum area. The cell with a white dot in the middle is inhibitory. (*F*) Same as in (*E*), but with two inhibitory cells. (From Van Ooyen et al., 1995.)

field of the inhibitory cell can remain small to get sufficient input.

Empirical

The model shows that excitatory cell loss will be accompanied by an increased neuritic (dendritic) field of the surviving neurons. In the human cortex, the dendritic extent per neuron increases steadily through old age (Coleman and Flood, 1986), which has been interpreted as a compensatory response to neuronal death (Coleman and Flood, 1986). This is consistent with the model and with the observation that no increase in dendritic extent occurs in brain regions that do not lose neurons with age (Coleman et al., 1986).

In the model, the neuritic fields of inhibitory cells tend to become smaller than those of excitatory cells. In the cerebral cortex, the dendritic (and axonal) fields of inhibitory neurons are indeed smaller, on the whole, than those of excitatory neurons. Pyramidal cells, which are excitatory, have large apical dendrites (which often cross several layers) and long axons. The nonpyramidal cells, most of which are inhibitory, usually have dendritic and axonal branches that extend only locally (e.g., Abeles, 1991).

Alternative Growth Function

In the growth function [Eq. (6.6)] used in the model, neurite retraction takes place only when neuronal activity is too high. If a growth function is used in which neurite retraction can also take place when neuronal activity is too low (see figure 6.1B), all the main results still hold, provided that the initial activity (as a result of initial connectivity and/or spontaneous activity) is high enough to stimulate outgrowth (Van Ooyen et al., 1996). A difference with this growth function is that not all cells will necessarily become part of the network; cells that receive too little excitation from their surrounding cells will completely retract their neuritic fields.

6.5 Discussion

It has previously been realized that activity-dependent neurite outgrowth could have considerable potential for controlling neuronal form and circuitry (e.g., Mattson, 1988). The model studies described in this chapter have begun to make this potential explicit. The way in which activity influences neurite outgrowth contributes to the homeostasis of neuronal activity, and our model studies have shown that this striving for homeostasis may underlie many of the seemingly unrelated phenomena observed in developing cultures of dissociated neurons. The phenomena that may be the consequence of activity-dependent neurite outgrowth, and that are observed both in culture and in the model, include (1) a transient phase of high connectivity during development, (2) the presence of stable end states of development at different connectivity levels, and (3) size differences in neurite length between excitatory and inhibitory cells. Without modeling studies, it would have been impossible to surmise that these three effects could be different aspects of the same underlying process.

With respect to effect (2), the model has shown that the state of the network (e.g., the average connectivity level) and the balance between excitatory and inhibitory elements can determine whether normal development will take place. Thus, *too much* inhibition prevents the normal pruning of exuberant connections and results in a network with highly oscillatory electrical activity (epileptiform activity).

With respect to effect (3), one must not draw the conclusion that intrinsic differences are unimportant in the development of size differences, but some aspects of differentiation can take place without them.

The results of the model do not depend critically on the neuritic fields being described as circles. Other approaches for modeling neurite outgrowth and connectivity give essentially the same results as described here (Van Ooyen and Van Pelt, 1994). Also, modeling axonal and dendritic fields separately, so that connectivity is no longer necessarily symmetrical, does not alter the main findings (see the section on alternative growth function and Van Ooyen et al., 1995).

Activity-dependent neurite outgrowth could play a role not only during development but also in the adult nervous system. Many studies have shown structural changes in axons and dendrites in response to lesions, aberrant activity, changes in environmental conditions, and behavioral training (for a review, see Woolley, 1999).

Future Modeling Studies

Our modeling studies have dealt with the implications of activity-dependent neurite outgrowth for neuronal morphology and gross network development. Possible roles of activity-dependent neurite outgrowth in learning and memory have not been studied (but see Raijmakers and Molenaar, 1999; see also section 6.3). A form of memory could occur when, after the application of external input, the network is pushed into a different stationary state with respect to neuritic field sizes and connectivity.

A first step toward a more detailed description of growing neurons will be to use, instead of circular neuritic fields, spatial functions describing the density of the neuron's axonal and dendritic branches. Connectivity between neurons can then be described as a convolution of the overlapping dendritic and axonal spatial density functions.

A second step toward a more detailed description could be to model the growing neurites themselves (for example, using a compartmental modeling approach; see chapter 4). The outgrowth of each neurite can then be made dependent on the (time-averaged)

local membrane potential at the tip of the neurite (the growth cone). This membrane potential will be influenced by both the firing rate at the soma and the local synaptic potentials at the neurite. In this approach, the effects of neuronal activity on neurite branching can also be incorporated.

Besides making the description of the growing neurites more detailed, the neuron model can be improved upon. Spiking neurons can be used, and the dynamics of intracellular calcium can be modeled explicitly (as in Abbott and Jensen, 1997). This has the advantages that outgrowth can be made directly dependent on intracellular calcium, the major mediator of activity-dependent neurite outgrowth (see section 6.2), and that the effects of different firing patterns on neurite outgrowth can be included. For example, Fields et al. (1990) found that phasic stimulation is more effective in inhibiting neurite outgrowth than stimulation with the same number of impulses at a constant frequency, an effect for which the calcium dynamics could be responsible.

Finally, activity-dependent neurite outgrowth can be studied in combination with other activity-dependent processes, such as changes in conductances of membrane currents (see chapter 8 and Abbott and Jensen, 1997).

Future Experimental Studies

Experiments are necessary for studying the extent to which activity-dependent neurite outgrowth indeed underlies the phenomena listed in section 6.4.4. Ideally, one would like to simultaneously record the morphological development and neuronal activity of individual cells in a developing network. Using a multielectrode setup, Van Pelt et al. (2003) can continuously record the electrical activity of up to sixty cells in a developing network for a period of at least 6 weeks in culture. In order to correlate activity history with morphology, steps are being undertaken to determine the morphology of the cells from which activity has been recorded.

Examples of the many specific questions that can now be investigated include the following:

• Do neurites retract or change their morphology when they become electrically active?

• Does synaptic input from inhibitory cells indeed influence the size of the dendrites of excitatory cells?

• Do excitatory cells in purely excitatory networks have less variation in the size of their neuritic extent than excitatory cells in a mixed network (as the model predicts; compare figures 6.7A and 6.7C)?

• Does the failure of synapse elimination to occur after chronically blocking electrical activity depend upon the length of the blockade? The model predicts that as long as the blockade does not exceed a critical time, and the connectivity level is therefore below a critical value, elimination of surplus synapses will still occur.

• Do the slow fluctuations observed in the level of electrical activity of individual cells correlate with periodic changes in neurite outgrowth?

It would be useful for further modeling studies if the relationship between intracellular calcium concentration and rate of outgrowth could be determined more quantitatively.

References

Abbott, L. F., and Jensen, O. (1997). Self-organizing circuits of model neurons. In *Computational Neuroscience, Trends in Research*, J. Bower, ed. pp. 227–230. New York: Plenum.

Abbott, L. F., and Nelson, S. B. (2000). Synaptic plasticity: Taming the beast. *Nat. Neurosci.*, Suppl. 3: 1178–1183.

Abeles, M. (1991). *Corticonics. Neural Circuits of the Cerebral Cortex*. Cambridge: Cambridge University Press.

Acebes, A., and Ferrús, A. (2000). Cellular and molecular features of axon collaterals and dendrites. *Trends Neurosci.* 23: 557–565.

Al-Mohanna, F. A., Cave, J., and Bolsover, S. R. (1992). A narrow window of intracellular calcium concentration is optimal for neurite outgrowth in rat sensory neurones. *Dev. Brain Res.* 70: 287–290.

Aspenstrom, P. (1999). The Rho GTPases have multiple effects on the actin cytoskeleton. *Exp. Cell Res.* 246: 20–25.

Berridge, M. J. (1998). Neuronal calcium signaling. *Neuron* 21: 13–26.

Bixby, J. L., Grunwald, G. B., and Bookman, R. J. (1994). Ca^{2+} influx and neurite outgrowth in response to purified N-cadherin and laminin. *J. Cell Biol.* 127: 1461–1475.

Buxbaum, R. E., and Heidemann, S. R. (1992). An absolute rate theory model for tension control of axonal elongation. *J. Theor. Biol.* 155: 409–427.

Cline, H. T. (1999). Development of dendrites. In *Dendrites*, G. Stuart, N. Spruston, and M. Häusser, eds. pp. 35–67. Oxford: Oxford University Press.

Cohan, C. S., and Kater, S. B. (1986). Suppression of neurite elongation and growth cone motility by electrical activity. *Science* 232: 1638–1640.

Cohan, C. S., Connor, J. A., and Kater, S. B. (1987). Electrically and chemically mediated increases in intracellular calcium in neuronal growth cones. *J. Neurosci.* 17: 3588–3599.

Coleman, P. D., and Flood, D. G. (1986). Dendritic proliferation in the aging brain as a compensatory repair mechanism. *Prog. Brain Res.* 70: 227–237.

Coleman, P. D., Buell, S. J., Magagna, L., Flood, D. G., and Curcio, C. A. (1986). Stability of dendrites in cortical barrels of C57B1/6N mice between 4 and 45 months. *Neurobiol. Aging* 7: 101–105.

Corner, M. A., and Ramakers, G. J. A. (1992). Spontaneous firing as an epigenetic factor in brain development—physiological consequences of chronic tetrodotoxin and picrotoxin exposure on cultured rat neocortex neurons. *Dev. Brain Res.* 65: 57–74.

Corner, M. A., Van Pelt, J., Wolters, P. S., Baker, R. E., and Nuytinck, R. H. (2002). Effects of sustained blockade of excitatory synaptic transmission on spontaneously active developing neuronal networks—an inquiry into the reciprocal linkage between intrinsic biorythms and neuroplasticity in early ontogeny. *Neurosci. Biobehav. Rev.* 26: 127–185.

Chen, H. J., Rojas, S. M., Oguni, A, and Kennedy, M. B. A. (1998). A synaptic Ras-GTPase activating protein (p135 SynCap) inhibited by CaM kinase II. *Neuron* 20: 895–904.

Davenport, R. W., and Kater, S. B. (1992). Local increases in intracellular calcium elicit local filopodial responses in *Helisoma* neuronal growth cones. *Neuron* 9: 405–416.

Eglen, S. J., Van Ooyen, A., and Willshaw, D. J. (2000). Lateral cell movement driven by dendritic interactions is sufficient to form retinal mosaics. *Network: Comput. Neural Syst.* 11: 103–118.

Fields, R. D., Neale, E. A., and Nelson, P. G. (1990). Effects of patterned electrical activity on neurite outgrowth from mouse sensory neurons. *J. Neurosci.* 10: 2950–2964.

Forscher, P. (1989). Calcium and polyphosphoinositide control of cytoskeletal dynamics. *Trends Neurosci.* 12: 468–474.

Golowasch, J., Casey, M., Abbott, L. F., and Marder, E. (1999). Network stability from activity-dependent regulation of neuronal conductances. *Neur. Comput.* 11: 1079–1096.

Goldberg, D. J., and Grabham, P. W. (1999). Braking news: Calcium in the growth cone. *Neuron* 22: 423–425.

Gomez, T. M., and Spitzer, N. C. (1999). *In vivo* regulation of axon extension and pathfinding by growth-cone calcium transients. *Nature* 397: 350–355.

Gorgels, T. G. M. F., De Kort, E. J. M., Van Aanholt, H. T. H., and Nieuwenhuys, R. (1989). A quantitative analysis of the development of the pyramidal tract in the cervical spine cord in the rat. *Anat. Embryol.* 179: 377–385.

Grossberg, S. (1988). Nonlinear neural networks: Principles, mechanisms, and architectures. *Neur. Net.* 1: 17–61.

Gu, X., and Spitzer, N. C. (1995). Distinct aspects of neuronal differentiation encoded by frequency of spontaneous Ca^{2+} transients. *Nature* 375: 784–787.

Guthrie, P. B., Mattson, M. P., Mills, L., and Kater, S. B. (1988). Calcium homeostasis in molluscan and mammalian neurons: Neuron-selective set-point of calcium rest concentrations. *Soc. Neurosci. Abstr.* 14: 582.

Habets, A. M. M. C., Van Dongen, A. M. J., Van Huizen, F., and Corner, M. A. (1987). Spontaneous neuronal firing patterns in fetal rat cortical networks during development in vitro: A quantitative analysis. *Exp. Brain Res.* 69: 43–52.

Hentschel, H. G. E., and Fine, A. (1996). Diffusion-regulated control of dendritic morphogenesis. *Proc. Roy. Soc. London B* 263: 1–8.

Horn, D., Levy, N., and Ruppin, E. (1998). Neuronal regulation versus synaptic unlearning in memory maintenance mechanisms. *Network: Comput. Neural Syst.* 9: 577–586.

Kater, S. B., and Guthrie, P. B. (1990). Neuronal growth cone as an integrator of complex environmental information. *Cold Spring Harbor Symp. Quant. Biol.* LV: 359–370.

Kater, S. B., and Mills, L. R. (1991). Regulation of growth cone behaviour by calcium. *J. Neurosci.* 11: 891–899.

Kater, S. B., Mattson, M. P., Cohan, C., and Connor, J. (1988). Calcium regulation of the neuronal growth cone. *Trends Neurosci.* 11: 315–321.

Kuhn, T. B., Williams, C. V., Dou, P., and Kater, S. B. (1998). Laminin directs growth cone navigation via two temporally and functionally distinct calcium signals. *J. Neurosci.* 18: 184–194.

Lankford, K. L., and Letourneau, P. (1991). Roles of actin filaments and three second messenger systems in short term regulation of chick dorsal root ganglion neurite outgrowth. *Cell Motil. Cytoskel.* 20: 7–29.

LeMasson, G., Marder, E., and Abbott, L. F. (1993). Activity-dependent regulation of conductances in model neurons. *Science* 259: 1915–1917.

Letourneau, P. C., Kater, S. B., and Macagno, E. R. (eds.) (1991). *The Nerve Growth Cone.* New York: Raven Press.

Li, Z., Van Aelst, L., and Cline, H. T. (2000). Rho GTPases regulate distinct aspects of dendritic arbor growth in *Xenopus* central neurons in vivo. *Nature Neurosci.* 3: 217–225.

Lin, C. H., and Forscher, P. (1995). Growth cone advance is inversely proportional to retrograde F-actin flow. *Neuron* 14: 763–771.

Maccioni, R., and Cambiazo, V. (1995). Role of microtubule associated proteins in the control of microtubule assembly. *Physiol. Rev.* 75: 835–864.

Marder, E., Abbott, L. F., Turrigiano, G. G., Liu, Z., and Golowasch, J. (1996). Memory from the dynamics of intrinsic membrane currents. *Proc. Natl. Acad. Aci. U.S.A.* 93: 13481–13486.

Marom, S., and Shahaf, G. (2002). Development, learning and memory in large random networks of cortical neurons: Lessons beyond anatomy. *Quart. Rev. Biophys.* 35: 63–87.

Mattson, M. P. (1988). Neurotransmitters in the regulation of neuronal cytoarchitecture. *Brain Res. Rev.* 13: 179–212.

Mattson, M. P., and Kater, S. B. (1987). Calcium regulation of neurite elongation and growth cone motility. *J. Neurosci.* 7: 4034–4043.

Mattson, M. P., and Kater, S. B. (1989). Excitatory and inhibitory neurotransmitters in the generation and degeneration of hippocampal neuroarchitecture. *Brain Res.* 478: 337–348.

Mattson, M. P., Dou, P., and Kater, S. B. (1988). Outgrowth-regulation actions of glutamate in isolated hippocampal pyramidal neurons. *J. Neurosci.* 8: 2087–2100.

McAllister, A. K. (2000). Cellular and molecular mechanisms of dendrite growth. *Cereb. Cortex* 10: 963–973.

Meinecke, D. L., and Peters, A. (1987). GABA immunoreactive neurons in rat visual cortex. *J. Comp. Neurol.* 261: 388–404.

O'Kusky, J. R. (1985). Synapse elimination in the developing visual cortex: A morphometric analysis in normal and dark-reared cats. *Dev. Brain Res.* 22: 81–91.

Raijmakers, M. E. J., and Molenaar, P. C. M. (1999). A biologically plausible maturation of an ART network. *Lect. Notes Comput. Sci.* 1606: 730–736.

Ramakers, G. J. A., Winter, J., Hoogland, T. M., Lequin, M. B., Van Hulten, P., Van Pelt, J., and Pool, C. W. (1998). Depolarization stimulates lamellipodia formation and axonal but not dendritic branching in cultured rat cerebral cortex neurons. *Dev. Brain Res.* 108: 205–216.

Ramakers, G. J. A., Avci, B., Van Hulten, P., Van Ooyen, A., Van Pelt, J., Pool, C. W., and Lequin, M. B. (2001). The role of calcium signaling in early axonal and dendritic morphogenesis of rat cerebral cortex neurons under non-stimulated growth conditions. *Dev. Brain Res.* 126: 163–172.

Romijn, H. J., Van Marle, J., and Janszen, A. W. J. W. (1993). Permanent increase of the GAD_{67}/synaptophysin ratio in rat cerebral cortex nerve endings as a result of hypoxic ischemic encephalopathy sustained in early postnatal life: A confocal laser scanning microscopic study. *Brain Res.* 630: 315–329.

Romijn, H. J., Voskuyl, R. A., and Coenen, A. M. L. (1994). Hypoxic ischemic encephalopathy sustained in early postnatal life may result in permanent epileptic activity and an altered cortical convulsive threshold in rat. *Epilepsy Res.* 17: 31–42.

Sanes, D. H., and Takács (1993). Activity-dependent refinement of inihibitory connections. *Eur. J. Neurosci.* 5: 570–574.

Sánchez, C., Díaz-Nido, J., and Avila, J. (2000). Phosphorylation of microtubule-associated protein 2 (MAP2) and its relevance for the regulation of the neuronal cytoskeleton function. *Prog. Neurobiol.* 61: 133–168.

Schilling, K., Dickinson, M. H., Connor, J. A., and Morgan, J. I. (1991). Electrical activity in cerebellar cultures determines Purkinje cell dendritic growth patterns. *Neuron* 7: 891–902.

Schilstra, M. J., Bayley, P. M., and Martin, S. R. (1991). The effect of solution composition on microtubule dynamic instability. *Biochem. J.* 277: 839–847.

Shepherd, G. M. (1990). *The Synaptic Organization of the Brain*. Oxford: Oxford University Press.

Stoop, R., and Poo, M. M. (1996). Synaptic modulation by neurotrophic factors: Differential and synergistic effects of brain-derived neurotrophic factor and ciliary neurotrophic factor. *J. Neurosci.* 16: 3256–3264.

Stuart, G. J., and Sakmann, B. (1994). Active propagation of somatic action potentials into neocortical pyramidal cell dendrites. *Nature* 367: 69–72.

Tapon, N., and Hall, A. (1997). Rho, Rac and Cdc42 GTPases regulate the organization of the actin cytoskeleton. *Curr. Opin. Cell Biol.* 9: 86–92.

Torreano, P. J., and Cohan, C. S. (1997). Electrically induced changes in Ca^{2+} in *Helisoma* neurons: Regional and neuron-specific differences and implications for neurite outgrowth. *J. Neurobiol.* 32: 150–162.

Turrigiano, G. G. (1999). Homeostatic plasticity in neuronal networks: The more things change, the more they stay the same. *Trends Neurosci.* 22: 221–227.

Turrigiano, G. G., Abbott, L. F., and Marder, E. (1994). Activity-dependent changes in the intrinsic properties of cultured neurons. *Science* 264: 974–977.

Turrigiano, G. G., Le Masson, G., and Marder, E. (1995). Selective regulation of current densities underlies spontaneous changes in the activity of cultured neurons. *J. Neurosci.* 15: 3640–3652.

Turrigiano, G. G., Leslie, K. R., Desai, N. S., Rutherford, L. C., and Nelson, S. B. (1998). Activity-dependent scaling of quantal amplitude in neocortical neurons. *Nature* 391: 892–896.

Uylings, H. B. M., Van Eden, C. G., Parnavelas, J. G., and Kalsbeek, A. (1990). The prenatal and postnatal development of rat cerebral cortex. In *The Cerebral Cortex of the Rat*, B. Kolb and R. C. Trees, eds. Cambridge, Mass.: MIT Press.

Van Huizen, F., and Romijn, H. J. (1987). Tetrodotoxin enhances initial neurite outgrowth from fetal rat cerebral cortex cells in vitro. *Brain Res.* 408: 271–274.

Van Huizen, F., Romijn, H. J., and Habets, A. M. M. C. (1985). Synaptogenesis in rat cerebral cortex is affected during chronic blockade of spontaneous bioelectric activity by tetrodotoxin. *Dev. Brain Res.* 19: 67–80.

Van Huizen, F., Romijn, H. J., Habets, A. M. M. C., and Van den Hooff, P. (1987a). Accelerated neural network formation in rat cerebral cortex cultures chronically disinhibited with picrotoxin. *Exp. Neurol.* 97: 280–288.

Van Huizen, F., Romijn, H. J., and Corner, M. A. (1987b). Indications for a critical period for synapse elimination in developing rat cerebellar cortex cultures. *Dev. Brain Res.* 13: 1–6.

Van Ooyen, A. (1994). Activity-dependent neural network development. *Network: Comput. Neural Syst.* 5: 401–423.

Van Ooyen, A., and Van Pelt, J. (1994). Activity-dependent outgrowth of neurons and overshoot phenomena in developing neural networks. *J. Theor. Biol.* 167: 27–43.

Van Ooyen, A., and Van Pelt, J. (1996). Complex periodic behaviour in a neural network model with activity-dependent neurite outgrowth. *J. Theor. Biol.* 179: 229–242.

Van Ooyen, A., Van Pelt, J., and Corner, M. A. (1995). Implications of activity-dependent neurite outgrowth for neuronal morphology and network development. *J. Theor. Biol.* 172: 63–82.

Van Ooyen, A., Pakdaman, K., Houweling, A. R., Van Pelt, J., and Vibert, J.-F. (1996). Network connectivity changes through activity-dependent neurite outgrowth. *Neur. Proc. Lett.* 3: 123–130.

Van Oss, C., and Van Ooyen, A. (1997). Effects of inhibition on neural network development through activity-dependent neurite outgrowth. *J. Theor. Biol.* 185: 263–280.

Van Pelt, J., Wolters, P. S., Willems, W. P. A., Rutten, W. L. C., Corner, M. A., Van Hulten, P., and Ramakers, G. J. A. (2003). Developmental changes in firing patterns in neocortical networks in vitro: Longitudinal recordings on multi-electrode arrays. *Exp. Brain Res.*, submitted.

Woolley, C. S. (1999). Structural plasticity of dendrites. In *Dendrites*, G. Stuart, N. Spruston, and M. Häusser, eds. pp. 339–364. Oxford: Oxford University Press.

Theoretical Models of Retinal Mosaic Formation

Stephen J. Eglen, Lucia Galli-Resta, and Benjamin E. Reese

7

The spatial distribution of neurons affects the efficiency of neural circuitry. This is particularly evident in the retina, where cells are regularly arranged so that the visual world can be efficiently sampled. In this chapter we outline several developmental mechanisms (cell fate, lateral cell movement, and cell death) by which postmitotic cells become organized into a regular spatial pattern. Here, we focus on two models and review other relevant models. In the first model, the d_{min} model, cellular positioning is subject only to the constraint that no cell can be closer than some minimal distance to any other cell of the same type. This simple model can account for the distribution of many classes of retinal cells. The second model, the lateral movement model, investigates the suggestion that dendritic interactions control cell positioning. This model shows that lateral movements are sufficient for achieving regularity. We conclude by comparing simulated mosaics produced by these two models and highlight areas of future experimental and theoretical research.

7.1   Neurobiological Background

The vertebrate retina contains six classes of retinal neuron, distributed in several layers. Each class of neuron typically occupies just one of these layers, with the neurons regularly distributed throughout a layer to ensure that the visual world is efficiently sampled and that each cell receives roughly the same number of inputs as its neighbors. Owing to the way that the cell bodies (and their dendrites) seem to tile

the retina (figure 7.1), these spatial distributions are often termed retinal mosaics. What developmental mechanisms help form these regular distributions?

The formation of regular cell distributions may not be a phenomenon specific to the retina; many other brain regions may assemble their functional circuits in an efficient manner, simplifying the task of making appropriate connections by virtue of regular arrays. In contrast, if cells are not regularly arranged, it may be more difficult for cells to find sufficiently close postsynaptic partners. Subpopulations of cells in cortical areas may be regularly distributed, but may simply have not yet been examined. The flat, layered structure of the retina makes it relatively easy to examine the distributions of cells compared with other neural tissue. Observing regular distributions of cells in other regions is also complicated by the large number of cell types that exist and the difficulty in correctly classifying them. Despite these problems, Cook and Chalupa (2000) describe examples of neuronal mosaics discovered in the rat cerebellum and the avian tectum. Mosaics are also likely to be found in non-neural tissue, such as skin.

7.1.1   Possible Mechanisms Underlying Mosaic Formation

Several developmental processes are likely to be involved in regulating the distribution of retinal cells throughout development, starting from the time that cells acquire their identity via fate determination mechanisms. Postmitotic cells leave the ventricular

(a)

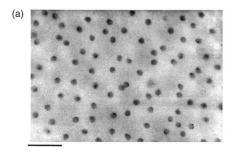

(b)

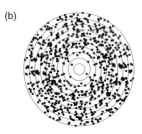

Figure 7.1
(*a*) Mosaic of adult mouse cholinergic amacrine cells (scale bar: 50 μm). The regularity index is 4.1. (*b*) Autocorrelation plot (defined in section 2 of the appendix) of this sample of cells; annuli are spaced 5 μm apart.

surface (near the future photoreceptor layer) and migrate (radially) toward their destination layer. Cells can also move tangentially (parallel to the retinal layers). Morphological differentiation begins while cells migrate and continues once they arrive at their destination layer. Throughout this period, there is massive cell death within the retina that could further sculpt retinal distributions. Each of these mechanisms is now described; for further details, see recent experimental reviews (Cook and Chalupa, 2000; Galli-Resta, 1998; Reese and Tan, 1998).

Cell Fate Mechanisms
The different classes of retinal cells are generated in two main phases of cell birth (reviewed in Robinson, 1991). In the first phase, retinal ganglion cells (RGCs) and cone photoreceptors are produced, followed by amacrine and horizontal cells. In the second phase, bipolar cells and rod photoreceptors are born. However, postmitotic cells in the mammalian retina are not predestined purely by their birth date to become one particular class, although their fate appears to become restricted over time (Cepko et al., 1996). The fate of cells to become a particular class may be influenced by neighboring cells. Early specification of one cell, the founder cell, could induce neighboring cells to adopt a different fate, as suggested for rods and cones in primates and goldfish (Wikler and Rakic, 1991; Stenkamp et al., 1997). The RGCs also secrete factors that inhibit neighboring cells in acquiring the same fate (Waid and McLoon, 1998). Much more is known about cell fate determination in invertebrate eye development. Genetic and molecular advances have shown that the differentiation of photoreceptors in each ommatidium of the *Drosophila* eye follows a sequence of fate determination signals once the founder cell, the photoreceptor R8, is determined (Freeman, 1997).

A combination of genetic and environmental influences is therefore likely to be involved in regulating the density of each type of retinal cell. However, the role of these processes in mosaic formation is unclear. Lateral inhibitory mechanisms should prevent primary fate cells from being close to each other, creating an initial regularity in the spatial distribution. In favor of this suggestion, early differentiating RGCs appear regularly distributed in chicks (McCabe et al., 1999). However, while cholinergic amacrine cells (Galli-Resta et al., 1997) and horizontal cells (Galli-Resta, unpublished observations) migrate to their destination layer, they are often much closer to each other than the minimal spacing observed between cells in the destination layer. This suggests that subsequent developmental events improve upon any order established by fate-based mechanisms.

Lateral Cell Movement

Early analysis of retinal cell migration from various lineage-tracing studies indicated that most cells moved radially from the site of their final mitotic division at the ventricular surface to yield columns of clonally related cells (Turner and Cepko, 1987). By using a transgenic mouse model enabling the marking of 50 percent of all clones, however, reanalysis of the migration patterns indicated that certain classes of retinal cell also move laterally away from their column of origin (Reese et al., 1995; Reese and Tan, 1998). Furthermore, the classes exhibiting lateral movement are the same classes that are regularly arranged. Cells moved typically less than 100 μm, and all cells within a given class moved (Reese et al., 1999).

Further evidence for lateral movement of retinal cells came from the discovery that cholinergic cells could be identified while they were still migrating to their destination layers (Galli-Resta et al., 1997; Galli-Resta, 2000). Migrating cholinergic cells were often found side by side, but once they arrived in their layer they acquired a regular spacing, which was maintained throughout the several days during which new cells entered the array. This suggests that the cells moved laterally within their destination layer to maintain regularity.

These two lines of evidence suggest that lateral movement of cells within a cell's destination layer is instructive for retinal mosaic formation. Indirect evidence in favor of lateral cell movement also comes from the observation that the positions of blue cone photoreceptors and their postsynaptic contacts, the blue cone bipolar cells, are positively correlated. Lateral cell movement is the most likely mechanism by which this patterning occurs (Kouyama and Marshak,

1997). The mechanisms driving cell movement have not yet been discovered, although one suggestion is that the cell body partially translocates within its processes (Reese and Tan, 1998; Cook and Chalupa, 2000).

Cell Death

Results from a range of mammalian species indicate that 50–90 percent of RGCs that are born will die during development (Finlay and Pallas, 1989; see also chapter 9). As well as refining the projection of RGCs to their targets (O'Leary et al., 1986), cell death might also influence mosaic formation by removing those neurons that are inappropriately placed among a population of cells (Galli-Resta, 1998; Cook and Chalupa, 2000). In favor of this hypothesis, early postnatal cell death, influenced by electrical activity, was implicated in the formation of ON- and OFF-center alpha RGC mosaics (Jeyarasasingam et al., 1998). It is possible that RGCs compete with their neighbors for trophic support or contacts from their afferents (Wässle and Rieman, 1978), and those cells that receive insufficient support will die (Kirby and Steineke, 1996). However, retinal cell death does not always guide mosaic formation. In rats, 20 percent of cholinergic amacrine cells die between postnatal day 4 (P4) and P12 without increasing mosaic regularity (measured using the regularity index, as defined in section 1 in the appendix to this chapter) (Galli-Resta and Novelli, 2000). Computer simulations showed that random deletion of 20 percent of these cells from P4 retina is sufficient to account for the distributions found at P12, implying that cell death occurs randomly across the retina without increasing mosaic regularity. Furthermore, cell death alone cannot create the spatial dependence observed between blue cones and the blue cone bipolar cells because the magnitude of cell death is too small to remove all the bipolar cells that are too far from the blue cones (Kouyama and Marshak, 1997).

7.2   The d_{min} Rule and Lateral Movement Model

Theoretical approaches to mosaic formation are instructive for investigating how local developmental rules can generate global properties of retinal mosaics. In this section we focus on two different approaches to modeling mosaic formation. Other relevant models are discussed in section 7.3. Theoretical models can be classified as either phenomenological or mechanistic (Nathan and Muller-Landau, 2000). Most models mentioned in this review are phenomenological models, since they replicate statistical properties of experimental mosaics without concern about the underlying developmental processes. In contrast, mechanistic models simulate potential developmental processes to see if they reproduce relevant aspects of development. Although a mechanistic model is usually preferred over a phenomenological model, both types of model are informative, as shown, for example, in models of dendritic tree formation (see chapter 4). In this section, we discuss the d_{min} rule (a phenomenological model) and the lateral movement model (a mechanistic model).

7.2.1   The d_{min} Rule

Autocorrelation plots of retinal mosaics typically have an empty space around the origin (figure 7.1b; section 2 in the appendix), suggesting that each cell has its own "exclusion zone"; i.e., each cell prevents other cells of the same type from getting too close to it. The d_{min} model (named after the minimal distance parameter in the model) (Galli-Resta et al., 1997) mimics this exclusion zone by constraining the addition of

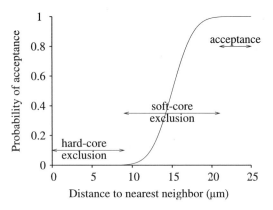

Figure 7.2
Example of a probability of acceptance function for an exclusion zone model (in this case, for a d_{min} model with a Gaussian distribution of $15 \pm 3.0\ \mu m$).

new cells into a mosaic. The model generates a distribution of cells by sequentially adding new cells into a (initially empty) two-dimensional array. A trial cell is randomly positioned within the array and given a d_{min} value from a Gaussian distribution of fixed mean and standard deviation. The trial cell is accepted if the distance to the nearest neighbor of previously accepted cells is greater than the d_{min} of the trial cell; otherwise the trial cell is discarded. This process continues until the desired number of cells has been placed in the array, or until no more cells can be positioned within the array.

The d_{min} model is one example of a general class of exclusion zone models in which the acceptance of a cell into the array is based upon a probability of acceptance function (figure 7.2). This function determines the probability of accepting a new cell according to the distance to the nearest neighboring cell. The exact shape of the function varies for different classes of retinal cell, but it has three regions. First, there is a hard-core exclusion zone that always prevents any two cells being closer to each other than some minimum distance. This area is partially, but not totally, accounted for by the restriction that somas within a layer cannot overlap. Second, there is a soft-core (or elastic) exclusion zone; the probability of accepting the new cell typically increases with distance. Third, after the nearest neighbor is farther than some upper distance, the trial cell is always accepted.

The d_{min} model requires only three parameters: (1) the density of cells to generate and (2) the mean and (3) standard deviation for the Gaussian distribution of d_{min} values. To fit the d_{min} model to a data set, the density of cells is the same as the density of the data set. A range of values for the mean and standard deviation of d_{min} is tested to see which values produce the closest fits to the experimental data. The quality of the match between simulated and experimental mosaics is measured by comparing distributions, such as nearest-neighbor distances and Voronoi-based features (section 3 in the appendix), and can be used by principled approaches to find the best model parameters (Diggle and Gratton, 1984).

Performance of the d_{min} Model

The d_{min} rule has successfully replicated the distribution of several developing and adult retinal mosaics, including rat cholinergic amacrine cells (Galli-Resta et al., 1997; Galli-Resta and Novelli, 2000), rod and cone photoreceptors in ground squirrels (Galli Resta et al., 1999; see figure 7.3), nicotinamide adenine dinucleotide phosphate (NADPH)-diaphorase active RGCs in chicks (Cellerino et al., 2000), and mouse horizontal cells (our unpublished results). The quality of these fits indicates that local, homotypic (i.e., interactions occur only between cells of the same class) exclusion zones are sufficient to recreate the spatial distributions of experimental mosaics. Furthermore, the d_{min} model has revealed additional features about retinal mosaics.

In the ground squirrel study, retinal mosaics were collected from a range of different cell densities. In these mosaics, the regularity index increased with cell density (figure 7.3). The d_{min} model fitted all rod and cone photoreceptor mosaics using the same Gaussian distribution (6.5 ± 1.0 μm for rods, 16.5 ± 3.5 μm for cones) across all cell densities. As cell density increases, the d_{min} model rejects more cells as it becomes harder to place trial cells that do not overlap with other exclusion zones, hence making more regular distributions. Also, when fitting cone photoreceptors at cell densities close to the packing density limit for the d_{min} rule, allowing less than 10 percent of the cells to ignore the d_{min} constraint was sufficient to fit the high-density mosaics.

In contrast, the regularity index of RGCs with NADPH-diaphorase activity in adult chicks (Cellerino et al., 2000), 6.3 ± 0.8 μm, did not vary with cell density across the retina. In this case, it was suggested that the initial population of RGCs across the entire retina was created at a constant density with one d_{min} value. Differential expansion of the retina then varied cell density across the retina while preserving regularity. This is a variant of the "little bang" hypothesis (Rodieck and Marshak, 1992), which suggests that cell bodies are initially close packed and then space out to different degrees during development because of nonuniform stretching of the retina (Mastronarde et al., 1984).

What Does the d_{min} Rule Tell Us about Development?

Although the d_{min} rule is a phenomenological model, it could inform us about some aspects of the development of retinal mosaics. It is a simple, robust model that requires only local information (over the range of the exclusion zone) when deciding whether to accept new cells into the mosaic. The model predicts that the cellular interactions underlying exclusion zones are

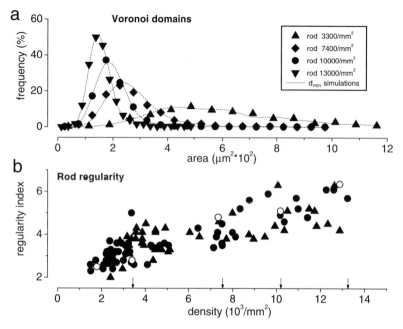

Figure 7.3
Comparison of the distribution of rod photoreceptors in the ground squirrel with the d_{min} model (using a Gaussian distribution of 6.5 ± 1.0 μm). (*a*) Distribution of Voronoi polygon areas for rod mosaics at four different densities (filled symbols). Dotted lines indicate the closely matching distribution of Voronoi polygon areas for mosaics produced by the d_{min} model at those four densities. (*b*) Regularity index of rod mosaics taken at different densities from two different animals (filled circles and triangles). Open circles indicate the regularity index of mosaics produced by the d_{min} model. The arrows along the abscissa indicate the four densities examined in (*a*). (Reprinted from figure 5 of Galli-Resta et al., 1999, with permission from Blackwell Science Ltd.)

homotypic and that the exclusion zones of neighboring cells are nonadditive (i.e., two closely neighboring cells do not make a stronger exclusion zone than one cell). In a recent experimental study prompted by these model predictions, both these features of the exclusion zone were verified for cholinergic amacrine cells (Galli-Resta, 2000). The d_{min} model is therefore a good example of theoretical work that has suggested new experiments to increase our knowledge of developmental mechanisms.

What cellular mechanisms effectively produce an exclusion zone around each cell? One possibility is that cell death removes new cells that are too close to preexisting cells within the mosaic (Galli-Resta, 1998). Although the number of rejected cells from the d_{min} model for low densities is small enough to be plausibly accounted for by cell death, at higher densities the number of rejected cells is much higher than the amount of cell death. In light of the studies showing cellular movement in the plane of the retina, cells may move laterally to observe the exclusion zone constraint rather than dying (Galli-Resta, 1998). The sufficiency of lateral cell movement to generate retinal mosaics is discussed further in section 7.2.2.

Related Exclusion Zone Models

Several other exclusion zone models have been used to model retinal mosaics. In the first reported model (Diggle and Gratton, 1984), the probability of accepting a new cell depended on the distance to all neighboring cells and successfully fitted the mosaic of displaced amacrine cells in the rabbit. The distribution of macaque blue cone photoreceptors was fitted by a d_{min}-like model using the distribution of nearest-neighbor distances measured from one experimental sample as the probability of acceptance function (Shapiro et al., 1985). Later work using the same "elastic ball" model found that it fitted the distribution of blue cones in the "foveal slope," but not the retinal periphery, of the macaque (Curcio et al., 1991).

Ammermüller et al. (1993) used a Boltzmann distribution as the probability of acceptance function to successfully fit the distribution of turtle horizontal cells; this fit was better than a lattice-based model (see section 7.3.1). The d_{min} rule was also used to fit embryonic rabbit horizontal cells (Scheibe et al., 1995). Outside of the retina, exclusion zone models have replicated many other spatial distributions, including gull's nests (Bartlett, 1975) and pine trees (Ripley, 1981). Owing to their wide use and slight variations, exclusion zone models have several names, including Strauss processes (Ripley, 1981), serial sequential inhibition, and Matérn processes (Diggle, 1983).

7.2.2 *Lateral Movement Mediated by Dendritic Interactions*

The d_{min} model has shown that local homotypic interactions are sufficient to replicate the spatial distribution of retinal cells. However, the biological mechanisms that might implement the d_{min} constraints are as yet unclear. One of the early mechanisms hypothesized for retinal mosaic formation was that each cell has a repulsive force so that an initial random distribution of cells can transform itself into a regular pattern (Wässle and Riemann, 1978). In this section we review a model investigating if dendritic interactions could mediate the repulsive force driving lateral cell movement and mosaic formation (Eglen et al., 2000).

Description of the Lateral Movement Model

The lateral movement model (Eglen et al., 2000) is based upon a previous model of dendritic outgrowth (Van Ooyen and Van Pelt, 1994; see also chapter 6). A number of n cells are positioned within a square region of tissue of side length l. Each cell i has three variables: $\mathbf{C}_i$, R_i, and X_i, where $\mathbf{C}_i$ (the bold denotes a 2-D vector) is the position of the cell body, R_i is the radius of the circular dendritic field around the cell body, and X_i is the mean membrane potential (although other physiological interpretations of X_i are equally plausible). Each cell is given a random initial position $\mathbf{C}_i$, and both R_i and X_i are zero initially. The following equations [Eqs. (7.1), (7.3), and (7.5)] then update each variable.

The mean membrane potential X_i is a weighted function of inputs from neighboring cells, along with a decay term with a time constant τ:

$$\frac{dX_i}{dt} = -\frac{X_i}{\tau} + (1 - X_i)\sum_{j=1}^{n} W_{ij}F(X_j), \qquad (7.1)$$

where

$$F(X_j) = \frac{1}{1 + \exp[(\theta - X_j)/\alpha]} \qquad (7.2)$$

and $W_{ij} = cA_{ij}$. The mean firing rate of cell j, $F(X_j)$, is thus a sigmoidal function with a threshold θ and steepness α. The function A_{ij} is the area of overlap between the dendritic fields of cells i and j, with $A_{ii} = 0$. The input from cell j to cell i, W_{ij}, is then cA_{ij}, where c is a constant.

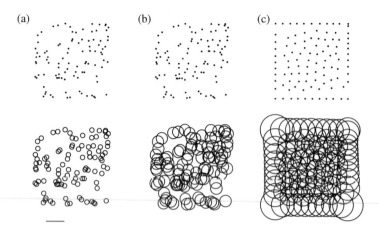

Figure 7.4
Typical lateral movement model at three stages during development ($n = 100$ cells). In the top row, each dot represents a cell body; in the bottom row, each circle represents the position of one cell, with its radius equal to the dendritic extent. (*a*) At the start of development, dendrites begin to grow outward, but cells have not yet moved very much and the mosaic is still random [regularity index (RI) = 2.1)]. (*b*) After a little while, dendrites have grown enough to cause some small movement and a slight increase in mosaic regularity (RI = 2.5). (*c*) At the end of development, dendrite sizes are uniform and the mosaic is highly regular (RI = 10.9). Scale bar: 100 μm.

The rate of dendritic outgrowth of each cell is controlled by the mean activation of the cell via the sigmoidal function G:

$$\frac{dR_i}{dt} = \rho G[F(X_i)], \tag{7.3}$$

where

$$G(x) = 1 - \frac{2}{1 + \exp[(\varepsilon - x)/\beta]}, \tag{7.4}$$

with β a constant that determines the steepness of the function. When the cell's firing rate is below threshold ε, $G[F(X_i)]$ is positive, causing outgrowth. Conversely, when it is above threshold, $G[F(X_i)]$ is negative, and the dendritic field retracts.

Cell bodies repel each other in proportion to their dendritic overlap:

$$\frac{d\mathbf{C}_i}{dt} = \eta \sum_{j=1}^{n} u(\mathbf{C}_i - \mathbf{C}_j) W_{ij}, \tag{7.5}$$

where $u(\mathbf{V})$ is the vector $\mathbf{V}$ normalized to a unit of length, except $u(\mathbf{0}) = \mathbf{0}$. Both elements of $\mathbf{C}_i$ were bounded to keep each cell within the fixed region of tissue.

In contrast to the relatively simple $d_{\min}$ model with three parameters, Eqs. (7.1)–(7.5) contain ten parameters that affect development (n, l, τ, θ, α, c, ε, β, ρ, η); typical values are given in Eglen et al. (2000).

Results from the Lateral Movement Model
Figure 7.4 shows a typical sequence of development in the model. In the early stages of development, since most dendrites do not overlap, they receive little input from neighboring cells and so the dendrites first expand. Once neighboring dendrites overlap, cells

then begin to repel each other and gradually reorganize into a regular mosaic while dendritic field sizes still change. Long after dendritic field sizes have stabilized, there is usually some small movement of a few cells, which continues to increase mosaic regularity. The small distances moved by each retinal cell are within the range observed experimentally (Reese et al., 1999). All cells within the model also move, which is in agreement with experimental results. Mosaic regularity is unaffected by cell density (figure 7.5a), although the values are higher than those typically found in experimental mosaics. Dendritic field size is inversely proportional to the cell density. Furthermore, if the overlap function A measures the relative rather than absolute amount of dendritic contact between cells, the coverage factor (dendritic area multiplied by cell density) is constant across retinal space.

Various components of the model were investigated to see how they affect development. First, different forms of the overlap function were considered. In the model, it is assumed that the dendrites of each cell are circular and that pairs of cells can calculate exactly by how much they overlap. If the estimate of the amount of overlap is made more qualitative by reducing the precision of the calculation, the regularity index decreases down to experimentally observed levels (see table 1 of Eglen et al., 2000).

Another contribution to mosaic regularity is whether dendritic fields change size during mosaic formation. If dendritic field sizes are fixed, regularity of the developed mosaic is no longer constant across cell densities, but instead increases with cell density (figure 7.5a). This might occur if dendrites grow on a much slower time scale than the rate at which cells move. Furthermore, when dendrites are fixed, cell movement is not universal. Instead, cells simply repel each other until they no longer overlap. The degree of cell movement is therefore proportional to the

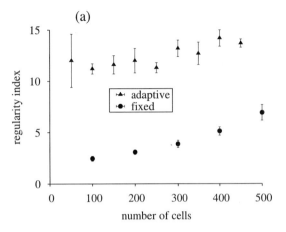

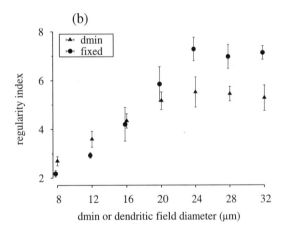

Figure 7.5

Regularity of the lateral movement models and comparison with the d_{min} model. (*a*) Regularity indexes in the adaptive-dendrite (triangles; five simulations per condition) and fixed-dendrite model (circles; ten simulations per condition) for different cell densities. (*b*) Comparison of the d_{min} model with the fixed-dendrite model. For each mean value of d_{min}, ten d_{min} mosaics (triangles) were simulated ($n = 200$ cells) using a Gaussian distribution with a standard deviation of 20 percent of d_{min}. Likewise, ten simulations of the fixed-dendrite lateral model (circles) were run using the same Gaussian distribution for the dendritic field diameters. In each simulation, the simulated region of the retina was $400 \times 400 \ \mu m^2$, and in both plots, error bars denote 1 standard deviation of the mean.

cell density and the size of the exclusion radius. For example, at low cell densities ($n = 100$ cells, figure 7.5a), about 25 percent of the cells moved; at higher densities ($n = 500$, figure 7.5a), about 80 percent of the cells moved. Regularity is proportional to cell density here because more cells move at higher densities to fit the model's constraints.

The lateral movement therefore behaves differently according to whether the dendrites adapt (the adaptive-dendrite model) or remain fixed (the fixed-dendrite model) during development. In the adaptive-dendrite model, cells move and dendrites adapt sizes until each cell receives a fixed amount of input (see chapter 6), and cells move until the movement forces cancel each other out. In the fixed-dendrite model, at low densities, cells can repel each other until their dendrites do not overlap.

These two variants of the model produce different results as cell density varies. In agreement with the adaptive-dendrite model, the regularity index of mammalian cone photoreceptors, cholinergic amacrine cells, and chick NADPH-diaphorase active RGCs (Jeffery et al., 1994; Rodieck and Marshak, 1992; Cellerino et al., 2000) did not increase with cell density. In contrast, and in agreement with the fixed-dendrite model, the regularity of rod photoreceptors in the ground squirrel increased with cell density (Galli-Resta et al., 1999; see also figure 7.3). Corresponding experimental data for the percentage of cells that move are not yet available. Measurements of cell movement can currently be made only in transgenic models, namely, the mouse retina, where there is a shallow center-to-periphery gradient in cell density. Experimental manipulations to alter the density of cells in the mouse retina are therefore necessary to test if the amount of cell movement is proportional to, or independent of, cell density.

The lateral movement model assumes that dendritic interactions produce repulsion between cells. To date,

the only evidence implicating dendrites in lateral cell movement is correlational. The morphology of horizontal cells changes from radial to tangential processes during the period of lateral cell movement (Reese et al., 1999). The role of dendritic interactions therefore awaits experimental investigation. However, nondendritic interpretations of the fixed-dendrite model are possible. Each cell could produce a chemorepellant that is effective over a fixed radius R to attain some minimal spacing between cells (Eglen et al., 2000), although no such chemorepellants have yet been identified. Furthermore, at least for cholinergic amacrine cells, the exclusion zone around each cell is not additive and so probably is not mediated by diffusible substances (Galli-Resta, 2000).

7.2.3 Comparison of d_{min} and Lateral Movement Model

The fixed-dendrite version of the lateral movement model prevents dendrites of neighboring cells from overlapping when possible. The lateral movement model could therefore be interpreted as implementing a d_{min} rule by repelling cells that lie within each other's exclusion zones (Eglen et al., 2000) if we equate the d_{min} value of a cell from the d_{min} model with the diameter of a cell's dendritic field in the lateral movement model. To test if the two models are functionally equivalent, we simulated a set of mosaics using the same Gaussian distribution for both the d_{min} values and dendritic field diameters (figure 7.5b). For each value of d_{min}, the regularity indexes of mosaics produced by the two models are significantly different ($p < 0.02$ in each comparison, Wilcoxon rank sum test) except when $d_{min} = 16$ μm ($p = 0.25$). Hence the two models are not functionally equivalent, although their regularity index curves follow the same general shape. In both models, regularity increases with d_{min} up to about 20 μm, but increases slightly faster in the fixed-dendrite model. For values of $d_{\min}$ greater than 20 µm, regularity in each model reaches its own plateau value because it is close to the packing density limit. No significant difference in regularity was found in this range. (When comparing the regularity of mosaics at 24 µm, 28 µm, and 32 µm with the Kruskal-Wallis test, $p = 0.43$ for the $d_{\min}$ model and $p = 0.29$ for the fixed-dendrite model.)

One source of the difference between the two models may be because the exclusion zones in the $d_{\min}$ model are nonadditive, whereas the repulsive forces in the lateral movement model are additive. Despite this difference, the two models may agree more closely after the repulsive force is altered in the lateral movement model. First, the force repels the two cells along a straight line between the cells. The direction of any repulsive force is likely to be much less accurate, perhaps just pushing the two cells away in different directions. Second, repulsion stops once two cells no longer overlap, whereas it could continue over longer distances. Third, if cells migrate into the destination layer over a long time, the "arrival time" of a cell into the layer might be important. For two overlapping cells, the newer cell could be the one that must translocate. This would be like the $d_{\min}$ model, in which only the later-positioned cell moves.

7.3 Related Theoretical Models of Mosaic Formation

Here we describe the main findings of other theoretical models, retinal and nonretinal, that are of relevance to retinal mosaic formation.

7.3.1 Lattice-Based Models

Exclusion zone models incrementally generate mosaics on a cell-by-cell basis. In contrast, lattice-based models generate the whole sample of cells in parallel by placing the cells on a predefined regular (hexagonal or rectangular) lattice. Each cell is then independently moved some random direction and distance, as long as cell bodies do not overlap. This method was used to fit the distribution of macaque blue cone photoreceptors (Shapiro et al., 1985) and turtle horizontal cells (Ammermüller et al., 1993). In both cases, better fits to the experimental data were obtained using exclusion zone models. Lattice models were also used to investigate the distribution of ON- and OFF-center cat beta RGCs (Zhan and Troy, 2000). In this distribution, the nearest neighbor of an ON-center cell is normally an OFF-center cell (and vice versa), and the ON (or OFF) mosaic is more regular than the combined (ON and OFF) mosaic. Zhan and Troy showed that just the ON- or OFF-center cells alone could be fit using a hexagonal lattice. Furthermore, to match the statistical properties of the overall population of beta RGCs, one array needed to be spatially offset from the other in a precise manner. If this result can be generalized across many samples of mosaics, it would indicate that there are positional constraints among the different subclasses, rather than the ON- and OFF-center mosaics being spatially independent, as originally suggested (Wässle et al., 1981).

Lattice-based models seem inappropriate for describing the biological processes underlying mosaic formation because they assume that cells are initially perfectly arranged in a lattice before some noisy process disturbs them. However, they are sometimes preferred by theorists because of their analytical tractability compared with exclusion zone models (Diggle, 1983).

7.3.2 Cell Fate Models

Models of cell fate processes have investigated how an initial population of undifferentiated cells could divide into several differentiated cell classes (see also chapters 1 and 2). Early cell fate models investigated whether the relative sizes of cell populations of primary and secondary fates could be explained by nearest-neighbor interactions within an initial undifferentiated population of cells (Honda et al., 1990). Undifferentiated cells were positioned on a surface using an exclusion zone model and then chosen either systematically (e.g., from left to right across the surface) or at random to become primary fate cells, forcing all the nearest neighboring cells to become secondary fate cells. This simple cell fate process produced about a 3:1 ratio of secondary to primary fate cells. A biologically realistic process for deciding cell fate among neighboring cells, simulating lateral inhibition mediated via Delta-Notch signaling, produced a similar ratio of secondary to primary fate cells (Collier et al., 1996). In this model, however, cells were arranged in a regular hexagonal grid so that each cell always had six neighbors. When the Delta-Notch process was applied to an array of undifferentiated cells generated using an exclusion zone model, the ratio of secondary to primary fate cells dropped to about 2.6:1, regardless of the regularity of the undifferentiated cell mosaic (Eglen and Willshaw, 2002). Furthermore, the regularity of the primary fate mosaic was always higher than that of the secondary fate mosaic.

Cell fate models were used to investigate the specific problem of the development of four classes of cone photoreceptor in zebrafish retina (Takesue et al., 1998; Tohya et al., 1999). Undifferentiated cells were placed in a rectangular grid. The potential differentiated state of each cell changed probabilistically according to its affinity with the current state of its neighbors. The final stable state of each cell was dependent on the relative affinities between different classes of photoreceptors. By exploring different affinity values, Tohya et al. (1999) predicted the relative size of affinities between different cell classes that were required to match the experimental distribution of photoreceptors.

7.3.3   Cell Sorting Models

A general property of embryonic cells, not just neurons, is that they normally show preferential grouping toward other cells of the same type. When two different types of cells are randomly mixed, they tend to eventually separate so that cells of the same type adhere to each other—a phenomenon known as cell sorting. It is suggested that cells have differential affinity for different classes of cells, and that cell movement is influenced by these affinities (see also chapter 1). For example, two cell classes will segregate if cells of the same type have a high affinity and cells of an opposite type have a low affinity. Different patterns can emerge according to the relative value of within and between-class affinities. Theoretical models of cell sorting can be used to discover the relative affinities (adhesive strengths) between the different classes of cells (Mochizuki et al., 1996), or how changes in the aggregate tissue shape depend on these affinities (Sulsky et al., 1984).

To date, cell sorting models have rarely been applied to the development of retinal mosaics. This could be because the distribution of one class of retinal cells is mostly independent of other classes of retinal cells (Galli-Resta, 2000; Rockhill et al., 2000; Cameron and Carney, 2000). The only known example is recent work describing a cell sorting model of zebrafish cone photoreceptors, similar to the cell fate model mentioned previously (Mochizuki, 2002). There is currently no experimental evidence, however, suggesting if the pattern of cone photoreceptors is due to either cell fate mechanisms operating among undifferentiated cells or to differentiated cells undergoing cell sorting.

7.4 Discussion

Exclusion zone models, such as the d_{min} model, are simple phenomenological models that can replicate the statistical distribution of many different classes of retinal cells. These models suggest that mosaics develop from purely local, homotypic interactions. This agrees with recent experimental findings, prompted by the modeling studies, that interactions are homotypic, since the regularity of cholinergic amacrine cells was unaffected by either up- or down-regulating the number of synaptic partners (Galli-Resta, 2000). This experimental evidence is supported by correlational studies of adult mosaics, which found no evidence for heterotypic interactions (Cameron and Carney, 2000; Rockhill et al., 2000).

Complementary mechanistic models, such as the lateral movement model described here, have shown that universal, small movements by retinal cells are one mechanism by which the exclusion zones can be enforced. Other recent theoretical work is beginning to investigate the importance of other developmental mechanisms, such as cell fate and cell sorting, in the development of retinal mosaics.

Future Modeling Studies

Given that the d_{min} model is so successful at replicating retinal mosaics, one potential area of future work is to obtain closer agreement between the d_{min} model and the mechanistic models, perhaps by making the changes to the lateral movement model described in section 7.2.3. The fixed-dendrite model could then be used to evaluate the effect of different degrees of dendritic overlap upon the resulting dispersion distances and regularity indexes. These model results could then be compared with dispersion distances observed experimentally for cells with different degrees of dendritic overlap, such as horizontal and cholinergic amacrine cells (Reese et al., 1999). These results might also suggest whether the relative differences in dendritic overlap observed in adult retina are also present at the time of mosaic formation.

However, in addition to lateral cell movement, many other mechanisms, such as cell death and the local regulation of cell genesis (Galli-Resta, 1998), might be involved in mosaic formation. Hence the translation from the phenomenological d_{min} model to a mechanistic model may require the simulation of several different mechanisms. Another challenge is if the mechanistic models can be extended to account for the distribution of those cells that are thought to be cross-correlated with other classes (Kouyama and Marshak, 1997; Ahnelt et al., 2000).

Theoretical research on retinal mosaic formation could also benefit from other areas of mathematical biology. Mathematical models of regular spatial distributions in ecological situations, such as territory formation (Tanemura and Hasegawa, 1980; Adams, 1998), have close parallels to the models proposed for retinal development. Models of gene networks (see chapter 2) are also now being built to investigate the formation of regularly spaced patterns (Von Dassow et al., 2000). Another relevant area is to consider models in which neighboring cells compete for limited resources, in analogy with trees in a forest that compete for sunlight and nutrients from the ground (Wässle and Riemann, 1978). Such competitive processes have long been proposed to explain the reorganization of dendrites in response to retinal lesions (Perry and Linden, 1982). Ongoing theoretical work into competition for neurotrophins (see chapter 10) may provide useful insights, especially since neurotrophins can modulate the outgrowth of retinal neurites (Bosco and Linden, 1999).

Future Experimental Studies

The theoretical work mentioned here raises several interesting questions to be investigated experimentally. Some of these questions may require experimental techniques that are not yet available, but may soon become possible. They include the ability to identify different cell types at early stages of development using suitable markers. Advances in imaging techniques may also allow us to follow the radial and tangential movements made by cells over sufficiently long periods of time.

One major outstanding question is how exclusion zones are mediated biologically. Candidate mechanisms to be explored include cell death and lateral cell movement (Galli-Resta, 1998). To investigate the role of cell death, the effect of altering the magnitude of cell death, perhaps using transgenic models (Galli-Resta, 2000), upon mosaic regularity should be evaluated. To test the importance of lateral cell movement in mosaic formation, the mechanisms driving such movement will first need to be found. Obvious mechanisms to be tested include diffusible repellants and dendritic interactions.

The current models also raise other questions best tested directly by experiments. The regularity index of some retinal mosaics does not vary with density (Rodieck and Marshak, 1992; Jeffery et al., 1994; Cellerino et al., 2000); theoretical models cannot discriminate between the possibility that these mosaics were created by differential expansion of a mosaic created with one exclusion zone (the "little bang" hypothesis) or by exclusion zones that vary in size depending on local cell density (Eglen et al., 2000). Assuming that markers are available to label certain cell types early enough in development, this could be investigated by seeing if an immature mosaic is uniformly regular across the retina before differential growth begins (Mastronarde et al., 1984). These markers may also be used to see if cell fate mechanisms initiate any spatial regularity, as predicted by models of lateral inhibition (Eglen and Willshaw, 2000). Finally, the cell fate models predict relative affinity levels between different classes of cone photoreceptors (Tohya et al., 1999) that need to be measured.

Acknowledgments

This work was supported by funding from the Wellcome Trust (051282 to SJE), the National Council of Research and European Commission DGXII Biotechnology Program (to LG-R), and the National Institutes of Health (EY-11087 to BER).

Appendix: Evaluating Mosaic Regularity

Here we outline the main methods used to quantify the degree of regularity of spatial distributions. These techniques are complementary and can describe different aspects of mosaic regularity (Galli-Resta et al., 1999). Cells at the border of the sample area are specially treated (either by excluding them or by using a weighting factor) to minimize biasing the distributions. Statistical methods, including ranking and bootstrap methods (Shapiro et al., 1985), can then be used to compare distributions of experimental and model mosaics.

1   Nearest-Neighbor Analysis

The simplest estimate of mosaic regularity is based upon measuring the distance from each cell to its nearest neighboring cell. For randomly arranged cells (assuming a negligible soma size), these distances match a Rayleigh distribution (Wässle and Riemann, 1978), whereas regular mosaics tend to have Gaussian-shaped nearest-neighbor distributions. One useful

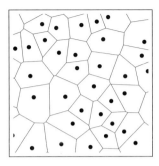 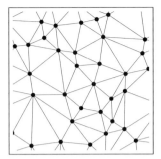

Figure 7.6
Voronoi (*left*) and Delaunay tessellations (*right*) of a central region of an example data set (cells marked with circles).

measure of this distribution, the regularity index (RI; Wässle and Riemann, 1978) or conformity ratio (Cook, 1996), is the mean divided by the standard deviation of the nearest-neighbor distances. For random distributions, the regularity index is about 1.9, whereas a conservative baseline for nonrandomness is about 3.0, depending on sample size and tissue geometry (Cook, 1996). Retinal mosaics typically have regularity indexes of about 4–9 (Wässle and Riemann, 1978), whereas for a perfectly regular pattern, the index is infinity.

This method has two main drawbacks (Cook, 1996). First, since it considers only the distance to the nearest neighboring cell, a 1-D line and 2-D array of cells can have the same regularity index, despite their different coverage of space. Second, the regularity index is sensitive to undersampling of cells.

2   *Auto- and Cross-Correlation Measures*

One improvement on nearest-neighbor analysis is to graphically observe the relative position of a cell to all other cells in an autocorrelation plot (see figure 7.1b). Taking each cell in turn, the relative position of all other cells is plotted. The size of the exclusion zone is proportional to the size of the central hole in the plot. This method can be generalized to the cross-correlation plot to check for correlations between two cell classes by taking each cell of class one in turn and plotting the relative positions of all class two cells (Rodieck, 1991). The autocorrelation plot is quantified using the density recovery profile (Rodieck and Marshak, 1992), which measures the expected number of cells within a certain distance of each other. Autocorrelation methods have the advantage of being more robust to undersampling than nearest-neighbor methods, and clearly show the size of any exclusion zone or deviation from randomness (Rodieck and Marshak, 1992).

3   *Voronoi-Based Methods*

The Voronoi tessellation divides the plane into nonoverlapping polygons; the Voronoi polygon of a cell encloses all points of the plane that are closer to the cell than to any other cell (figure 7.6). The Delaunay triangulation is an equivalent representation in which lines, or segments, are drawn between cells that share an edge of a Voronoi polygon, and the neighboring segments form triangles. From these tessellations, several useful distributions can be measured (Galli-Resta et al., 1999; Zhan and Troy, 2000), including the area of Voronoi polygons and Delaunay triangles, the length of Delaunay segments, the number of Voronoi polygon edges, and the internal angles in Delaunay triangles or Voronoi polygons. Area-based distributions measure the uniformity of coverage of visual space, whereas angle-based distributions indicate if cells are arranged in a distorted lattice (Shapiro et al., 1985). The extent to which these measures covary, or whether some are more discriminative than others, is not yet known. From experience, the distribution of Delaunay segment lengths is a reliable measure to use (Galli-Resta et al., 1999).

References

Adams, E. S. (1998). Territory size and shape in fire ants: A model based on neighborhood interactions. *Ecology* 79: 1125–1134.

Ahnelt, P. K., Fernández, E., Martinez, O., Bolea, J. A., and Kübber-Heiss, A. (2000). Irregular S-cone mosaics in felid retinas. Spatial interaction with axonless horizontal cells, revealed by cross correlation. *J. Opt. Soc. Am. A* 17: 580–588.

Ammermüller, J., Möckel, W., and Rugan, P. (1993). A geometrical description of horizontal cell networks in the turtle retina. *Brain Res.* 616: 351–356.

Bartlett, M. S. (1975). *The Statistical Analysis of Spatial Pattern.* New York: Wiley.

Bosco, A., and Linden, R. (1999). BDNF and NT-4 differentially modulate neurite outgrowth in developing retinal ganglion cells. *J. Neurosci. Res.* 57: 759–769.

Cameron, D. A., and Carney, L. H. (2000). Cell mosaic patterns in the native and regenerated inner retina of zebrafish: Implications of retinal assembly. *J. Comp. Neurol.* 416: 356–367.

Cellerino, A., Novelli, E., and Galli-Resta, L. (2000). Retinal ganglion cells with NADPH-diaphorase activity in the chick form a regular mosaic with a strong dorsoventral asymmetry that can be modeled by a minimal spacing rule. *Eur. J. Neurosci.* 12: 613–620.

Cepko, C. L., Austin, C. P., Yang, X., Alexiades, M., and Ezzeddine, D. (1996). Cell fate determination in the vertebrate retina. *Proc. Natl. Acad. Sci. U.S.A.* 93: 589–595.

Collier, J. R., Monk, N. A. M., Maini, P. K., and Lewis, J. H. (1996). Pattern formation by lateral inhibition with feedback: A mathematical model of Delta-Notch intracellular signaling. *J. Theor. Biol.* 183: 429–446.

Cook, J. E. (1996). Spatial properties of retinal mosaics: An empirical evaluation of some existing measures. *Vis. Neurosci.* 13: 15–30.

Cook, J. E., and Chalupa, L. M. (2000). Retinal mosaics: New insights into an old concept. *Trends Neurosci.* 23. 26–34.

Curcio, C. A., Allen, K. A., Sloan, K. R., Lerea, C. L., Hurley, J. B., Klock, I. B., and Milam, A. H. (1991). Distribution and morphology of human cone photoreceptors stained with anti-blue opsin. *J. Comp. Neurol.* 312: 610–624.

Diggle, P. J. (1983). *Statistical Analysis of Spatial Point Patterns.* Orlando, Fla.: Academic Press.

Diggle, P. J., and Gratton, R. J. (1984). Monte Carlo methods of inference for implicit statistical models. *J. Roy. Stat. Soc. B* 46: 193–227.

Eglen, S. J., and Willshaw, D. J. (2002). Influence of cell fate mechanisms upon retinal mosaic formation: A modelling study. *Development* 129: 5399–5408.

Eglen, S. J., Van Ooyen, A., and Willshaw, D. J. (2000). Lateral cell movement driven by dendritic interactions is sufficient to form retinal mosaics. *Network: Comput. Neural Syst.* 11: 103–118.

Finlay, B. L., and Pallas, S. L. (1989). Control of cell number in the developing mammalian visual system. *Prog. Neurobiol.* 32: 207–234.

Freeman, M. (1997). Cell determination strategies in the *Drosophila* eye. *Development* 124: 261–270.

Galli-Resta, L. (1998). Patterning the vertebrate retina: The early appearance of retinal mosaics. *Semin. Cell Dev. Biol.* 9: 279–284.

Galli-Resta, L. (2000). Local, possibly contact-mediated signaling restricted to homotypic neurons controls the regular spacing of cells within the cholinergic arrays in the developing rodent retina. *Development* 127: 1509–1516.

Galli-Resta, L., and Novelli, E. (2000). The effects of natural cell loss on the regularity of the retinal cholinergic arrays. *J. Neurosci.* 20(RC60): 1–5.

Galli-Resta, L., Resta, G., Tan, S.-S., and Reese, B. E. (1997). Mosaics of Islet-1-expressing amacrine cells assembled by short-range cellular interactions. *J. Neurosci.* 17: 7831–7838.

Galli-Resta, L., Novelli, E., Kryger, Z., Jacobs, G. H., and Reese, B. E. (1999). Modelling the mosaic organization of rod and cone photoreceptors with a minimal-spacing rule. *Eur. J. Neurosci.* 11: 1461–1469.

Honda, H., Tanemura, M., and Yoshida, A. (1990). Estimation of neuroblast numbers in insect neurogenesis using the lateral inihibition hypothesis of cell differentiation. *Development* 110: 1349–1352.

Jeffery, G., Darling, K., and Whitmore, A. (1994). Melanin and the regulation of mammalian photoreceptor topography. *Eur. J. Neurosci.* 6: 657–667.

Jeyarasasingam, G., Snider, C. J., Ratto, G. M., and Chalupa, L. M. (1998). Activity-regulated cell death contributes to the formation of on and off alpha ganglion cell mosaics. *J. Comp. Neurol.* 394: 335–343.

Kirby, M. A., and Steineke, T. C. (1996). Morphogenesis of retinal ganglion cells: A model of dendritic, mosaic, and foveal development. *Perspect. Dev. Neurobiol.* 3: 177–194.

Kouyama, N., and Marshak, D. W. (1997). The topographical relationship between two neuronal mosaics in the short wavelength-sensitive system of the primate retina. *Vis. Neurosci.* 14: 159–167.

Mastronarde, D. N., Thibeault, M. A., and Dubin, M. W. (1984). Non-uniform postnatal growth of the cat retina. *J. Comp. Neurol.* 228: 598–608.

McCabe, K. L., Gunther, E. C., and Reh, T. A. (1999). The development of the pattern of retinal ganglion cells in the chick retina: Mechanisms that control differentiation. *Development* 126: 5713–5724.

Mochizuki, A. (2002). Pattern formation in the zebrafish retina: A cell rearrangement model. *J. Theor. Biol.* 215: 345–361.

Mochizuki, A., Iwasa, Y., and Takeda, Y. (1996). A stochastic model for cell sorting and measuring cell-cell adhesion. *J. Theor. Biol.* 179: 129–146.

Nathan, R., and Muller-Landau, H. C. (2000). Spatial patterns of seed dispersal, their determinants and consequences for recruitment. *Trends Ecol. Evol.* 15: 278–285.

O'Leary, D. D. M., Fawcett, M., and Cowan, W. M. (1986). Topographic targeting errors in the retinocollicular projection and their elimination by ganglion cell death. *J. Neurosci.* 6: 3692–3705.

Perry, V. H., and Linden, R. (1982). Evidence for dendritic competition in the developing retina. *Nature* 297: 683–685.

Reese, B. E., and Tan, S.-S. (1998). Clonal boundary analysis in the developing retina using X-inactivation transgenic mosaic mice. *Semin. Cell Dev. Biol.* 9: 285–292.

Reese, B. E., Harvey, A. R., and Tan, S.-S. (1995). Radial and tangential dispersion patterns in the mouse retina are cell-class specific. *Proc. Natl. Acad. Sci. U.S.A.* 92: 2494–2498.

Reese, B. E., Necessary, B. B., Tam, P. P. L., Faulkner-Jones, B., and Tan, S.-S. (1999). Clonal expansion and cell dispersion in the developing mouse retina. *Eur. J. Neurosci.* 11: 2965–2978.

Ripley, B. D. (1981). *Spatial Statistics.* New York: Wiley.

Robinson, S. R. (1991). Development of the mammalian retina. In *Neuroanatomy of the Visual Pathways and their Development*, B. Dreher and S. R. Robinson, eds. pp. 69–128. Houndmills, UK: Macmillan Press.

Rockhill, R. L., Euler, T., and Masland, R. H. (2000). Spatial order within but not between types of retinal neurons. *Proc. Natl. Acad. Sci. U.S.A.* 97: 2303–2307.

Rodieck, R. W. (1991). The density recovery profile: A method for the analysis of points in the plane applicable to retinal studies. *Vis. Neurosci.* 6: 95–111.

Rodieck, R. W., and Marshak, D. W. (1992). Spatial density and distribution of choline acetyltransferase immunoreactive cells in human, macaque and baboon retinas. *J. Comp. Neurol.* 321: 46–64.

Scheibe, R., Schnitzer, J., Röhrenbeck, J., Wohlrab, F., and Reichenbach, A. (1995). Development of A-type (axonless) horizontal cells in the rabbit retina. *J. Comp. Neurol.* 354: 438–458.

Shapiro, M. B., Schein, S. J., and deMonasterio, F. M. (1985). Regularity and structure of the spatial pattern of blue cones of macaque retina. *J. Am. Stat. Assoc.* 80: 803–812.

Stenkamp, D. L., Barthel, L. K., and Raymond, P. A. (1997). Spatiotemporal coordination of rod and cone photoreceptor differentiation in goldfish retina. *J. Comp. Neurol.* 382: 272–284.

Sulsky, D., Childress, S., and Percus, J. K. (1984). A model of cell sorting. *J. Theor. Biol.* 106: 275–301.

Takesue, A., Mochizuki, A., and Iwasa, Y. (1998). Cell-differentiation rules that generate regular mosaic patterns: Modeling motivated by cone mosaic formation in fish retina. *J. Theor. Biol.* 194: 575–586.

Tanemura, M., and Hasegawa, M. (1980). Geometrical models of territory. I. Models for synchronous and asynchronous settlement of territories. *J. Theor. Biol.* 82: 477–496.

Tohya, S., Mochizuki, A., and Iwasa, Y. (1999). Formation of cone mosaic of zebrafish retina. *J. Theor. Biol.* 200: 231–244.

Turner, D. L., and Cepko, C. L. (1987). A common progenitor for neurons and glia persists in rat retina late in development. *Nature* 328: 131–136.

Van Ooyen, A., and Van Pelt, J. (1994). Activity-dependent outgrowth of neurons and overshoot phenomena in developing neural networks. *J. Theor. Biol.* 167: 27–43.

Von Dassow, G., Meir, E., Munro, E. M., and Odell, G. M. (2000). The segment polarity network is a robust developmental module. *Nature* 406: 188–192.

Waid, D. K., and McLoon, S. C. (1998). Ganglion cells influence the fate of dividing retinal cells in culture. *Development* 125: 1059–1066.

Wässle, H., and Riemann, H. J. (1978). The mosaic of nerve cells in the mammalian retina. *Proc. Roy. Soc. London B* 200: 441–461.

Wässle, H., Boycott, B. B., and Illing, R. B. (1981). Morphology and mosaic of on-beta and off-beta cells in the cat retina and some functional considerations. *Proc. Roy. Soc. London B* 212: 177–195.

Wickler, K. C., and Rakic, P. (1991). Relation of an array of early-differentiating cones to the photoreceptor mosaic in the primate retina. *Nature* 351: 397–400.

Zhan, X. J., and Troy, J. B. (2000). Modeling of cat retinal beta-cell arrays. *Vis. Neurosci.* 17: 23–39.

Activity-Dependent Modification of Intrinsic and Synaptic Conductances in Neurons and Rhythmic Networks

8

L. F. Abbott, Kurt A. Thoroughman, Astrid A. Prinz, Vatsala Thirumalai, and Eve Marder

How do neurons and networks develop and maintain stable physiological properties despite a continuous turnover of the ion channels and receptors that underlie neuronal signaling? In this chapter we present a series of computational models based on the premise that neurons have mechanisms to monitor their own patterns of activity and use these sensors to control the strengths of their voltage-dependent and synaptic conductances. Single neurons and small networks with these properties can self-assemble and recover from perturbations. These models predict that individual neurons of the same type, or identical networks in different animals, can produce similar activity patterns using different mixtures of conductances, all consistent with the same behavior.

8.1 Introduction

During development, neurons and neural circuits construct the response characteristics they need to function properly, and this process is regulated, at least in part, by activity (see also chapter 6). It seems likely that many of the activity-dependent processes that guide development remain active throughout adult life to maintain cells and circuits. The ion channels, synaptic receptors, vesicle trafficking proteins, and numerous other components that give neurons and neural circuits their functional properties have lifetimes much shorter than the animals in which they are found. As a result, continuous maintenance and renewal of these components is needed to maintain stable functional and behavioral performance

throughout the lifetime of an animal. Homeostatic maintenance of stability in adult circuits may require continuous implementation of many of the same processes used during development.

To function properly, neural circuits must generate appropriate patterns of electrical activity, so it seems almost unavoidable that activity plays a vital role in guiding the development and maintenance of neuronal characteristics. The role of activity in the development of functional circuits has been studied extensively in sensory systems with defined topographic maps (Miller, 1996), and in motor systems where the resulting behavior can be used as an assay for when the circuit is producing appropriate output (Fénelon et al., 1998; O'Donovan, 1999; Marder and Bucher, 2001). An interesting common feature of embryonic motor systems is that they are spontaneously active, although they generate movements that are not directly used for behavior (Bekoff et al., 1975; Bekoff, 1992; Casasnovas and Meyrand, 1995; Richards et al., 1999; Saint-Amant and Drapeau, 2000). This suggests that early rhythmic activity may provide a signal that guides the tuning of neuronal and circuit properties.

In this chapter we describe a series of models in which the properties of neurons and synapses are dynamically maintained and modified by processes that are sensitive to and regulated by patterns of neuronal activity. Much of this work focuses on the crustacean stomatogastric ganglion (STG), a small motor pattern generator that controls a number of digestive rhythms in crustaceans. The STG is an excellent preparation

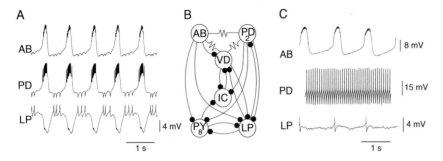

Figure 8.1

Pyloric rhythm in decapod crustaceans. (*A*) When modulatory inputs to the STG are intact, the pyloric neurons produce a triphasic motor pattern. The electrically coupled AB and PD neurons fire together, followed by the LP neurons first and then by the PY neurons (not shown). (*B*) Synaptic connections between neurons of the pyloric network. Chemical synaptic connections are represented by lines ending in black circles. In the STG, all of the chemical synapses are inhibitory. Electrical synapses are represented by resistor symbols. There are two PD neurons and 8 PY neurons, and one neuron each for the other cell types. (*C*) Activity patterns produced in pyloric neurons when each cell type is isolated from its presynaptic partners by photoinactivating them. An isolated AB neuron continues to produce bursts, whereas the PD and LP neurons fire tonically.

for these studies because it generates stereotypical patterns of activity that can act as targets for activity-dependent processes during circuit development and maintenance. For modeling purposes, the STG rhythms provide an easily identifiable circuit output that indicates when a simulation has achieved a desired goal. Furthermore, neurons within the STG can themselves have interesting rhythmic properties, and these can be used to study activity-dependent regulation of characteristics at the single neuron level. Because the STG plays a central role in the studies we discuss, we begin with an introduction to its relevant features.

8.2   The Stomatogastric Ganglion: Neurobiological Background

In vivo, the STG controls the movements of the crustacean foregut and produces rhythms that are necessary for feeding and processing food. When the STG of an adult lobster or crab is removed from the animal and placed in a saline solution, it generates a triphasic motor pattern known as the pyloric rhythm (figure 8.1) (Marder and Calabrese, 1996; Abbott and Marder, 1998).

Although the frequency of the rhythm and the phase relations among the constituent neurons are modified on short time scales by the behavioral state of an animal (Clemens et al., 1998b) and by sensory (Sigvardt and Mulloney, 1982; Hooper and Moulins, 1989; Combes et al., 1999) and modulatory inputs (Nagy and Dickinson, 1983; Nusbaum and Marder, 1989; Coleman et al., 1995; Marder and Calabrese, 1996), in vivo recordings show that the pyloric rhythm is always active (Rezer and Moulins, 1983; Clemens et al., 1998a), and under stable recording conditions the in vitro pyloric rhythms recorded from different adult animals are remarkably consistent. This argues that the STG in adult animals is always tuned to produce a stereotypical pyloric rhythm.

The pyloric rhythm (figure 8.1A) is produced by fourteen neurons that fall into six cell types (figure

8.1B). The neurons of the pyloric network are connected by electrical and inhibitory synapses, and in general the neurons fire on rebound from inhibition. Each cell type in the pyloric network displays characteristic intrinsic properties (Harris-Warrick et al., 1992) that depend on particular combinations of ionic conductances expressed in the cell membrane. For example, the anterior burster (AB) neuron bursts when isolated, while the lateral pyloric (LP) and pyloric dilator (PD) neurons do not burst, but fire at a constant rate when isolated (figure 8.1C).

As is the case with many motor systems, the STG is already active in embryos, although the animal is not yet feeding (Casasnovas and Meyrand, 1995; Le Feuvre et al., 1999; Richards et al., 1999; Richards and Marder, 2000). Recordings from embryonic and larval animals show that these early motor patterns are less frequent and less regular than in adults (Casasnovas and Meyrand, 1995; Richards et al., 1999; Richards and Marder, 2000). In contrast, once the animals have metamorphosed into juveniles, their motor patterns are statistically identical to those seen in the adult (figure 8.2A). The juvenile data shown in figure 8.2 were taken from 18-month-old animals that were approximately one-eighth the size of adult animals, and the STG itself was about one-quarter the size of the adult STG (figure 8.2B). An adult lobster of sufficient size to reach the dinner table is commonly 5 to 7 years old.

In summary, the development of the STG reveals two stages. Early in development, the circuit, while active, is irregular and does not produce a reliable set of motor patterns. Later in development, the circuit is active and produces motor patterns identical to those in the adult, although all of the neurons that participate in the circuit are still experiencing an enormous amount of structural growth. At this stage, circuit behavior remains constant despite the fact that all of the constituent neurons change size. Thus, activity-dependent regulatory processes that tune circuits to achieve stable behavior must maintain this behavior not only over extended periods of time but also during periods of significant growth.

8.3 Models of Activity-Dependent Regulation of Conductances in Single Neurons

All nervous systems contain neurons with a wide range of intrinsic properties that arise from particular combinations of ionic conductances. Some neurons are silent unless driven by synaptic input; some neurons fire action potentials spontaneously; other neurons fire action potentials in rhythmic bursts; while still others display spike-frequency adaptation or other characteristic activity signatures. Appropriate function in a neural circuit relies on the specific intrinsic properties of its constituent neurons. The models we discuss in this chapter address the question of how neurons develop and maintain the arrangement of conductances they need to perform properly.

Models of activity-dependent regulation of neuronal conductances require two critical features: sensors of activity that monitor changes in neuronal output and identify when neuronal activity is appropriate, and mechanisms that allow these sensors to control changes in intrinsic properties when activity patterns are inappropriate. A great deal of experimental work indicates that intracellular Ca^{2+} concentrations are a good indicator of neuronal activity. Specifically, intracellular Ca^{2+} concentrations become elevated in response to activity and fall during inactive periods (Ross, 1989). Finally, during development, some neurons produce slow Ca^{2+} spikes before they acquire the ability to generate fast action potentials (Baccaglini and Spitzer, 1977; O'Dowd et al., 1988; Ribera and Spitzer, 1998), and an important function of these early Ca^{2+} spikes is to provide signals

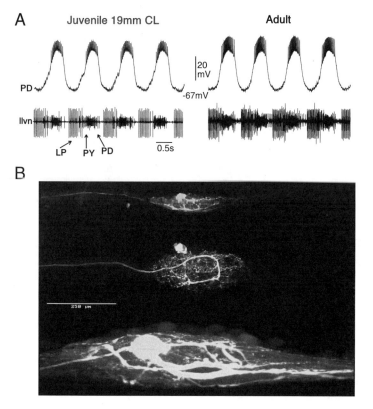

Figure 8.2

Maintenance of the pyloric rhythm during growth. (*A*) The STG of a juvenile lobster produces pyloric motor patterns that are similar to those seen in the adult. This figure shows an intracellularly recorded PD neuron and an extracellular nerve recording from the lower lateral ventricular nerve (llvn). The nerve recording shows the spiking patterns of the three pyloric motor neurons, LP, PY, and PD. (*B*) Growth of a pyloric neuron between juvenile and adult stages. The LP neuron was identified, filled with Lucifer yellow, and imaged on a confocal microscope during different stages of growth. The cell shown at the top is an LP neuron from a juvenile lobster with a carapace length of 12 mm. The cell in the middle is from a juvenile lobster with a carapace length of 20 mm, and the cell at the bottom is an LP neuron from an adult lobster (carapace length 110 mm). (From unpublished work of V. Thirumalai and E. Marder.)

for developmental tuning of excitability (Spitzer, 1994a,b; Gu and Spitzer, 1995; Spitzer and Ribera, 1998). Therefore, in these models, the intracellular Ca^{2+} concentration, or a filtered version of the inward Ca^{2+} current (representing a Ca^{2+}-dependent molecular pathway), is used as the feedback element that monitors activity. In the first-generation models (Abbott and LeMasson, 1993; LeMasson et al., 1993; Siegel et al., 1994), the intracellular Ca^{2+} concentration itself was used in this role, while in a second-generation model, three Ca^{2+}-current filtering sensors (fast, slow, and steady-state) were used (Liu et al., 1998).

The Ca^{2+} sensors in any such models are used to modify and control the maximal conductances of different membrane currents. This corresponds to Ca^{2+}-dependent processes that insert, remove, and modify ion channels within the cell membrane. The Ca^{2+} sensors are constructed using the same formalism as in standard models of voltage-dependent conductances, with Ca^{2+} playing the role that the membrane potential would normally play in those models. In other words, the sensors are described in terms of Ca^{2+}-dependent activation and inactivation variables. The integration time for each sensor is set by adjusting the time constants of the activation and inactivation processes. The sensors then drive changes in the maximal conductances of the active membrane currents in the model. Specifically, the maximal conductance $\bar{g}_i$ for current i is modified in terms of a set of calcium sensors S_a according to the equation

$$\tau \frac{d\bar{g}_i}{dt} = \sum_a B_{ia}(\bar{S}_a - S_a)\bar{g}_i,$$ (8.1)

where τ is a time constant that reflects the slow dynamics of the activity-dependent processes. The parameter B_{ia} is a coupling coefficient that determines how sensor a affects conductance i. The parameter $\bar{S}_a$ is the target value for Ca^{2+} sensor a, and these target values collectively determine the equilibrium point of the model. When constructing the model, these are set by hand so that the equilibrium point generates a desired pattern of activity.

In general, when neuronal activity is high, the activity-dependent rules are set to modify conductances so that excitability is decreased. When neuronal activity is low, they increase excitability. If these models are to represent homeostatic regulation and maintenance of neuronal properties, they must achieve a stable steady-state configuration of conductances that produces a desired pattern of activity. We first show how an activity-regulated model can self-assemble the conductances needed to generate such a steady-state configuration. To analyze stability, we then show the response to a perturbation.

Figure 8.3 shows examples of a model neuron with Ca^{2+} sensors that control the maximal values of its membrane conductances self-assembling to produce a bursting pattern of activity (Liu et al., 1998). The two examples show the model spontaneously developing sets of maximal conductances that produce bursting behavior starting from two different initial conditions. Although the final activity shown in the middle panels is similar, the maximal conductances established by the model in the two cases (figures 8.3A and 8.3B) are different, as are the trajectories followed by the model as it self-assembles (bottom panel). A study of these models indicates that there is a non-unique map between maximal conductances and activity (Goldman et al., 2001). The final set of conductances attained by the model depends on initial conditions and is highly variable even though the pattern of activity ultimately produced is not.

A similar development of intrinsic properties was seen in adult STG neurons that were placed in dissociated cell culture (Turrigiano et al., 1995). In the intact adult network, STG neurons are always rhythmically active, but when they are acutely isolated

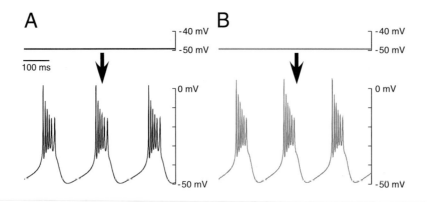

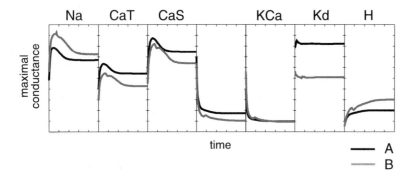

Figure 8.3
Self-assembly of an activity-regulated model neuron. The top traces in A and B show the activity of the model neuron in two silent initial states with different sets of ionic conductances. Over time, the model adjusted the maximal conductances of its ionic currents until the activity shown in the middle traces was obtained. The values of the maximal conductances over a period of 1000 sec are shown in the bottom row of plots. The vertical axes extend from 0 to 2 mS/nF for CaT, CaS, and H and from 0 to 50 mS/nF for all other currents. In both A and B, the model neuron achieves a bursting activity pattern, but the final equilibrium values of the maximal conductances are different. (Figures 8.3–8.5 were constructed using the model of Liu et al., 1998.)

from the presynaptic drive, most of them are not intrinsic bursters. In contrast, after several days in dissociated cell culture, the vast majority of isolated STG neurons fire in bursts when depolarized. This suggests that in response to removal of rhythmic drive, STG neurons rebuild the conductances they need to produce, by themselves, the rhythmic bursting activity that they normally achieve as part of a network (Turrigiano et al., 1994). As is seen in embryonic de-

velopment, the adult neurons follow a sequence of developmental changes—in this case, first becoming tonically active and then bursting. This sequence is also seen in specific temporal changes in membrane conductances as the neurons move to their final target activity levels (Turrigiano et al., 1995).

Figure 8.4 shows another example of a self-regulating model neuron. In this case, the model started out in a spiking mode, while its destined target

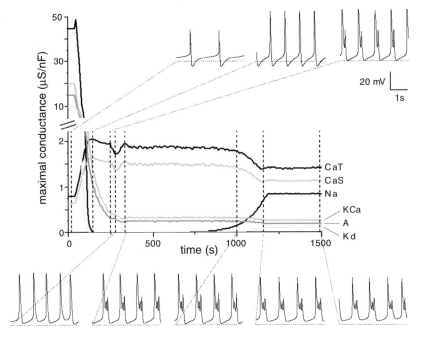

Figure 8.4
Activity and conductances of a regulating model neuron during self-tuning. The central graph shows the development of the seven activity-dependent ionic conductances of a regulating model neuron during its transition from spiking to a bursting activity pattern. The insets illustrate the activity of the model at different times during the approach to equilibrium. Similar sets of conductances can result in different activity patterns, while similar bursting patterns can be obtained from very different ionic conductances.

activity was bursting. Here we have plotted the changes in activity and conductances as the model self-tuned. Note that early in the tuning process all of the conductances were significantly altered. The model very rapidly moved into a bursting mode, but then it moved away from this activity and back into a spiking mode for a short time as it continued to tune. Close examination of this figure shows areas where large changes in conductance produce relatively little change in activity, while at other times during the tuning process, relatively modest changes in conductance are associated with more dramatic changes in the firing properties of the neuron.

Figure 8.5 shows an example of the stability of the activity in such a model when perturbations are applied. The model was initially in a bursting mode, but was then perturbed by simulating the application of high K^+ extracellular saline. Immediately after the perturbation, the model neuron changed its firing pattern from bursting to steady firing. Then, over time, the Ca^{2+}-dependent processes responded to this change in activity by modifying the parameters controlling the different membrane conductances in the model. Ultimately, this returned the activity to something quite close to the initial bursting pattern. At this point, the Ca^{2+} signal indicated that the desired

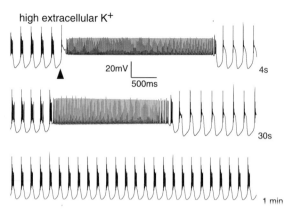

high extracellular K+

Figure 8.5

Response of a self-regulating model neuron to a change in extracellular K+ concentration. The electrical activity of a model neuron before and at different times after the reversal potential for K+ was shifted from −80 mV to −65 mV to simulate an increase in extracellular K+. Immediately after the change (indicated by the triangle) the model neuron spiked rapidly, but later it managed to return to its initial bursting activity by adjusting its conductances to the new ionic environment. Note that the activity switches between spiking and bursting several times before an equilibrium is reached.

activity pattern had been achieved, and further modification was halted. In the end, the model has intrinsic properties that are different from those when the simulation started, which allows it to produce a similar activity pattern in a different environment. Experimental support for the types of adaptations seen in figure 8.4 has been obtained in cultured neurons from the STG (Turrigiano et al., 1994).

Models of this type predict that neuronal conductances are likely to depend on the past history of activity. Consequently, individual examples of the same neuronal type could easily have quite different membrane conductances, and membrane conductances measured in a given neuron might vary over time as a consequence of activity. Figure 8.6 shows

that both of these are true for individual neurons of the STG. Figure 8.6A shows that individual neurons of the same cell type show a considerable range of measured conductance densities for three different K+ conductances (Golowasch et al., 1999b). Figure 8.6, panels B and C, shows that the strengths of these K+ conductances are altered by prolonged stimulation (Golowasch et al., 1999a). A similar result was found in experiments on cultured rat cortical neurons (Desai et al., 1999). In these experiments, the cultures were silenced for 48 hr by the application of tetrodotoxin, and, as predicted by these models, the Na^1 conductance increased and the delayed rectifier K+ conductance decreased.

The models discussed to this point have a single compartment describing both the membrane potential and Ca^{2+} concentration of the entire cell. It is possible to construct multicompartment models in which local Ca^{2+} concentrations control the membrane conductances within each compartment (Siegel et al., 1994). This introduces the idea that morphological features of neuronal structure might affect the distribution of conductances across the surface of the cell membrane. Indeed, in such models, conductances typically are not distributed uniformly across the cell. Instead, somatic and axonal regions, where Ca^{2+} concentrations are typically low, develop high concentrations of Na+ conductances. Dendritic regions, where synaptic activation results in higher Ca^{2+} concentrations, develop lower levels of Na+ and higher levels of K+ conductances (Siegel et al., 1994).

8.4 Activity-Dependent Conductance Regulation at the Network Level

Thus far, we have discussed models and experimental data that deal with the ability of single neurons to self-tune and regulate their own properties. However,

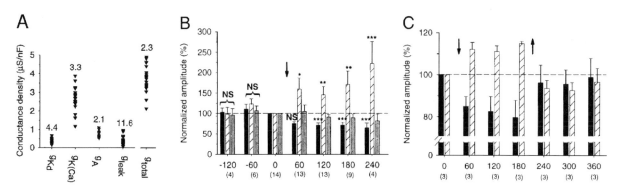

Figure 8.6

Ranges and activity-dependent shifts in conductances. (*A*) Conductance densities measured from eighteen different IC (inferior cardiac) neurons (see figure 8.1) in the crab STG for delayed rectified K^+ (K_d), Ca^{2+}-dependent K^+, A-type K^+, and leak conductances. Each plotted point corresponds to a different preparation. The numbers above the points show the ratios of the maximum to minimum values. (*B*) Patterned stimulation changes K^+ conductances. Conductances were measured in IC neurons every 60 min, before (to the left of the downward arrow) and during (to the right of the downward error) patterned stimulation applied continuously to the neurons. The amplitudes shown are for the sum of delayed rectified and Ca^{2+}-dependent K^+ conductances (black bars), A-type K^+ conductances (striped bars), and leakage conductances (gray bars). The stars indicate statistically significant changes in conductance that are due to the patterned stimulation. NS indicates a difference that is not statistically significant. The numbers in parentheses below the *x*-axis indicate the numbers of neurons measured in each case. (*C*) Same as (*B*), except for three neurons measured before (to the left of the downward arrow), during (between the two arrows), and after (to the right of the upward arrow) patterned stimulation. Patterned stimulation decreased the sum of delayed rectified and Ca^{2+}-dependent K^+ conductances (black bars) and increased A-type K^+ conductances (striped bars), and this effect reversed after the stimulation stopped. (Adapted from Golowasch et al., 1999a,b.)

what really matters for an animal is not how individual neurons fire, but how the circuits in which these neurons are embedded function. How might stability of circuit output be ensured? It might seem that circuit stability would require regulation by a mechanism that senses the output of the entire circuit, but it is difficult to imagine how this would be achieved. Instead, we have found that circuit stability can be achieved by cell-autonomous regulation that only involves activity sensing at the single neuron level (Golowasch et al., 1999b).

In this work, a three-cell model of the triphasic pyloric rhythm of the STG was constructed. Each model neuron had a simple activity-dependent con-

ductance modification mechanism, and the synaptic connectivity was fixed to match that of the STG. Even when the initial intrinsic properties of the individual neurons are set randomly, the network develops a triphasic activity pattern similar to the pyloric motor pattern. Moreover, as in the case of the biological STG, when the individual neurons are acutely isolated from their presynaptic inputs, their intrinsic activity patterns are different from those they display within the network, indicating that activity-regulated neurons change their properties in response to synaptic inputs (Golowasch et al., 1999b).

The canonical pyloric rhythm depends on the presence of modulatory inputs to the STG, so that when

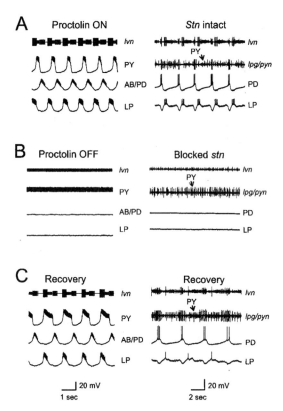

Figure 8.7
Effect of blocking neuromodulators on model and biological pyloric rhythms. In the experiments, recordings were made before, immediately after, and 24 hr after blocking modulatory inputs, indicated in the figure by "Stn intact" and "Blocked stn" (the stn is a nerve carrying modulatory input to the STG). (*A*) The triphasic pyloric rhythm in model and biological networks when modulator inputs are intact. (*Left*) "Extracellular" recordings (lvn nerve) from the model and "intracellular" recordings of model PY, AB/PD, and LP neurons in steady-state control conditions. (*Right*) Extracellular recordings from two motor nerves (lvn and lpg/pyn) and intracellular recordings of the PD and LP neurons. Activity of the PY neuron appears as the small unit on the lvn and lpg/pyn recordings (arrow). (*B, left*) Immediately after the proctolin conductance was set to zero in both the AB/PD and the LP neurons, the PY neuron remained depolarized and tonically fired action potentials at high

the STG is isolated from descending modulatory inputs, the pyloric rhythm usually slows considerably or stops. Nonetheless, after several days of silence, rhythmicity returns and individual neurons become more excitable (Thoby-Brisson and Simmers, 1998; Golowasch et al., 1999b; Thoby-Brisson and Simmers, 2000; Mizrahi et al., 2001). Figure 8.7 shows that a cell-autonomous model can replicate these findings. On the left side of the figure, a three-cell model is shown. The presence of the modulatory inputs is modeled by adding a "proctolin current" (Golowasch and Marder, 1992; Swensen and Marder, 2000) to the AB/PD and LP neurons. In figure 8.7A, the triphasic pattern depends on the presence of this modulatory current, as is seen when the proctolin current is removed in figure 8.7B. However, over time the individual neurons in the circuit sense the lack of rhythmic activity and regulate their conductances accordingly, resulting in a restoration of the triphasic pattern, as seen in figure 8.7C. It is important to stress that the triphasic motor patterns seen in figure 8.7A and C, although quite similar in appearance, result from different configurations because the circuit elements have retuned themselves so that the circuit output is maintained. The right panels in figure 8.7 show the same sequence of events occurring in a cultured STG circuit when it was isolated from its modulatory inputs for a prolonged period.

frequency, while the AB/PD and LP neurons were silent. (*Right*) Soon after modulatory input was blocked, the PY units (recorded extracellularly on the lpg/pyn and lvn) showed high-frequency firing, and the LP and PD neurons were silent. (*C, left*) The pyloric activity in the model has recovered after prolonged deactivation of the proctolin current owing to changes in intrinsic conductances. (*Right*) The activity of the biological network also recovers after prolonged block of modulatory inputs. (Adapted from Golowasch et al., 1999b.)

8.5 Activity-Dependent Tuning of Synaptic Inhibition

Pharmacological perturbations, dynamic clamp manipulations, and computational modeling have demonstrated that properties of the triphasic pyloric rhythm depend critically on proper tuning of both the intrinsic and synaptic conductances within the STG (Eisen and Marder, 1984; Johnson and Harris-Warrick, 1990; Harris-Warrick et al., 1995; Abbott and Marder, 1998; Kloppenburg et al., 1999; Swensen and Marder, 2001). Thus, a discussion of the role of activity in the development and maintenance of this circuit is not complete without an analysis of the role of activity-dependent modification of synapses. As stated earlier, all the chemical synapses within the STG are inhibitory. Across neurobiological preparations, many experimental and theoretical studies have elucidated aspects of activity-dependent modification of excitatory synapses. Although inhibitory synapses are present in virtually all neural systems, much less attention has been paid to how activity affects their long-term strength (Marty and Llano, 1995). Furthermore, theoretical work has not identified sensible and useful rules for inhibitory synaptic tuning similar to what the Hebb rule and theories of LTP and LTD have done for excitatory synaptic modification. We have modeled small networks with inhibitory synapses to determine whether activity-dependent modification of inhibitory synapses allows them to self-assemble to produce particular forms of rhythmic activity (Soto-Treviño et al., 2001). In this work, the intrinsic conductances of the neurons were held fixed in order to focus on the effects of regulating synaptic strength.

We first modeled two-neuron circuits with mutual inhibition. Two LP neurons, modeled by Morris-Lecar equations and connected by instantaneous, voltage-dependent inhibitory synapses, oscillate at frequencies that depend nonlinearly on the values of both synaptic strengths. The intracellular Ca^{2+} concentration in the postsynaptic neuron, however, smoothly and monotonically decreases when inhibitory synaptic strength increases. This suggests that intracellular Ca^{2+} can serve as an effective activity sensor, and it led to a modification rule that changes the strength of synapses in proportion to the difference between the current postsynaptic Ca^{2+} concentration and a fixed target value (Soto-Treviño et al., 2001). Such a rule did indeed drive all initial synaptic strengths to a single fixed point (figure 8.8A), at which the two neurons oscillated in the target 1:1 pattern of entrainment (figure 8.8B).

We then modeled a three-cell network (LP, PY, and one cell combining the electrically coupled AB and PD neurons) that, when properly connected with inhibitory synapses, mimicked the triphasic pyloric rhythm of the STG. This model network has five inhibitory synapses; therefore, two cells each received two inhibitory synapses. Because each cell possesses only one Ca^{2+} concentration, such a neuron must increase or decrease the strengths of all the synapses it receives proportionally, using the Ca^{2+} concentration rule. It cannot differentially regulate their strengths. Despite this limitation, the Ca^{2+} concentration rule successfully drives the majority of networks with randomly selected initial synaptic strengths to configurations that produce the target triphasic rhythm. However, if the network encounters synaptic strengths outside this majority of values, it requires a supplemental mechanism during either development or adult homeostasis because differential regulation of synaptic strengths is required. We hypothesized that such a synapse-specific mechanism (as opposed to the global mechanism that modifies all synapses onto a given neuron on the basis of its intracellular Ca^{2+} concentration) could alter a synapse only when it is

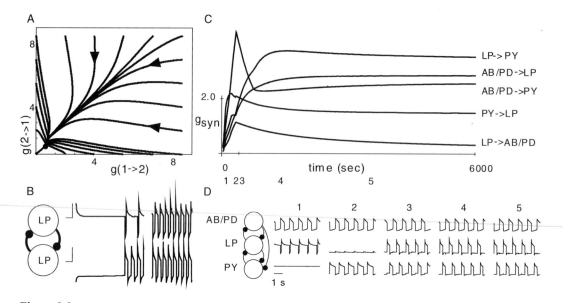

Figure 8.8

(*A*) Trajectories of synaptic strengths in LP-LP circuits initiated from many initial values, as modified by a Ca^{2+}-dependent rule. The dot at the lower left indicates a global fixed point of the dynamics. (*B*) Example voltage traces of two coupled LP cells as synaptic strengths are modified. (*C*) Time series of synaptic strengths in a three-cell model of the pyloric network starting from a single initial condition, as modified by a combination of global and synapse-specific rules. (*D*) Voltage traces for all three model neurons taken at the five different times indicated by the numbers underneath the synaptic strength time series in (*C*). (Adapted from Soto-Treviño et al., 2001.)

active. Thus, the synapse-specific synaptic modification rule depends on the presynaptic membrane potential, as well as on postsynaptic Ca^{2+} (Soto-Treviño et al., 2001).

For the synapse-specific rule, we based this latter dependence on the Ca^{2+} current, rather than on the Ca^{2+} concentration, to increase temporal specificity. Because local Ca^{2+} concentrations near the cell membrane are proportional to the Ca^{2+} current, this dependence can be construed as reflecting a dependence on a local, rather than on the bulk, Ca^{2+} concentration. The combination of global, Ca^{2+} concentration-dependent and synapse-specific synaptic modification rules successfully drove networks with any set of initial synaptic strengths to a fixed-

point configuration (figure 8.8C) at which the three neurons produced the target triphasic rhythm (figure 8.8D).

This modeling work suggests that reasonably simple, activity-dependent modification rules are sufficient to drive inhibitory synapses in small networks to values appropriate for generating specific patterns of activity. These results also suggest that Ca^{2+} concentrations and currents can drive sensors that monitor activity to produce this tuning. We do not expect that our proposed rule is uniquely appropriate for all cases of inhibitory synaptic tuning, nor do we know whether biological networks use these mechanisms. Nevertheless, these simulations show that individual synapses in networks coupled through inhibition can autonomously control their strengths on the basis of pre- and postsynaptic activity to assemble and maintain a network that exhibits a particular emergent behavior.

8.6 Discussion

The modeling work described here raises several novel issues. All neural circuits function as a consequence of both the intrinsic properties of their neurons and the strengths of the synapses among those neurons, but there is not a one-to-one map between activity patterns and configurations of intrinsic and synaptic conductances. We therefore predict that all neurons of the same type need not have, nor would be expected to have, identical sets of conductances, and that the same neuron might display similar patterns of activity over its lifetime produced by different conductances. Likewise, at the network level, because both synaptic and intrinsic conductances can be coordinately tuned, the same network in different animals, or in the same animal at different times, is likely to produce similar behavior by different combinations of synaptic strengths and intrinsic membrane conductances. The essential feature of this paradigm is that tuning rules must be continuously operational over development and in adult life to maintain stable network dynamics and behavior.

Future Modeling Studies

A future goal of this research is to implement plasticity of both intrinsic and synaptic conductances in the same model. Synaptic plasticity is thought to be a major mechanism of learning and memory, but changing synapses without adjusting intrinsic neuronal excitability can lead to severe instabilities in network activity, and it may not allow optimal plasticity in

network function. It is important to understand how intrinsic and synaptic plasticity might interact constructively to maximize learning and adaptation to novel situations.

Future Experimental Studies

The experimental challenge of this work is to understand more fully the nature of the rules governing these types of activity-dependent plasticity and to discover the mechanisms by which such plasticity is generated. Theoretical work should serve as an important guide both by suggesting lines of inquiry and in the interpretation of experimental results.

Acknowledgments

This research was supported by the National Institute of Mental Health (MH46742), the Sloan Center for Theoretical Neurobiology at Brandeis University, and the W. M. Keck Foundation.

References

Abbott, L. F., and LeMasson, G. (1993). Analysis of neuron models with dynamically regulated conductances in model neurons. *Neur. Comput.* 5: 823–842.

Abbott, L. F., and Marder, E. (1998). Modeling small networks. In *Methods in Neuronal Modeling: From Ions to Networks*, 2nd ed. C. Koch and I. Segev, eds. pp. 361–410. Cambridge, Mass.: MIT Press.

Baccaglini, P. I., and Spitzer, N. C. (1977). Developmental changes in the inward current of the action potential of Rohon-Beard neurones. *J. Physiol. (London)* 271: 93–117.

Bekoff, A. (1992). Neuroethological approaches to the study of motor development in chicks: Achievements and challenges. *J. Neurobiol.* 23: 1486–1505.

Bekoff, A., Stein, P. S., and Hamburger, V. (1975). Coordinated motor output in the hindlimb of the 7-day chick embryo. *Proc. Natl. Acad. Sci. U.S.A.* 72: 1245–1248.

Casasnovas, B., and Meyrand, P. (1995). Functional differentiation of adult neural circuits from a single embryonic network. *J. Neurosci.* 15: 5703–5718.

Clemens, S., Combes, D., Meyrand, P., and Simmers, J. (1998a). Long-term expression of two interacting motor pattern-generating networks in the stomatogastric system of freely behaving lobster. *J. Neurophysiol.* 79: 1396–1408.

Clemens, S., Massabuau, J. C., Legeay, A., Meyrand, P., and Simmers, J. (1998b). In vivo modulation of interacting central pattern generators in lobster stomatogastric ganglion: Influence of feeding and partial pressure of oxygen. *J. Neurosci.* 18: 2788–2799.

Coleman, M. J., Meyrand, P., and Nusbaum, M. P. (1995). A switch between two modes of synaptic transmission mediated by presynaptic inhibition. *Nature* 378: 502–505.

Combes, D., Meyrand, P., and Simmers, J. (1999). Dynamic restructuring of a rhythmic motor program by a single mechanorecptor neuron in lobster. *J. Neurosci.* 19: 3620–3628.

Desai, N. S., Rutherford, L. C., and Turrigiano, G. G. (1999). Plasticity in the intrinsic excitability of cortical pyramidal neurons. *Nat. Neurosci.* 2: 515–520.

Eisen, J. S., and Marder, E. (1984). A mechanism for production of phase shifts in a pattern generator. *J. Neurophysiol.* 51: 1375–1393.

Fénelon, V. S., Casasnovas, B., Simmers, J., and Meyrand, P. (1998). Development of rhythmic pattern generators. *Curr. Opin. Neurobiol.* 8: 705–709.

Goldman, M. S., Golowasch, J., Marder, E., and Abbott, L. F. (2001). Global structure, robustness, and modulation of neuronal models. *J. Neurosci.* 21: 5229–5238.

Golowasch, J., and Marder, E. (1992). Proctolin activates an inward current whose voltage dependence is modified by extracellular Ca^{2+}. *J. Neurosci.* 12: 810–817.

Golowasch, J., Abbott, L. F., and Marder, E. (1999a). Activity-dependent regulation of potassium currents in an identified neuron of the stomatogastric ganglion of the crab *Cancer borealis*. *J. Neurosci.* 19: RC33.

Golowasch, J., Casey, M., Abbott, L. F., and Marder, E. (1999b). Network stability from activity-dependent regulation of neuronal conductances. *Neur. Comput.* 11: 1079–1096.

Gu, X., and Spitzer, N. C. (1995). Distinct aspects of neuronal differentiation encoded by frequency of spontaneous Ca^{2+} transients. *Nature* 375: 784–787.

Harris-Warrick, R. M., Marder, E., Selverston, A. I., and Moulins, M. (1992). *Dynamic Biological Networks. The Stomatogastric Nervous System*. Cambridge, Mass.: MIT Press.

Harris-Warrick, R. M., Coniglio, L. M., Levini, R. M., Gueron, S., and Guckenheimer, J. (1995). Dopamine modulation of two subthreshold currents produces phase shifts in activity of an identified motoneuron. *J. Neurophysiol.* 74: 1404–1420.

Hooper, S. L., and Moulins, M. (1989). Switching of a neuron from one network to another by sensory-induced changes in membrane properties. *Science* 244: 1587–1589.

Johnson, B. R., and Harris-Warrick, R. M. (1990). Aminergic modulation of graded synaptic transmission in the lobster stomatogastric ganglion. *J. Neurosci.* 10: 2066–2076.

Kloppenburg, P., Levini, R. M., and Harris-Warrick, R. M. (1999). Dopamine modulates two potassium currents and inhibits the intrinsic firing properties of an identified motor neuron in a central pattern generator network. *J. Neurophysiol.* 81: 29–38.

Le Feuvre, Y., Fénelon, V. S., and Meyrand, P. (1999). Unmasking of multiple adult neural networks from a single embryonic circuit by removal of neuromodulatory inputs. *Nature* 402: 660–664.

LeMasson, G., Marder, E., and Abbott, L. F. (1993). Activity-dependent regulation of conductances in model neurons. *Science* 259: 1915–1917.

Liu, Z., Golowasch, J., Marder, E., and Abbott, L. F. (1998). A model neuron with activity-dependent conductances regulated by multiple calcium sensors. *J. Neurosci.* 18: 2309–2320.

Marder, E., and Bucher, D. (2001). Central pattern generators and the control of rhythmic movements. *Curr. Biol.* 11: R986–R996.

Marder, E., and Calabrese, R. L. (1996). Principles of rhythmic motor pattern generation. *Physiol. Rev.* 76: 687–717.

Marty, A., and Llano, I. (1995). Modulation of inhibitory synapses in the mammalian brain. *Curr. Opin. Neurobiol.* 5: 335–341.

Miller, K. D. (1996). Receptive fields and maps in the visual cortex: Models of ocular dominance and orientation columns. In *Models of Neural Networks*, E. Domany, V. L. van Hemmen, and K. Schulten, eds. pp. 55–78. New York: Springer-Verlag.

Mizrahi, A., Dickinson, P. S., Kloppenburg, P., Fénelon, V., Baro, D. J., Harris-Warrick, R. M., Meyrand, P., and Simmers, J. (2001). Long-term maintenance of channel distribution in a central pattern generator neuron by neuromodulatory inputs revealed by decentralization in organ culture. *J. Neurosci.* 21: 7331–7339.

Nagy, F., and Dickinson, P. S. (1983). Control of a central pattern generator by an identified modulatory interneurone in crustacea. I. Modulation of the pyloric motor output. *J. Exp. Biol.* 105: 33–58.

Nusbaum, M. P., and Marder, E. (1989). A modulatory proctolin-containing neuron (MPN). I. Identification and characterization. *J. Neurosci.* 9: 1591–1599.

O'Donovan, M. J. (1999). The origin of spontaneous activity in developing networks of the vertebrate nervous system. *Curr. Opin. Neurobiol.* 9: 94–104.

O'Dowd, D. K., Ribera, A. B., and Spitzer, N. C. (1988). Development of voltage-dependent calcium, sodium, and potassium currents in *Xenopus* spinal neurons. *J. Neurosci.* 8: 792–805.

Rezer, E., and Moulins, M. (1983). Expression of the crustacean pyloric pattern generator in the intact animal. *J. Comp. Physiol.* A 153: 17–28.

Ribera, A. B., and Spitzer, N. C. (1998). Development of electrical excitability: Mechanisms and roles. *J. Neurobiol.* 37: 1–2.

Richards, K. S., and Marder, E. (2000). The actions of crustacean cardioactive peptide on adult and developing stomatogastric ganglion motor patterns. *J. Neurobiol.* 44: 31–44.

Richards, K. S., Miller, W. L., and Marder, E. (1999). Maturation of the rhythmic activity produced by the stomatogastric ganglion of the lobster, *Homarus americanus. J. Neurophysiol.* 82: 2006–2009.

Ross, W. N. (1989). Changes in intracellular calcium during neuron activity. *Annu. Rev. Physiol.* 51: 491–506.

Saint-Amant, L., and Drapeau, P. (2000). Motoneuron activity patterns related to the earliest behavior of the zebrafish embryo. *J. Neurosci.* 20: 3964–3972.

Siegel, M., Marder, E., and Abbott, L. F. (1994). Activity-dependent current distributions in model neurons. *Proc. Natl. Acad. Sci. U.S.A.* 91: 11308–11312.

Sigvardt, K. A., and Mulloney, B. (1982). Sensory alteration of motor patterns in the stomatogastric nervous system of the spiny lobster *Panulirus interruptus. J. Exp. Biol.* 97: 137–152.

Soto-Treviño, C., Thoroughman, K. A., Marder, E., and Abbott, L. F. (2001). Activity-dependent modification of inhibitory synapses in models of rhythmic neural networks. *Nat. Neurosci.* 4: 297–303.

Spitzer, N. C. (1994a). Development of voltage-dependent and ligand-gated channels in excitable membranes. *Prog. Brain Res.* 102: 169–179.

Spitzer, N. C. (1994b). Spontaneous Ca^{2+} spikes and waves in embryonic neurons: Signaling systems for differentiation. *Trends Neurosci.* 17: 115–118.

Spitzer, N. C., and Ribera, A. B. (1998). Development of electrical excitability in embryonic neurons: Mechanisms and roles. *J. Neurobiol.* 37: 190–197.

Swensen, A. M., and Marder, E. (2000). Multiple peptides converge to activate the same voltage-dependent current in a central pattern-generating circuit. *J. Neurosci.* 20: 6752–6759.

Swensen, A. M., and Marder, E. (2001). Modulators with convergent cellular actions elicit distinct circuit outputs. *J. Neurosci.* 21: 4050–4058.

Thoby-Brisson, M., and Simmers, J. (1998). Neuromodulatory inputs maintain expression of a lobster motor pattern-generating network in a modulation-dependent state: Evidence from long-term decentralization in vitro. *J. Neurosci.* 18: 212–2225.

Thoby-Brisson, M., and Simmers, J. (2000). Transition to endogenous bursting after long-term decentralization requires de novo transcription in a critical time window. *J. Neurophysiol.* 84: 596–599.

Turrigiano, G., Abbott, L. F., and Marder, E. (1994). Activity-dependent changes in the intrinsic properties of cultured neurons. *Science* 264: 974–977.

Turrigiano, G. G., LeMasson, G., and Marder, E. (1995). Selective regulation of current densities underlies spontaneous changes in the activity of cultured neurons. *J. Neurosci.* 15: 3640–3652.

Models of Neuronal Death in Vertebrate Development: From Trophic Interactions to Network Roles

9

Peter G. H. Clarke

Large numbers of neurons die in development at the time when their connections are being formed, and the survival of individual neurons during this period depends on their integrated response to anterograde signals received from their afferents as well as retrograde signals received from their efferent targets. In central neurons, both the anterograde and the retrograde signals are multiple. Both involve neurotrophic factors and both include a component that is dependent on electrical activity. The roles of the neuronal death are unclear, but the most widely invoked hypotheses involve the regulation of neuronal number or the refinement of connectivity.

While the signals controlling neuronal death appear to be moderately well understood, we have only a minimal conception of how they combine to affect the developing networks of the nervous system. To achieve such a network-level understanding, it will be necessary to incorporate the principles of neuron-to-neuron signaling into network-level models. The few models so far available are reviewed here and placed in their biological context.

9.1 Neurobiological Background

9.1.1 The When and Where of Neuronal Death in the Developing Nervous System

The fact that large numbers of neurons die during development is now a commonplace, and large numbers of papers, coasting on the wave of the boom in research on apoptosis, are currently devoted to unraveling the underlying molecular mechanisms. Yet the basic question of what role the neuronal death serves in development has received less attention.

Two Phases of Cell Death Affecting Developing Neurons

Cell death during development is a widespread phenomenon, affecting most kinds of cells. In the nervous system, it affects glia as well as neurons, but this review is limited to neurons.

Even with this restriction, there are multiple phenomena. During proliferation and migration, cells with the potential to become neurons are eliminated (Lewis, 1975; Carr and Simpson, Jr., 1981). This early phase of neuronal death is followed by a period without cell death, after which there is a second phase of neuronal death, occurring at the time when the neurons are making and receiving connections.

The early phase of neuronal death was long thought to be a relatively minor phenomenon, but its numerical importance was recently suggested by the fact that a highly sensitive adaptation (ISEL+) of a standard method for in situ end labeling (ISEL) of fragmented DNA (considered to be a marker of cell death) showed labeling of up to 70 percent of cells in many cortical and subcortical regions, including even higher percentages among proliferative cells (Blaschke et al., 1996, 1998). The interpretation of this result is not straightforward, because DNA fragmentation may not be an infallible criterion of cell death (Gilmore et al., 2000), but the importance of the early phase of neuronal death is supported by analyses of the effects of preventing it (Kuida et al., 1996, 1998; Frade et al.,

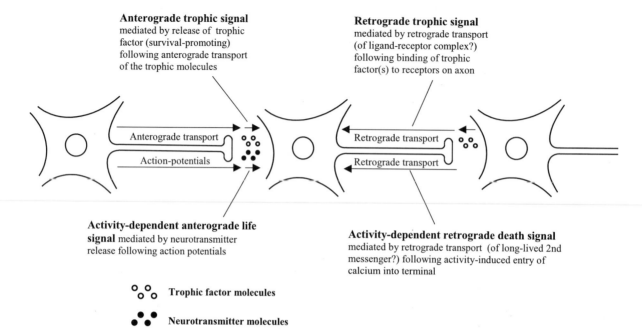

Anterograde trophic signal mediated by release of trophic factor (survival-promoting) following anterograde transport of the trophic molecules

Retrograde trophic signal mediated by retrograde transport (of ligand-receptor complex?) following binding of trophic factor(s) to receptors on axon

Activity-dependent anterograde life signal mediated by neurotransmitter release following action potentials

Activity-dependent retrograde death signal mediated by retrograde transport (of long-lived 2nd messenger?) following activity-induced entry of calcium into terminal

Trophic factor molecules

Neurotransmitter molecules

Figure 9.1
The different anterograde and retrograde signals so far shown to affect neuronal survival. The diagram shows three different signals promoting the survival of the center neuron and one promoting its death. While all four signals have been shown to exist, it is not clear that all operate in all neurons. The existence of additional signals cannot be excluded.

1997). However, very little is known concerning its role, and this chapter concentrates on the second phase of neuronal death.

This second phase has been studied in many neural systems and occurs roughly at the time when the neurons in question are making and receiving connections. It involves the death of large numbers of postmitotic, postmigratory neurons that are already well differentiated, being endowed with dendrites and an axon. The extent of loss varies from region to region, but in most neuronal populations accounts for 25–75 percent of the initial number (Clarke, 1985). The timing of this naturally occurring death suggests a relationship between it and the establishment of connectivity.

9.1.2   The Multiple Signals Controlling Neuronal Death

Since the neuronal death occurs when afferent (input) and efferent (output) connections are being formed, the question arises whether these are instrumental in regulating the phenomenon, and the answer is that they are. Both anterograde and retrograde signals play a role. In each direction, electrical activity and neurotrophic factors both contribute to the signaling, as discussed later (figure 9.1).

Retrograde Survival Signals and Death Signals
The textbook view is that during a critical period in development neurons need to receive one or more

"trophic" (literally, "nourishing") substances, taken up by their axon terminals and transported retrogradely to the cell body, and that natural neuronal death is due to failure in the competition for such substances. This view is based on considerable experimental evidence in the peripheral nervous system (Fariñas et al., 1994), including the fact that nerve growth factor (NGF) is available in the skin in limited amounts that depend on competition (Davies et al., 1987; Korsching and Thoenen, 1985), as well as the fact that the death of sensory and sympathetic neurons is reduced in transgenic mice that overproduce NGF specifically in skin.

The textbook view is supported also in the central nervous system by the facts that target-specific injections of brain-derived neurotrophic factor (BDNF) reduce neuronal death in the centrifugally projecting isthmo-optic nucleus (ION) of chick embryos (Von Bartheld et al., 1994; Primi and Clarke, 1996) and in retinal ganglion cells of neonatal rats (Ma et al., 1998). These responses to exogenous BDNF do not on their own prove that endogenous BDNF is exerting a survival role in normal development. However, BDNF is known to be present normally in the injected target areas, and in the case of the isthmo-optic projection, the survival role of an endogenous ligand to TrkB (the high-affinity receptor to BDNF and neurotrophin-4, NT-4) in the isthmo-optic target area (i.e., the retina) has been shown by the fact that intraocular injection of anti-TrkB Fab fragments reduces the survival of ION neurons (Von Bartheld et al., 1996b).

The analysis of knockout mice lacking particular neurotrophic factors has given substantial support to the textbook view in the peripheral nervous system, but not in the central nervous system. Thus, deletion of the genes for NGF, BDNF, NT-3, and NT-4/5 (all members of the neurotrophin family of neurotrophic factors), or of their high-affinity receptors, leads to the death of particular classes of sensory neurons, but the central effects are much milder, and indeed minimal during the period of naturally occurring neuronal death (Conover and Yancopoulos, 1997). To explain this unexpected absence of neuronal death, it is sometimes suggested that neurons may switch their neurotrophic factor dependence in the knockout mice, or that individual central neurons may depend on several different neurotrophic factors, necessitating the knockout of several genes in order to cause massive neuronal death centrally. These alternative explanations have yet to be thoroughly tested, but the double knockout of BDNF and NT-4 does not significantly increase neuronal death (Conover et al., 1995).

In view of these uncertainties, we should be open to other possible mechanisms, and there is in fact evidence for a second retrograde signal that is much less conventional—a retrograde death signal—in the chick embryo's isthmo-optic pathway, which terminates in the retina. This death signal is mediated (or perhaps modulated) by electrical activity in the isthmo-optic terminals (Primi and Clarke, 1997a,b) owing to calcium entry through N-type channels and probably to the subsequent, calcium-mediated activation of nitric oxide synthase in the terminals (Posada and Clarke, 1999a,b). This signal acts very rapidly, promoting the death of ION neurons within as little as 3–6 hr after the arrival of an action potential in their terminals.

Afferent Survival Signals

Neuronal survival is also influenced by afferents. Despite an early suggestion by Levi-Montalcini (1949), afferents were for many years believed to influence neuronal death only much later than the neuronal death period. It is now, however, known that when a major proportion of the afferents are prevented from arriving, this greatly enhances neuronal death during its natural period in many neuronal populations. These include spinal motor neurons, the isthmo-optic nucleus, and the ciliary ganglion in chick embryos; and the parabigeminal nucleus in baby rats (reviewed in Linden, 1994; Sherrard and Bower, 1998). The anterograde signals involved appear to be multiple, involving separate components dependent on and independent of activity. Thus, a blockade in activity in optic afferents to the developing tectum causes neurons there to die within a few hours (Catsicas et al., 1992; Galli-Resta et al., 1993). The activity-dependent interneuronal signal is probably the transmitter itself rather than a coreleased neurotrophic agent because in parasympathetic ganglia, cholinergic receptor blockade is sufficient to increase neuronal death (Meriney et al., 1987).

However, an activity-independent survival signal has also been demonstrated. In the chick embryo's tectum, neuronal death is greatly enhanced from about 12 hr after the blockade of axonal transport in the optic afferents (Catsicas et al., 1992). The nature of the activity-independent signal is not known, but neurotrophins are candidates because the intraocular injection of radiolabeled NT-3 or BDNF is followed by their anterograde transport along optic axons, release, and uptake by tectal neurons (Von Bartheld et al., 1996a).

The fact that neurons need afferents in order to survive does not prove this to be critical in the regulation of naturally occurring neuronal death. Although massive elimination of afferents does increase neuronal death, removing a small proportion of them often has little effect. The normal variability in afferent supply to individual neurons might therefore be too small to affect their survival, or the neurons might be saturated with afferents. The latter possibility would be ruled out if reducing competition between afferents could be shown to reduce neuronal death. A report

that increased optic innervation of rodent visual centers reduces neuron death (Cunningham et al., 1979) was not confirmed by another group (Raabe et al., 1986). However, there is evidence that the retinal ganglion cells may compete with each other for inputs from retinal interneurons onto the ganglion cell dendrites (Linden and Serfaty, 1985).

9.2   Review of Models

Models are needed to go from the above data on afferents and efferents to an understanding of the roles of neuronal death in development. Many suggestions have been made concerning these roles, but the two most widely considered are the control of neuronal number and the elimination of axonal targeting errors. A third possibility, one that has appealed to modelers, is the optimization of learning. Throughout the rest of this chapter, we group the different models of neuronal death according to the postulated role of the death: number control, error elimination, or optimization of learning. Yet another possibility, the refinement of neuronal mosaics, as in the retina (Cook and Chalupa, 2000), is not discussed here, but is mentioned in chapter 7.

It is perhaps worth pointing out that in the present stage of knowledge it is impossible to produce detailed, fully realistic models of so complex a phenomenon as neuronal death. All existing models of neuronal death are open to criticism, and I will attempt to provide it, but that does not imply a negative value judgment about the work, merely a reminder that more needs to be done. Moreover, the aim of the modeling is not generally the formulation of a quantitative replica, but the construction of aids to conceptualization, and this can sometimes be achieved even with a biologically unrealistic model.

9.2.1 Models of Neuronal Number Control by Neuronal Death

The Numerical Matching Hypothesis and the Notion of Competition for Retrograde Support

The numerical matching hypothesis is the most widely held interpretation of neuronal death. It states that the role or purpose of neuronal death is to match the number of neurons in a given population to the size of their axonal target territory or, in recent versions, to the number of their afferents as well. The underlying assumption is that many, even most, of the neurons that die are not aberrant, but merely numerically superfluous.

This hypothesis dates back to the classic experiments of Hamburger and Levi-Montalcini on spinal ganglia and motor neurons. They reported that before the period of neuron death, the spinal ganglia were of a similar size at all levels (Hamburger and Levi-Montalcini, 1949) and that the motor column (which contains the motor neurons) was likewise homogeneous in size (Levi-Montalcini, 1950). They also claimed that the subsequently larger numbers of spinal ganglion cells and motor neurons at brachial and lumbar levels were due to a retrograde influence from the periphery on neuronal proliferation and neuronal death. It soon became clear that there is no such effect of the periphery on the proliferation of neurons (although neurotrophic factors from a more local source may indeed affect proliferation; Geffen and Goldstein, 1996), but the notion that axial-level related differences in neuronal number are due to sculpting through regulation by the periphery of neuronal death became widely accepted and still exerts a major influence.

Axial Level-Dependent Differences Are Not Sculpted by Neuronal Death

However, several of the early claims of Hamburger and Levi-Montalcini turned out to be false. Whereas they had thought neuronal death occurred almost exclusively at upper cervical and thoracic levels, it is now known to occur in substantial amounts at all levels, in both the spinal ganglia and the motor columns. Moreover, even before the period of neuronal death, axial-level related differences occur both in sensory ganglia, where they seem to be imposed by local mesoderm (Goldstein et al., 1995), and in the motor column (Hollyday and Hamburger, 1977). Thus, sculpting of regional differences in neuronal number seems to be at most a minor purpose of neuronal death in peripheral projections.

In the central nervous system, there is no evidence for the sculpting by neuronal death of regional differences, but, as in the peripheral system, neuronal survival is reduced by reductions in the number of afferent or target neurons and enhanced by their increase (e.g., Herrup and Sunter, 1987).

Numerical Matching in an Evolutionary Context

An influential version of the numerical matching hypothesis casts it in the context of evolution (Katz and Lasek, 1978). It is argued that since the nervous system is organized in sets of "matching populations" of neurons, a mutation that changed the production of neurons in a given population would disturb the match, which might be nonadaptive. However, an initial overproduction of neurons would permit the match to be restored by reduced death in the partners, giving the species buffer capacity and enabling it to tolerate mutations and hence evolve more readily. I do not find this argument compelling. It fails to explain the proportionally massive overproduction (typically, 50 percent) that normally occurs. I know of no evidence for viable mutations causing such large changes. Moreover, a mathematical analysis of motor neuron counts made by the same first author concluded that the initial excess in their numbers is far greater than that required to buffer the variability in their numbers (Katz and Grenander, 1982).

Two Early Models Predicting the Time Course of Neuronal Death

The earliest mathematical approach to neuronal death was the study by Rager (1978), who integrated into a single model a consideration of the proliferation, axonal growth, and terminal competition of retinal ganglion cells, and then formulated differential equations predicting the time course of several variables, including neuronal number. All the differential equations resembled the Verhulst equation and were integrated directly. By a suitable choice of parameters, it was possible to make the resulting curves fit the author's own extensive counts of retinal ganglion cell axons with remarkable precision.

Another early model derived the time course of neuronal loss from a consideration of the dynamics of competition among neurons (Borsellino, 1980). The author drew on the mathematics of competition for light among plants and managed to fit reasonably well the time course of neuronal death in the isthmo-optic nucleus. However, this model incorporated none of the available data on neuronal competition and arbitrarily assumed that the competing neurons occurred in pairs, artificially ensuring a death rate of 50 percent. Another implausible characteristic was the assumption that the stronger member of each pair could deprive the weaker one of resources, but that the weaker could not affect the stronger. This is plausible for trees competing for light, but not for neurons competing for neurotrophic factors.

A Model Incorporating Mutual Neurotrophic Interactions between Input and Output

A substantial modeling contribution to neuronal death theory focused on the mutuality of neurotrophic interactions (Galli-Resta and Resta, 1992).

Until the late 1970s, neurotrophic theory emphasized almost exclusively competition among neurons for a unidirectional trophic influence from their efferent target, which was considered to imply a positive, almost linear, correlation between the final neuron number and the size of the target. This implication was contradicted by Lamb (1980), who showed that when both right and left hind-limb motor neurons of the *Xenopus* frog were made to project to a single limb at an early stage of development, the number of motor neurons surviving death was close to normal. Lamb concluded that this refuted the notion of peripheral competition, but an alternative reaction was to complicate the competition hypothesis without rejecting it, by assuming that the motor axons regulate the target property that they seek (Purves, 1980). A further consideration is that cell death may occur in the target territory as well as in the innervating population, and that the afferents may influence, not only the properties of the target neurons, but also their very survival. Both these considerations are taken into account in the model of Galli-Resta and Resta (figure 9.2).

This model deals with the interaction between two groups of neurons containing, at time t, $N_1(t)$ and $N_2(t)$ neurons, each of which is subject to both neurogenesis and cell death, the timing and rates of both phenomena being determined partly by prespecified Gaussian functions of time, and partly by trophic influences. It is assumed that group 1 innervates group 2 and that the trophic influence exerted by group 2 on group 1 is given by

$$F = \alpha_{12}N_1(t)N_2(t), \tag{9.1}$$

where α_{12} is a constant. The trophic term F is assumed to linearly decrease the rate of cell death in population 1.

The fact that F depends on N_2 is intuitively reasonable because the amount of trophic factor pro-

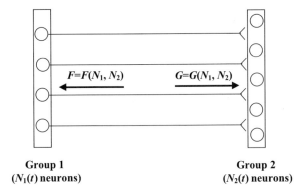

Group 1
($N_1(t)$ neurons)

Group 2
($N_2(t)$ neurons)

Figure 9.2
The model of Galli-Resta and Resta (1992). Group 1 innervates group 2 and provides an anterograde trophic influence G, whereas group 2 exerts a retrograde trophic influence F on group 1.

duced in group 2 will depend on the number of cells in it. The dependence of F on N_1 means that the cells in group 1 upregulate the trophic factor production in group 2, which is biologically reasonable given that electrical activity mediated by afferents has been shown to upregulate the production and release of neurotrophins in a variety of forebrain structures (e.g., Castren et al., 1992; Blochl and Thoenen, 1996). In the model, the term F reduces the rate of cell death. The model also incorporates an equation for the proliferation of the neurons in each group. These considerations are combined to yield a rather complicated differential equation expressing dN_1/dt as a function of N_1, N_2, and t.

The further consideration, that the afferents exert a trophic influence on target cell survival, is then taken to justify a symmetrical equation in which N_1 and N_2 are interchanged. Thus, dN_2/dt is expressed as a function of N_1, N_2, and t.

This set of two equations is simulated, and it is shown that a wide variety of results can be accounted for (with a suitable choice of constants). They include

• The changes in retinal ganglion cell numbers during normal development of the rat

• The changes in numbers of retinal ganglion cells and their target neurons in the lateral geniculate nucleus during normal development of the rhesus monkey

• And, most important in the present context, the final numbers of motor neurons in Lamb's (1980) experiment involving innervation of a single limb by the motor neurons on both sides of the spinal cord; the persistence of normal numbers on both sides can be accounted for

An important prediction of this model is that it is only the initial numbers of cells generated in groups 1 and 2 that determine how many in each group will survive to adulthood. If more are generated, more will survive. The model does not predict numerical matching in the simple sense of adjusting the number of input neurons to be essentially proportional to the number of target neurons, but implies a more complex kind of numerical adjustment between input and target.

I consider this to be the most useful model of neuronal death in development so far, but I nevertheless have some criticisms. First, I find the expression for trophic influence, $F = \alpha_{12} N_1(t) N_2(t)$, inappropriate. The authors do in fact attempt to justify it on the grounds that it is the lowest nonzero term of the Taylor series expansion for $F(N_1, N_2)$, but I find implausible its implication that the *relative* trophic influence (i.e., F/N_1 or F/N_2) will depend on the *absolute* number of neurons N_2 or N_1, respectively. For example, if $N_1 = N_2 = 100$, the trophic influence per neuron will be 10,000 times less than if $N_1 = N_2 = 1,000,000$. This counterintuitive situation could be avoided if the expression for trophic influence were changed to $F = \alpha_{12}[N_1(t) N_2(t)]^{0.5}$, although it is not known if the model works with this change. A second criticism is that the model ignores, or at least fails to make explicit, the role of electrical activity, which is known to exert an important influence on neuronal survival (see section 9.1.2). A third is that the model deals only with the global influence of the whole target on the whole input population and vice versa. It ignores the differences among the cells of a given population and does not explicitly deal with competition among the cells. None of these objections abrogates the importance of the model, which lies in its providing an intellectual handle to the conceptually elusive question of mutual trophic interactions and in its proposing a resolution to the seemingly paradoxical results from Lamb's dual innervation experiments.

9.2.2 Models of the Elimination of Axonal Targeting Errors by Neuronal Death

Evidence for and against Elimination of Targeting Errors by Neuronal Death

The main alternative hypothesis to numerical matching is that neuronal death might be involved in error elimination. That initial neural connections are refined in vertebrates by regressive events such as the loss of axonal branches is undeniable, but the role of neuronal death in the refinement is less clear. However, there is evidence from the retinopetal (brain-to-retina) and retinofugal (retina-to-brain) visual systems that here at least, neuronal death does play a significant role in eliminating neurons with connections that are aberrant by adult standards.

In the isthmo-optic projection of chick embryos, a very small proportion (about 0.2 percent) of the axons initially project to the ipsilateral retina, whereas the adult projection is entirely to the contralateral retina; the elimination of the transient ipsilateral projection involves the death of the parent cell bodies during the neuronal death period (Clarke and Cowan, 1976; O'Leary and Cowan, 1982). A much greater number of transient "aberrant" axons project to the "correct"

(contralateral) retina, but to the "wrong" part of it by the standard of adult topography; again their elimination involves the death of their parent cell bodies (Catsicas et al., 1987). The exact percentage of topographically aberrant axons is uncertain, but may be 25–50 percent. When isthmo-optic neuron death is reduced by intraocular injections of tetrodotoxin (a blocker of action potentials), this permits the survival of aberrantly projecting isthmo-optic neurons of both types ("wrong" laterality and "wrong" topography) (Péquignot and Clarke, 1992).

Similarly, in the retinocollicular projection of young rodents, neuronal death has been reported to eliminate preferentially retinal ganglion cells projecting to inappropriate regions of the superior colliculus, and here too the sculpting of the final precise projection by cell death is essentially blocked by intraocular tetrodotoxin (O'Leary et al., 1986). However, the importance of neuronal death for refining topography in the mammalian retinofugal projection is currently controversial (Chalupa and Dreher, 1991; Thompson and Cordery, 1997).

Model of Topographical Sculpting by Cell Death
A computational model (figure 9.3) was therefore elaborated for neuronal death-mediated changes in the topographical organization of a two-layered neural network (Posada and Clarke, 1999c). There is evidence that, at least in the retina, the spontaneous electrical activity occurring during the cell death period occurs in waves that sweep across the tangential extent of the retina (Wong, 1999), so this was modeled as an "activity wave" sweeping regularly through layer 1. There is further evidence that synaptic modification according to a Hebbian mechanism might be responsible for the topographical refinement, either through a secondary effect on neuronal death (O'Leary et al., 1986) or independently of neuronal death (Stollberg, 1995), but there were also some

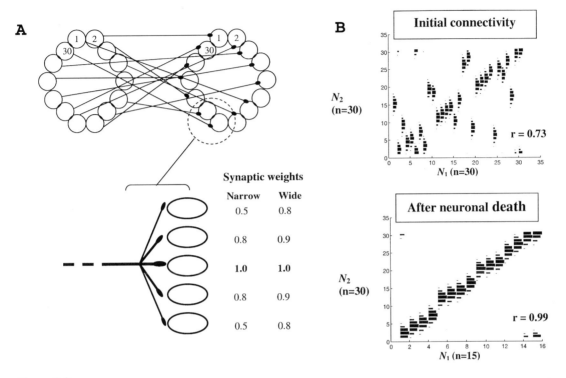

Figure 9.3
The model of Posada and Clarke. (*A*) Connectivity between the two layers. Each layer network is circular. An axon from layer 1 reaches a neuron in layer 2 and ramifies. The synaptic values are defined by a Gaussian function; narrow axonal ramification is given by a Gaussian variance (*V*) equal to 3; wide axonal ramification is given by *V* = 10. Each neuron from layer 2 receives only one axon with a synaptic weight equal to 1. The numbers 1, 2, ..., 30 represent topographical neuron positions. (*B*) Representative examples of topographical refinement involving removal of approximately half the neurons in layer 1. (*Top*) Initial connectivity between the two layers, 50 percent correct and 50 percent random. (*Bottom*) Remaining network after selective neuronal death. Rectangles represent synaptic weight values. After neuronal death, the network has been reconstructed (i.e., closed up to fill the gaps left by dead neurons) and the rectangles have been adapted to the new scale. The correlation coefficient *r* gives an indication of the precision of the topography. (Based on figures 1A and 2 of Posada and Clarke, 1999c.)

grounds for thinking that the activity-mediated retrograde death signal in the isthmo-optic pathway (see section 9.1.2) might likewise result from a Hebbian mechanism (Posada and Clarke, 1999a). For these reasons, in different versions of the model, transmission of activity to layer 2 was assumed to cause neuronal death in layer 1 or synaptic modification in layer

2 or a combination of the two. In the simulations involving neuronal death, transmission of activity to layer 2 was assumed to generate, according to a Hebbian rule, a retrograde death signal compensated for by a trophic survival signal from the target cells. The interaction between the two retrograde signals determined the death or survival of layer 1 neurons.

The main results were as follows: In the simulations involving neuronal death, with a suitable choice of parameters, about 50 percent of the neurons in layer 1 died, and this produced a substantial improvement in the topography. In simulations involving Hebbian synaptic modification without neuronal death, an equivalent reorganization occurred, but with less precision and efficiency. With the two mechanisms combined, synaptic modification provided no further improvement over that achieved by neuronal death alone.

This model thus supports the hypothesis that neuronal death could play a role in the refinement of topographic projections during neural development, but it is not a realistic model of what happens in detail. It ignores the trophic effects on the layer 2 neurons of the afferents from layer 1. It makes no provision for neuronal death in layer 2. It did not include inhibitory neurons, or lateral or feedback connections. And most of the parameters were chosen without biological justification.

9.2.3 Models Postulating Optimization of Learning by Neuronal Death

The two models just described dealt with relatively simple hypotheses of the role of neuronal death; both of them proposed that it was optimizing some aspect of the nervous system: neuronal number or connectivity. It remains possible that neuronal death may play a much more subtle role, one that involves the functioning of the nervous system in complex tasks. While most neuronal death occurs too early for this, the hippocampal dentate gyrus is an exception, since neuronal death there continues into adult life and has been shown to be reduced in juvenile rats exposed to an enriched environment (Young et al., 1999).

Such subtle roles will be all the more difficult to unravel, but computational models can at least provide clues as to the possibilities. Several such models have been produced, and all of these relate to some kind of optimization of learning (where "learning" can be interpreted broadly to mean almost any kind of adaptive plasticity, e.g., in development; see also chapter 14). I first mention briefly some studies that were concerned with the optimization of artificial neural networks without claiming to model the nervous system.

Optimization of Artificial Networks by Elimination of Units

Several authors have tested the effects of eliminating units in artificial networks. A major purpose of this has been to reduce the size of the network so as to optimize generalization (appropriate classification even of inputs not used to train the network). If the network is too large, it has too much freedom and may fit the intricacies in input space rather than extracting the underlying trends, so that when a novel input is presented, i.e., one that was not in the training set, it cannot make a correct classification.

On the other hand, having a large number of units tends to enhance the speed of learning. In an attempt to combine speed of learning with constrained generalization, Mozer and Smolensky (1989) trained a three-layer feedforward network until all output activities were within a specified range of the target value. They then measured a "relevance" parameter for each unit, a criterion of how important the unit was for the overall performance, and removed the least relevant units one by one, with retraining between the removals. This procedure improved performance (a combination of speed and generalization) on a variety of tasks. Other authors have adopted similar strategies, although the improvement in generalization was not always found (Sietsma and Dow, 1988, 1991). Since the aim of these artificial networks was not to simulate neuronal death, some authors used networks that could both increase and decrease the number of hidden units, and in this case the improvements were greater (Hirose et al., 1991). In this respect, it is interesting that neurogenesis in the adult may be involved in the formation of memories (Shors et al., 2001).

Optimization of Learning by Neuronal Death in a Neural Network Model

The neuron removal algorithms used in the papers described above are unlikely to be adopted by the brain because they require nonlocal computations (e.g., the decision of which neurons to remove is based on knowledge of the overall network performance or the behavior of all the neurons). Therefore, Brown et al. (1994) drew on the above ideas, as applied to a three-layer backpropagation network (figure 9.4), but developed an alternative criterion for neuron removal

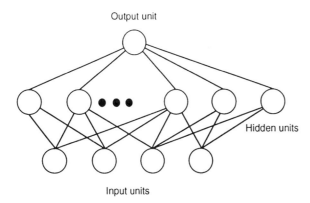

Figure 9.4
The backpropagation model of Brown et al. (1994). The activation of hidden units and the output unit is determined by the activation of all the units they are connected to in the previous layer, and by the strengths of these connections. The strengths are adjusted by a standard gradient-descent learning algorithm (backpropagation). (Reproduced from Brown et al., 1994, with permission.)

that relied solely on local computation. They assumed that each "neuron" has a constant death tendency K that is counteracted by a supply $P1$ of protective factor that is directly proportional to the average rate of change in the neuron's output activity. In order to ensure that the process would be self-limiting (all the neurons would not die), the authors further assumed a fixed supply $P2$ of a second protective factor that was equally divided among the neurons. Their simulations of a simple pattern-recognition task indicated that an initial oversupply of neurons in the hidden unit layer permitted the network to more efficiently learn to solve the task, and that although the performance was temporarily disrupted by the death of a neuron, the subsequent relearning was rapid (although in simulations performed after publication, the results turned out to be somewhat sensitive to weight initialization parameters; G. D. A. Brown, personal communication).

In my view, the main value of this paper is that it reminds us that neuronal death may play a much more subtle role than the regulation of neuronal number or long-range connectivity. Adjusting the parameter space of a learning network is one possibility among several. However, the model proposed is (inevitably) less than fully realistic biologically, for several reasons, including the following: It assumes without justification that retrograde signals can mediate error-sensitive gradient descent. It ignores the activity-independent component of afferent trophic signals. While the assumption that a neuron's activity will in some way affect its receipt of trophic factor is biologically supported, there is no evidence that the receipt will be proportional to the average rate of change in activity. The network modeled contains "neurons" that do not behave like biological neurons; for example, they are not either excitatory or inhibitory. And most of the parameters were chosen without biological justification.

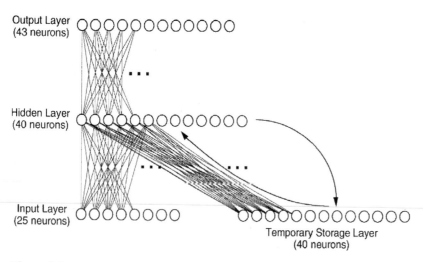

Figure 9.5

The speech perception neural network studied by Hoffman and McGlashan (1997, 1999) in relation to schizophrenia. The hidden layer receives "phonetic information," while the output layer codes for "words." Projections are unidirectional and flow upward in the figure except for those from the hidden layer to the temporary storage layer. The latter retains a copy of the hidden layer from the previous iteration. (Reproduced from Hoffman and McGlashan 1997, 1999, with permission.)

Model of the Developmental Basis of Schizophrenia

A still more complex neural network model (figure 9.5) employed 148 units organized in four layers ("input," "hidden," "output," and "temporary storage") to explore the possibility that abnormalities in synaptic elimination and neuronal death might contribute to the etiology of anomalous speech perception in schizophrenia (Hoffman and McGlashan, 1997, 1999). An interesting feature of the model is its capacity to acquire and use linguistic expectations, stored as activation patterns resonating between the hidden and temporary storage layers, to guide detection of words. The main finding was that low levels of synaptic elimination or excitotoxic neuron death (i.e., death that is due to intense electrical activation) improved perceptual ability (word detection), whereas excessive synaptic elimination (but not neuron death) led to hallucinated speech (detection of

nonexistent words). It is of course difficult to make a biologically realistic model of so complex a process as speech perception. I therefore refrain from detailed criticism, but would point out that while excitotoxicity is sometimes postulated as a cause of neuronal death in schizophrenia, the regulatory role of electrical activity in neuronal death during development is certainly more complex, as discussed in section 9.1.2.

9.3   Discussion

We need models primarily to bridge the gap between the wealth of data on trophic interactions and a network-level understanding of the role of the neuronal death. Models can do this in two ways. Simulations can demonstrate the diversity of network consequences that could arise from a given set of cell-to-cell mechanisms. And analytical mathematics can show definitively the inevitability of certain consequences over a wide range of functions, parameters, and initial conditions. However, it has to be admitted that so far the modeling of neuronal death is immature. It lacks the sophistication that has been achieved in the modeling of axonal competition and withdrawal (see chapters 10–12).

This may arise from three factors. First, the causation of neuronal death appears to be secondary to that of axonal competition, and more complex, since the retrograde signals that regulate it depend on all the events that affect axonal competition; however, neuronal death also depends on a further set of influences, the afferent signals. Second, the available biological data are not adequate, because a network-level analysis of neuronal death requires the integration of data concerning perikaryal survival with data on the axonal terminal fields, which is difficult to obtain. Third, the modeling of neuronal death has simply attracted less attention.

As the supply of biological data continues to increase, the usefulness of modeling neuronal death should increase substantially over the next few years.

Acknowledgments

I am grateful to Dr. Lucia Galli-Resta and the editor of this volume for their helpful comments on the manuscript. The experimental research of my own group was supported by the Swiss National Foundation for Scientific Research, grant 31-50598.97, and by the Swiss Federal Office for Education and Science (EU Bio4 CT 96-0649).

References

Blaschke, A. J., Staley, K., and Chun, J. (1996). Widespread programmed cell death in proliferative and postmitotic regions of the fetal cerebral cortex. *Development* 122: 1165–1174.

Blaschke, A. J., Weiner, J. A., and Chun, J. (1998). Programmed cell death is a universal feature of embryonic and postnatal neuroproliferative regions throughout the central nervous system. *J. Comp. Neurol.* 396: 39–50.

Blochl, A., and Thoenen, H. (1996). Localization of cellular storage compartments and sites of constitutive and activity-dependent release of nerve growth factor (NGF) in primary cultures of hippocampal neurons. *Mol. Cell. Neurosci.* 7: 173–190.

Borsellino, A. (1980). Neuronal death in embryonic development: A model for selective cell competition and dominance. In *Multidisciplinary Approach to Brain Development*, C. Di Benedetta, ed. pp. 495–550. Amsterdam: Elsevier/North-Holland.

Brown, G. D. A., Hulme, C., Hyland, P. D., and Mitchell, I. J. (1994). Cell suicide in the developing nervous system: A functional neural network model. *Cognit. Brain Res.* 2: 71–75.

Carr, V. M., and Simpson, S. B., Jr. (1981). Rapid appearance of labeled degenerating cells in the dorsal root ganglia after exposure of chick embryos to tritiated thymidine. *Brain Res.* 254: 157–162.

Castren, E., Zafra, F., Thoenen, H., and Lindholm, D. (1992). Light regulates expression of brain-derived neurotrophic factor mRNA in rat visual cortex. *Proc. Natl. Acad. Sci. U.S.A.* 89: 9444–9448.

Catsicas, M., Péquignot, Y., and Clarke, P. G. H. (1992). Rapid onset of neuronal death induced by blockade of either axoplasmic transport or action potentials in afferent fibers during brain development. *J. Neurosci.* 12: 4642–4650.

Catsicas, S., Thanos, S., and Clarke, P. G. H. (1987). Major role for neuronal death during brain development: Refinement of topographical connections. *Proc. Natl. Acad. Sci. U.S.A.* 84: 8165–8168.

Chalupa, L. M., and Dreher, B. (1991). High precision systems require high precision "blueprints": A new view regarding the formation of connections in the mammalian visual system. *J. Cognit. Neurosci.* 3: 209–219.

Clarke, P. G. H. (1985). Neuronal death during development in the isthmo-optic nucleus of the chick: Sustaining role of afferents from the tectum. *J. Comp. Neurol.* 234: 365–379.

Clarke, P. G. H., and Cowan, W. M. (1976). The development of the isthmo-optic tract in the chick, with special reference to the occurrence and correction of developmental errors in the location and connections of isthmo-optic neurons. *J. Comp. Neurol.* 167: 143–164.

Conover, J. C., and Yancopoulos, G. D. (1997). Neurotrophin regulation of the developing nervous system: Analyses of knockout mice. *Rev. Neurosci.* 8: 13–27.

Conover, J. C., Erickson, J. T., Katz, D. M., Bianchi, L. M., Poueymirou, W. T., McClain, J., Pan, L., Helgren, M., Ip, N. Y., Boland, P., et al. (1995). Neuronal deficits, not involving motor neurons, in mice lacking BDNF and/or NT4. *Nature* 375: 235–238.

Cook, J. E., and Chalupa, L. M. (2000). Retinal mosaics: New insights into an old concept. *Trends Neurosci.* 23: 26–34.

Cunningham, T. J., Huddelston, C., and Murray, M. (1979). Modification of neuron numbers in the visual system of the rat. *J. Comp. Neurol.* 184: 423–433.

Davies, A. M., Bandtlow, C., Heumann, R., Korsching, S., Rohrer, H., and Thoenen, H. (1987). Timing and site of nerve growth factor synthesis in developing skin in relation to innervation and expression of the receptor. *Nature* 326: 353–358.

Fariñas, I., Jones, K. R., Backus, C., Wang, X.-Y., and Reichardt, L. F. (1994). Severe sensory and sympathetic deficits in mice lacking neurotrophin-3. *Nature* 369: 658–661.

Frade, J. M., Bovolenta, P., Martínez-Morales, J. R., Arribas, A., and Rodríguez-Tébar, A. (1997). Control of early cell death by BDNF in the chick retina. *Development* 124: 3313–3320.

Galli-Resta, L., and Resta, G. (1992). A quantitative model for the regulation of naturally occurring cell death in the developing vertebrate nervous system. *J. Neurosci.* 12: 4586–4594.

Galli-Resta, L., Ensini, M., Fusco, E., Gravina, A., and Margheritti, B. (1993). Afferent spontaneous electrical

activity promotes the survival of target cells in the developing retinotectal system of the rat. *J. Neurosci.* 13: 243–250.

Geffen, R., and Goldstein, R. S. (1996). Rescue of sensory ganglia that are programmed to degenerate in normal development: Evidence that NGF modulates proliferation of DRG cells in vivo. *Dev. Biol.* 178: 51–62.

Gilmore, E. C., Nowakowski, R. S., Caviness, V. S., and Herrup, K. (2000). Cell birth, cell death, cell diversity and DNA breaks: How do they all fit together? *Trends Neurosci.* 23: 100–105.

Goldstein, R. S., Avivi, C., and Geffen, R. (1995). Initial axial level dependent differences in size of avian dorsal root ganglia are imposed by the sclerotome. *Dev. Biol.* 168: 214–222.

Hamburger, V., and Levi-Montalcini, R. (1949). Proliferation, differentiation and degeneration in the spinal ganglia of the chick embryo under normal and experimental conditions. *J. Exp. Zool.* 111: 457–501.

Herrup, K., and Sunter, K. (1987). Numerical matching during cerebellar development: Quantitative analysis of granule cell death in staggerer mouse chimeras. *J. Neurosci.* 7: 829–836.

Hirose, Y., Yamashita, K., and Hijiya, S. (1991). Back-propagation algorithm which varies the number of hidden units. *Neur. Net.* 4: 61–66.

Hoffman, R. E., and McGlashan, T. H. (1997). Synaptic elimination, neurodevelopment, and the mechanism of hallucinated "voices" in schizophrenia. *Am. J. Psychiatry* 154: 1683–1689.

Hoffman, R. E., and McGlashan, T. H. (1999). Using a speech perception neural network simulation to explore normal neurodevelopment and hallucinated "voices" in schizophrenia. *Prog. Brain Res.* 121: 311–325.

Hollyday, M., and Hamburger, V. (1977). An autoradiographic study of the formation of the lateral motor column in the chick embryo. *Brain Res.* 132: 197–208.

Katz, M. J., and Grenander, U. (1982). Developmental matching and the numerical matching hypothesis for neuronal cell death. *J. Theor. Biol.* 98: 501–517.

Katz, M. J., and Lasek, R. J. (1978). Evolution of the nervous system: Role of ontogenetic mechanisms in the evolution of matching populations. *Proc. Natl. Acad. Sci. U.S.A.* 75: 1349–1352.

Korsching, S., and Thoenen, H. (1985). Nerve growth factor supply for sensory neurons: Site of origin and competition with the sympathetic nervous system. *Neurosci. Lett.* 54: 201–205.

Kuida, K., Zheng, T. S., Na, S., Kuan, C., Yang, D., Karasuyama, H., Rakic, P., and Flavell, R. A. (1996). Decreased apoptosis in the brain and premature lethality in CPP32-deficient mice. *Nature* 384: 368–372.

Kuida, K., Haydar, T. F., Kuan, C. Y., Gu, Y., Taya, C., Karasuyama, H., Su, M. S., Rakic, P., and Flavell, R. A. (1998). Reduced apoptosis and cytochrome c-mediated caspase activation in mice lacking caspase 9. *Cell* 94: 325–337.

Lamb, A. H. (1980). Motoneurone counts in *Xenopus* frogs reared with one bilaterally-innervated hindlimb. *Nature* 284: 347–350.

Levi-Montalcini, R. (1949). The development of the acoustico-vestibular centers in the chick embryo in the absence of the afferent root fibers and of descending tracts. *J. Comp. Neurol.* 91: 209–242.

Levi-Montalcini, R. (1950). The origin and development of the visceral system in the spinal cord of the chick embryo. *J. Morphol.* 86: 253–283.

Lewis, P. D. (1975). Cell death in the germinal layers of the postnatal rat brain. *Neuropathol. Appl. Neurobiol.* 1: 21–29.

Linden, R. (1994). The survival of developing neurons: A review of afferent control. *Neuroscience* 58: 671–682.

Linden, R., and Serfaty, C. A. (1985). Evidence for differential effects of terminal and dendritic competition upon developmental neuronal death in the retina. *Neuroscience* 15: 853–868.

Ma, Y. T., Hsieh, T., Forbes, M. E., Johnson, J. E., and Frost, D. O. (1998). BDNF injected into the superior colliculus reduces developmental retinal ganglion cell death. *J. Neurosci.* 18: 2097–2107.

Meriney, S. D., Pilar, G., Ogawa, M., and Nuñez, R. (1987). Differential neuronal survival in the avian ciliary ganglion after chronic acetylcholine receptor blockade. *J. Neurosci.* 7: 3840–3849.

Mozer, M. C., and Smolensky, P. (1989). Skeletonization: A technique for trimming the fat from a network via relevance assessment. *Connect. Sci.* 1: 3–26.

O'Leary, D. D. M., and Cowan, W. M. (1982). Further studies on the development of the isthmo-optic nucleus with special reference to the occurrence and fate of ectopic and ipsilaterally projecting neurons. *J. Comp. Neurol.* 212: 399–416.

O'Leary, D. D. M., Fawcett, J. W., and Cowan, W. M. (1986). Topographic targeting errors in the retinocollicular projection and their elimination by selective ganglion cell death. *J. Neurosci.* 6: 3692–3705.

Péquignot, Y., and Clarke, P. G. H. (1992). Maintenance of targeting errors by isthmo-optic axons following the intra-ocular injection of tetrodotoxin in chick embryos. *J. Comp. Neurol.* 321: 351–356.

Posada, A., and Clarke, P. G. H. (1999a). Fast retrograde effects on neuronal death and dendritic organization in development: The role of calcium influx. *Neuroscience* 89: 399–408.

Posada, A., and Clarke, P. G. H. (1999b). Role of nitric oxide in a fast retrograde signal during development. *Brain Res.* 114: 37–42.

Posada, A., and Clarke, P. G. H. (1999c). The role of neuronal death during the development of topographically ordered projections: A computational approach. *Biol. Cybern.* 81: 239–247.

Primi, M.-P., and Clarke, P. G. H. (1996). Retrograde neurotrophin-mediated control of neurone survival in the developing central nervous system. *NeuroReport* 7: 473–476.

Primi, M.-P., and Clarke, P. G. H. (1997a). Early retrograde effects of blocking axoplasmic transport in the axons of developing neurons. *Dev. Brain Res.* 99: 259–262.

Primi, M.-P., and Clarke, P. G. H. (1997b). Presynaptic initiation by action potentials of retrograde signals in developing neurons. *J. Neurosci.* 17: 4253–4261.

Purves, D. (1980). Neuronal competition. *Nature* 287: 585–586.

Raabe, J. I., Windrem, M. S., and Finlay, B. L. (1986). Control of cell number in the developing visual system. III. Effects of visual cortex ablation. *Brain Res.* 393: 23–31.

Rager, G. (1978). Systems-matching by degeneration. II. Interpretation of the generation and degeneration of retinal ganglion cells in the chicken by a mathematical model. *Exp. Brain Res.* 33: 79–90.

Sherrard, R. M., and Bower, A. J. (1998). Role of afferents in the development and cell survival of the vertebrate nervous system. *Clin. Exp. Pharmacol. Physiol.* 25: 487–495.

Shors, T. J., Miesegaes, G., Beylin, A., Zhao, M., Rydel, T., and Gould, E. (2001). Neurogenesis in the adult is involved in the formation of trace memories. *Nature* 410: 372–375.

Sietsma, J., and Dow, R. J. F. (1988). Neural net pruning— Why and how. Proc. *IEEE Int. Conf. Neural Networks* 1: 325–333.

Sietsma, J., and Dow, R. J. F. (1991). Creating artificial neural networks that generalize. *Neur. Net.* 4: 67–79.

Stollberg, J. (1995). Synapse elimination, the size principle, and Hebbian synapses. *J. Neurobiol.* 26: 273–282.

Thompson, I. D., and Cordery, P. M. (1997). Cell death and the elimination of topographic errors in the hamster retinocollicular projection. *J. Physiol. (London).* 501: 95P.

Von Bartheld, C. S., Kinoshita, Y., Prevette, D., Yin, Q.-W., Oppenheim, R. W., and Bothwell, M. (1994). Positive and negative effects of neurotrophins on the isthmo-optic nucleus in chick embryos. *Neuron* 12: 639–654.

Von Bartheld, C. S., Byers, M. R., Williams, R., and Bothwell, M. (1996a). Anterograde transport of neurotrophins and axodendritic transfer in the developing visual system. *Nature* 379: 830–833.

Von Bartheld, C. S., Williams, R., Lefcort, F., Clary, D. O., Reichardt, L. F., and Bothwell, M. (1996b). Retrograde transport of neurotrophins from the eye to the brain in chick embryos: Roles of the p75[NTR] and trkB receptors. *J. Neurosci.* 16: 2995–3008.

Wong, R. O. L. (1999). Retinal waves and visual system development. *Annu. Rev. Neurosci.* 22: 29–47.

Young, D., Lawlor, P. A., Leone, P., Dragunow, M., and During, M. J. (1999). Environmental enrichment inhibits spontaneous apoptosis, prevents seizures and is neuroprotective. *Nat. Med.* 5: 448–453.

Competition in the Development of Nerve Connections

Arjen van Ooyen and Richard R. Ribchester

10

During development, neurons and other target cells are often initially innervated by more axons than ultimately remain into adulthood. The process that leads to elimination of connections is referred to as axonal or synaptic competition. This chapter reviews the models of competition that have been proposed for the neuromuscular and the visual system, and describes in detail a model that links competition in the development of nerve connections with the underlying actions and biochemistry of neurotrophic factors.

10.1 Competition

The establishment and refinement of neural circuits involve both the formation of new connections and the elimination of existing connections (e.g., Lohof et al. 1996). A well-studied case of this form of remodeling is the withdrawal of connections that takes place during development. Neurons, and other cell types, are initially innervated by more axons than they ultimately maintain into adulthood (Purves and Lichtman, 1980; Lohof et al., 1996). This is a widespread phenomenon in the developing nervous system and occurs, for example, in the development of connections between motor neurons and muscle fibers (reviewed in Jansen and Fladby, 1990; Sanes and Lichtman, 1999; Ribchester, 2001; see also section 10.2.2), the formation of ocular dominance columns (see chapter 12 and section 10.2.3), and the climbing fiber innervation of Purkinje cells (Crepel, 1982).

The process that reduces the amount of innervation onto a postsynaptic cell is often referred to as axonal or synaptic competition, although neither term describes the competitors adequately (Colman and Lichtman, 1992; Snider and Lichtman, 1996). Since a single axon can branch to innervate, and compete on, many postsynaptic cells simultaneously, competition is perhaps better described as occurring between axon branches rather than between axons. By further arborization, the contact between an axon branch and a postsynaptic cell can involve several synaptic boutons, so that competition occurs not between single synapses but between groups of synapses.

Defining synaptic competition has exercised a number of authors. In discussing the neuromuscular system, Van Essen et al. (1990) gave one of the most general definitions of competition: a process in which there are multiple participants whose behavior is governed by certain rules so that one or more of the participants emerge as victors. This definition leaves open the processes by which the victors arise. Based on whether or not there are interactions between the participants, Colman and Lichtman (1992) distinguished two ways by which victors can come about, leading to two types of competition:

1. In *independent competition*, victors do not arise as a result of interactions (either direct or indirect) between the participants, but are chosen (by "judges") based on a comparison of the performance or desirable features of the participants (e.g., as in a beauty contest). In this form of competition, one participant cannot influence the performance of the others during the competition. Lotteries are another example of this form of competition. Here the criteria for selection are random, and there is nothing a single ticket holder can do to influence the outcome. In axonal competition, this would mean that the axons innervating the same postsynaptic cell do not affect each other and that the postsynaptic cell would decide, on the basis of some performance or random criteria, which axon(s) would win. Since axons do affect each other (see section 10.2), and synapse elimination is nonrandom, this form of competition is unlikely.

2. In *interdependent competition*, victors emerge as a result of direct or indirect interactions between the participants, affecting their performance. This is the type of competition that is considered in population biology, where two species of organisms are said to compete if they exert negative effects on the growth of each other's population. Ribchester (1992) and Ribchester and Barry (1994) extended this definition to neurobiology; they defined competition as the negative effects that one neuron or its synapses have on others. Based on how the negative interactions come about, two types of interdependent competition can be distinguished (Yodzis, 1989; see also figure 10.1):

• In *consumptive competition*, in systems of consumers and resources, each consumer hinders the others solely by consuming resources that they might otherwise have consumed; in other words, consumers hinder each other because they share the same resources. In neurobiology, competition is commonly associated with this dependence on shared resources (Purves and Lichtman, 1985; Purves 1988, 1994). In particular, it is believed that axons compete for target-derived neurotrophic factors (see section 10.2.1).

• In *interference competition*, instead of hindrance through dependence on shared resources, there is direct interference between individuals, e.g., direct

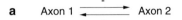

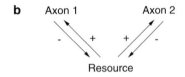

Figure 10.1
(*a*) Interference competition. (*b*) Consumptive competition. See the text for details. (Modified from Huisman, 1997.)

negative interactions, such as aggressive or toxic interactions. In axonal competition, nerve terminals could hinder each other by releasing toxins or proteases (see section 10.2.2). If some essential resource can be obtained only by occupying, more or less exclusively, some portion of space (competition for space), this is also primarily interference competition, because each consumer is seeking to monopolize a portion of space rather than to share resources (Yodzis, 1989).

Although the notion of competition is commonly used in neurobiology, there is little understanding of the type of competitive process or the underlying molecular mechanisms. In this chapter we discuss the different models of competition that have been proposed, both in the neuromuscular and in the visual system (for a more detailed review, see Van Ooyen, 2001). We classify the models according to the forms of (interdependent) competition that are distinguished in population biology (as described earlier). Before presenting the models, we briefly review the biology of neurotrophic factors—which play an important role in many models—and the development of the neuromuscular and the visual system, the two systems where competition is most widely studied.

10.2 Neurobiological Background

10.2.1 Neurotrophic Factors

During an early stage of development, when initial synaptic contacts are made, neurotrophic factors have a well-established role in the regulation of neuronal survival (see chapter 9). However, many studies now indicate that neurotrophic factors may also be involved in the later stages of development, when there is further growth and elimination of innervation (see sections 10.2.2 and 10.2.3; for a critical review, see Snider and Lichtman, 1996). For example, neurotrophic factors have been shown to regulate the degree of arborization of axons (e.g., Cohen-Cory and Fraser, 1995; Funakoshi et al., 1995; Alsina et al., 2001).

In addition to their decisive role in the fate of neurons and the disposition of their connections, neurotrophic factors have well-defined roles in modulating synaptic transmission. For instance, neurotrophins (i.e., neurotrophic factors of the NGF family, including BDNF, NT-3 and NT-4/5; Bothwell, 1995; Lewin and Barde, 1996), acting on their specific Trk receptors, may phosphorylate synapse-specific proteins and enhance transmitter release (Lohof et al., 1993). Similar effects are exerted by other neurotrophic factors, of the ciliary neurotrophic factor (CNTF) and glial cell line-derived neurotrophic factor (GDNF) classes (Ribchester et al., 1998; Stoop and Poo, 1996). It is of some interest that positive effects of neurotrophic factors on synaptic transmission and growth can be commuted to negative effects, depending on the relative levels of intracellular signaling molecules such as cyclic nucleotides (Boulanger and Poo, 1999; Poo, 2001).

10.2.2 Neuromuscular System

Adult System and Development

In adult mammals, each muscle fiber is innervated at the endplate—a discrete region near the midpoint of the muscle fiber—by the axon from a single motor neuron. This state is referred to as mononeuronal (μ) or "single" innervation (figure 10.2b). However, a single motor neuron, through its axonal branches, typically contacts many muscle fibers. The motor neuron and the group of muscle fibers it innervates is referred to as the motor unit, and the number of fibers contacted by a given motor neuron is called the motor unit size. Motor neurons with higher firing thresholds—which may therefore be less frequently activated—have progressively larger motor units (the size principle; Henneman, 1985).

During prenatal development, the axons of the motor neurons grow toward their target muscle, and near the muscle each axon arborizes to innervate a large number of muscle fibers. At birth, the endplate of each muscle fiber is contacted by axons from several different motor neurons, a state referred to as polyneuronal (π) or "multiple" innervation (figure 10.2a). During the subsequent few weeks, axonal branches are removed or withdrawn until the motor endplate of each muscle fiber is taken over by the synaptic boutons derived from a single motor axon collateral (Brown et al., 1976; Betz et al., 1979; Keller-Peck et al., 2001a; Walsh and Lichtman, 2003). Thus, during the elimination of polyneuronal innervation, the number and size of the synaptic boutons of the winning axon increase, while the synaptic boutons of the losing axon are either gradually retracted or nipped off from their parent neuron (Keller-Peck et al., 2001b). With contemporaneous addition and loss of synaptic boutons, the synaptic area on the endplate actually increases during the elimination

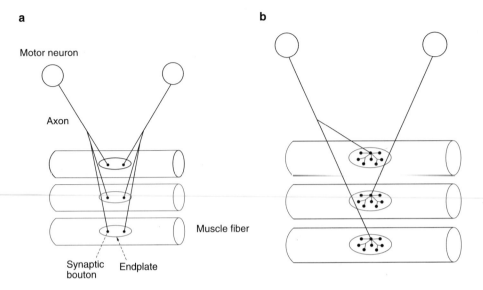

Figure 10.2

The development of connections between motor neurons and muscle fibers. (*a*) At birth, each fiber is innervated by axons from several different neurons. (*b*) In adulthood, each fiber is innervated by the axon from a single neuron. (From Van Ooyen, 2001.)

of polyneuronal innervation (Sanes and Lichtman, 1999). Motor unit sizes, as well as the range of sizes, decrease during elimination of polyneuronal innervation (Brown et al., 1976; Betz et al., 1979; Balice-Gordon and Thompson, 1988).

Competition

The elimination of polyneuronal innervation appears to be a competitive process. Following removal of some motor axons at birth, the average size of the remaining motor units after elimination of poly-neuronal innervation is larger than normal (Thompson and Jansen, 1977; Betz et al., 1979; Fladby and Jansen, 1987). This competition for the endplate (postsynaptic competition), however, cannot explain why larger motor units decrease in size more than smaller ones (thus reducing the range of motor unit sizes) and why sometimes branches at singly innervated fibers—where there is no competition—

apparently withdraw (Fladby and Jansen, 1987). This process of "intrinsic withdrawal" has not yet been observed directly, but Keller-Peck et al. (2001b) have argued that the asynchronous pattern of synapse loss observed within motor units precludes intrinsic withdrawal as an integral component of synapse elimination. It should now be possible to resolve this issue by repeating the earlier studies of Betz et al. (1979) and Fladby and Jansen (1987) using thy1-YFP transgenic mice, which have endogenously fluorescent motor axons and synapses, facilitating repeated visualization of identified neuromuscular junctions (Feng et al., 2000). If confirmed, the existence of intrinsic stimuli to synapse elimination would imply that there are also presynaptic constraints that restrict the number of axon branches each neuron can maintain.

What mediates the competition in the development of mononeuronal innervation? We discuss this for both consumptive and interference competition.

Consumptive Competition

Muscles might release diffusible neurotrophic factors for which axons compete (Snider and Lichtman, 1996). Several factors produced by muscles are capable of retarding elimination of polyneuronal elimination when applied to postnatal muscles (English and Schwartz, 1995; Kwon and Gurney, 1996; Jordan, 1996). For example, transgenic mice overexpressing the neurotrophic factor GDNF show extensive polyneuronal innervation at a relatively late postnatal stage (Nguyen et al., 1998). Mononeuronal innervation is eventually established, but about 2 weeks later than normal. Exogenous administration of GDNF to neonatal muscle also delays elimination of π-junctions, but the pattern of innervation, together with the decline in sensitivity, suggested that the predominant effect of this growth factor is to stimulate or maintain nerve branch points, rather than synaptic terminals per se (Keller-Peck et al., 2001a). It remains uncertain where the receptors for GDNF are located and how their expression is regulated. However, the observation that small but significant enhancements in neurotransmitter release occur in response to low concentrations of GDNF suggests that immature synaptic terminals at least express the receptor (Ribchester et al., 1998). It is not known whether nodes of Ranvier, or other sites of neural sprouting, also express GDNF receptors.

Interference Competition: Competition for Space

Until recently, the notion that competition occurs only for space at the endplate was controversial because some observations of developing neuromuscular junctions in vivo revealed that as one terminal is withdrawn, the space it occupied is left vacant rather than being taken over by another terminal (Balice-Gordon and Lichtman, 1994). However, a takeover of existing space clearly occurs during reinnervation of partially denervated muscle (Costanzo et al., 2000).

Moreover, very recent observations made by Lichtman and colleagues, utilizing transgenic thy1-YFP mice, suggest that both processes—takeover and withdrawal without takeover—occur at the same or different junctions during neonatal synaptic competition as well (Walsh and Lichtman, 2003). At present, it remains to be seen whether takeover will turn out to be the predominant mechanism, as in reinnervated muscle (Barry and Ribchester, 1995; Costanzo et al., 1999, 2000).

However, attempts to identity molecules that might mediate a spatial competition have so far been unsuccessful (Ribchester, 2001). For example, normal elimination of synapses occurs in various transgenic animals in which expression of cell surface or extracellular matrix molecules, such as neural cell adhesion molecules (N-CAMs), has been disrupted (Sanes et al., 1998). But since synapse elimination must, at some stage in the process, involve weakening of the adhesive bonds between synaptic membranes and molecules in the extracellular matrix, the notion of interference competition based on access to synaptic space should therefore still receive attention. Intraneuritic tension-adhesion mechanisms have been posited to account for morphogenesis in the brain (Van Essen, 1997), and such mechanisms may be accessible to experimental investigation at a cellular level, using the neuromuscular junction as a paradigm. Mechanical stimulation (stretch) of motor nerve endings regulates transmitter release at neuromuscular junctions, and this effect is mediated by integrins, receptors for adhesion molecules (Kashani et al., 2001). Integrins are implicated in synapse formation, growth, and specificity in *Drosophila* muscle (Beumer et al., 1999).

Interference Competition: Direct Negative Interactions

Another possibility is axon-derived or axon-stimulated release of interfering molecules. For instance, proteases might mediate direct negative interactions between axons (Sanes and Lichtman, 1999). Many proteases and protease inhibitors are located at the neuromuscular junction (Hantai et al., 1988), and various proteases have been proposed to play a role in synapse destabilization (e.g., Zoubine et al., 1996). Highly selective proteases could also work indirectly, mediating the kind of spatial competition indicated earlier.

Role of Electrical Activity

Does the overall level of activity affect the rate of synapse elimination? Blocking activity (by interfering with input activity, synaptic transmission, or muscle activity) delays or prevents synapse elimination (Thompson et al., 1979; Brown et al., 1982; Ribchester and Taxt, 1984; Callaway and Van Essen, 1989; Barry and Ribchester, 1995), while stimulating activity accelerates synapse elimination (O'Brien et al., 1978; Thompson, 1983; Zhu and Vrbova, 1992; Vyskocil and Vroba, 1993; for a review, see Ribchester, 2001).

Do differences in the activity of innervating axons confer competitive advantages on the more active axons? Here the findings are less clear-cut. Selectively stimulating motor neurons in neonates, Ridge and Betz (1984) found that the more active axons have a competitive advantage over the less active ones, whereas Callaway et al. (1987), using selective blocking, found the opposite. Experiments in tissue culture also show opposing results (Magchielse and Meeter, 1986; Nelson et al., 1993). Based on observations that synapse elimination begins with elimination of AChRs (the postsynaptic receptors for acetylcholine, the neurotransmitter in motor neurons) and that in adults partial but not complete paralysis of the endplate leads to the elimination of the terminals overlying the silent patches, Balice-Gordon and Lichtman (1993, 1994) suggested that electrically

active synapses are the stimulus for removing the AChRs underlying the less active synapses, which are then eliminated. However, when motor endplates are made completely silent by blocking nerve conduction and synaptic transmission during nerve regeneration, inactive terminals appear capable of competitively displacing other, active or inactive, terminals (Ribchester, 1988, 1993; Costanzo et al., 2000). Thus, differences in activity are not strictly necessary for synapse elimination.

Electrical activity also seems to be insufficient for synapse elimination. Barry and Ribchester (1995) found that following recovery from chronic nerve conduction block, many reinnervated muscle fibers in partially denervated muscles retain polyneuronal innervation, in spite of the resumption of normal neuromuscular activity. Following on from this, Costanzo et al. (1999) showed that the synaptic efficacy per unit area was similar in the coinnervating inputs to the muscle fibers, whatever the relative synaptic area covered by each motor nerve terminal.

In conclusion, activity is clearly influential in synaptic competition—particularly in regard to its effects on the rate of synapse elimination—but activity does not seem to be decisive (Costanzo et al., 2000; Ribchester, 2001). To reconcile the different findings, one possibility is that activity is just one of many influences in competition. Perhaps its main influence is restricted to critical periods during the competitive process, while the actual competition is governed by other factors, e.g., neurotrophic factors, adhesion molecules, and their receptors (Costanzo et al., 2000; see also section 10.4).

10.2.3   Visual System

Adult System and Development

In the adult visual system, the different layers of the lateral geniculate nucleus (LGN) receive axons from

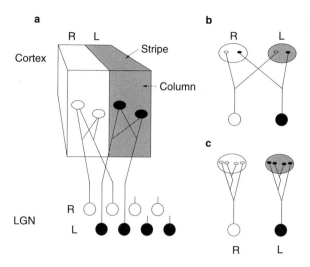

Figure 10.3
The development of ocular dominance columns. (*a*) The adult visual system. The lateral geniculate nucleus (LGN) of the thalamus is composed of two or more layers, each of which receives axons from either the left or the right eye. In the visual cortex, cells in layer IV respond preferentially to input from either the left or the right eye. (*b*) In the immature system, the arbors of the geniculate axons overlap extensively within layer IV. (*c*) During further development, remodeling of axonal arbors takes place so that each cortical cell receives axons from either the left-eye or right-eye geniculate neurons. (From Van Ooyen, 2001.)

either the left or the right eye (figure 10.3a). Like the different layers in the LGN, columns of cells in layer IV of the visual cortex (to which the axons from the LGN project) respond preferentially to input from either the left or the right eye (ocular dominance; see also chapter 12).

The formation of eye-specific layers and columns requires anatomical remodeling of axonal arbors during development (figures 10.3b and 10.3c). Initially, the retinal axons from the two eyes overlap extensively within the LGN. Similarly, the arbors of geniculate axons are initially evenly distributed within

layer IV. Just as in the elimination of polyneuronal innervation in the neuromuscular system, the refinement of connections to the LGN and cortex involves both the retraction of axonal side branches that project to the wrong region and the elaboration of branches that project to the correct region, and the total number of synapses onto a postsynaptic cell actually increases during the period in which elimination takes place.

Competition

As in the neuromuscular system, the formation of eye-specific layers and columns might involve competition between axons or axon branches for target-derived neurotrophic factors. Continuous infusion of the neurotrophins NT-4/5 or BDNF in the cat visual cortex prevents the formation of ocular dominance columns (Cabelli et al., 1995), presumably because the LGN axon branches fail to retract. In monocular deprivation experiments (see the following section) in cats and rats, excess neurotrophic factor mitigates or abolishes the relative increase of the ocular dominance stripes associated with the open eye (e.g., Yan et al., 1996; see also chapter 12).

Role of Electrical Activity

The process of segregation into eye-specific regions is influenced by neural activity, which arises not only from visual stimulation through photoreceptor activation but also from spontaneous activity in retinal ganglion cells. When all activity in both eyes of kittens is blocked by tetrodotoxin, ocular dominance columns do not form at all (Stryker and Harris, 1986). When only the visually driven activity is blocked, however, as in macaque monkeys reared in complete darkness, a normal pattern of ocular dominance columns is found (LeVay et al., 1980). In fact, in monkeys at least, ocular dominance columns are present prior to birth and eye opening (Horton and Hocking, 1996). Taken together, these observations suggest that spontaneous activity in the retina may instruct the formation of ocular dominance columns. This is inconsistent with the finding that eye removal in ferrets early in visual development does not prevent the development of ocular dominance columns (Crowley and Katz, 1999). Recently, Crowley and Katz (2000) showed that in ferrets, ocular dominance columns appear much earlier during development than previously thought, and that these early columns are unaffected by imbalances in retinal activity. They proposed that axon guidance cues are sufficient to initially establish columns.

Although activity might not be necessary for the initial formation of ocular dominance columns, the prevailing view is that it does play a decisive role in their later plasticity. For example, when vision through one eye is prevented by suturing the eyelids shut after birth, the stripes or patches formed by the sutured eye's input become smaller than those formed by the open eye's input (e.g., Shatz and Stryker, 1978; see also chapter 12). However, even this bastion of synaptic plasticity seems to be under renewed assault (Wickelgren, 2000; Crowley and Katz, 2000; Crair et al., 2001). In conclusion, just as activity has a role in the development of the neuromuscular system, activity is also influential in the development of the visual system, but it may not be overwhelmingly decisive.

10.3   Review of Models

Models in which competition plays an important role have been proposed for both the neuromuscular and the visual system. In the neuromuscular system, the main aim is to explain the change from polyneuronal to mononeural innervation of muscle fibers. In the visual system, the main aim is to explain the development of columnar organization of synaptic connec-

tivity, especially ocular dominance. The presentation of the various models here is structured on the basis of how competition is implemented: through synaptic normalization and modified Hebbian learning rules (section 10.3.1), dependence on shared resources (section 10.3.2), or interference (section 10.3.3). For each model, we identify its underlying positive feedback loop; this is what enables one or more competitors to outcompete the others. To show the differences and similarities in modeling approach, mathematical equations are given for one model of each type.

10.3.1   Competition Through Synaptic Normalization and Modified Hebbian Learning Rules

Many models—especially those of the formation of ocular dominance—enforce competition rather than implement its putative underlying mechanisms (for a review, see Miller, 1996). That is, these models explore the consequences of imposing certain rules that are introduced to ensure competition between axons. These models usually describe changes in synaptic strength (physiological plasticity) rather than changes in axonal arborization (anatomical plasticity). To see how competition can be enforced, consider n inputs with synaptic strengths $w_i(t)$ $(i = 1, \ldots, n)$ impinging on a given postsynaptic cell at time t. Simple Hebbian rules for the change $\Delta w_i(t)$ in synaptic strength in time interval Δt state that the synaptic strength should grow in proportion to the product of the postsynaptic activity level $y(t)$ and the presynaptic activity level $x_i(t)$ of the ith input:

$$\Delta w_i(t) \propto y(t)x_i(t)\Delta t. \tag{10.1}$$

According to Eq. (10.1), only increases in synaptic strength can take place, and if the activity levels of two inputs (e.g., two eyes) are both sufficient to achieve potentiation, then both pathways are strongly potentiated (and no ocular dominance can occur). To achieve the situation that when the synaptic strength of one input grows, the strengths of the other one shrinks (i.e., competition), $\sum_i^n w_i(t)$ should be kept constant (synaptic normalization). At each time interval Δt—following a phase of Hebbian learning, in which $w_i(t + \Delta t) = w_i(t) + \Delta w_i(t)$—the new synaptic strengths are forced to satisfy the normalization constraint, either by multiplying each synaptic strength by a certain amount (multiplicative normalization; Willshaw and Von der Malsburg, 1976) or by subtracting a certain amount from each synaptic strength (subtractive normalization; Miller et al., 1989). The final outcome of development may depend on whether multiplicative or subtractive normalization is used (Miller and MacKay, 1994). Multiplicative, but not subtractive, normalization prevents the development of ocular dominance if there are positive between-eye correlations (which are likely to be present when the two eyes are open). Experimental evidence for multiplicative normalization has been found in cultures of cortical neurons (Turrigiano et al., 1998; see also chapter 8).

Another approach for achieving competition is to modify Eq. (10.1) so that both increases in synaptic strength (long-term potentiation, or LTP) and decreases in synaptic strength (long-term depression, or LTD) can take place. Assume that $y(t)$ and $x_i(t)$ must be above some thresholds θ_y and θ_x, respectively, to achieve LTP, and otherwise yield LTD (Miller, 1996); i.e.,

$$\Delta w_i(t) \propto [y(t) - \theta_y][x_i(t) - \theta_x]\Delta t. \qquad (10.2)$$

A stable mechanism for ensuring that when some synaptic strengths increase others must correspondingly decrease is to make one of the thresholds variable. If θ_x^i increases sufficiently as $y(t)$ or $w_i(t)$ (or both) increases, conservation of synaptic strength can

be achieved (Miller, 1996). Similarly, if θ_y increases faster than linearly with the average postsynaptic activity, then the synaptic strengths will adjust to keep the postsynaptic activity near a set point value (Bienenstock et al., 1982; see also chapter 12).

Yet another mechanism that can balance synaptic strengths is based on (experimentally observed) spike timing-dependent plasticity (STDP; reviewed in Bi and Poo, 2001). Presynaptic action potentials that precede postsynaptic spikes strengthen a synapse, whereas presynaptic action potentials that follow postsynaptic spikes weaken it. Subject to a limit on the strengths of individual synapses, STDP keeps the total synaptic input to the neuron roughly constant, independent of the presynaptic firing rates (Song et al., 2000).

10.3.2 Consumptive Competition: Competition for a Target-Derived Resource

Keeping the total synaptic strength on a postsynaptic cell constant (synaptic normalization) is a biologically unrealistic way of modeling competition during development. In both the neuromuscular and the visual system, the total number of synapses on a postsynaptic cell increases during competition as the winning axons elaborate their branches and the losing axons retract branches (see section 10.2). In models that implement consumptive competition, competition between input connections does not have to be enforced, but comes about naturally through their dependence on the same target-derived resource. There are two ways in which this can be modeled:

1. In *fixed-resource models*, the total amount of postsynaptic resource is kept constant. The total amount of resource is the amount taken up by the input connections (i.e., the total synaptic strength if the resource is "converted" into synaptic strength) plus the amount left at the target. Thus, the total synaptic strength is not kept constant and can increase during development when the resource becomes partitioned among the input connections.

2. In *variable-resource models*, it is not imposed that even the total amount of resource should remain constant. In these models, there is continuous production of neurotrophin and continuous uptake or binding of neurotrophin. Continuous uptake or binding ("consumption") of neurotrophin is needed to sustain the axonal arbors and synapses. This view of the way in which the resource exerts its effects is closer to the biology of neurotrophins, and is also closer to other consumer–resource systems in biology; organisms need a continuous supply of food (resource) to sustain themselves.

Fixed-Resource Models

Dual Constraint Model (Bennett and Robinson, 1989; Rasmussen and Willshaw, 1993)

Based on experimental results that suggest a role for both a postsynaptic and a presynaptic resource in the development of neuromuscular connections (see section 10.2.2), the dual constraint model combines competition for both these types of resources. Each muscle fiber m has a postsynaptic resource B (in amount B_m), and each motor neuron n has a presynaptic resource A, which is located in its cell soma (in amount A_n) and in all its terminals nm (in amount A_{nm}). In the synaptic cleft, a reversible reaction takes place between A and B to produce binding complex C:

$$A_{nm} + B_m \rightleftarrows C_{nm}, \tag{10.3}$$

with

$$\frac{dC_{nm}}{dt} = \alpha A_{nm} B_m C_{nm}^\mu - \beta C_{nm}, \tag{10.4}$$

where α and β are rate constants. The size of the terminal is assumed to be proportional to C_{nm}. Including C_{nm}^μ (with $\mu > 0$) in Eq. (10.4) incorporates a positive feedback and is needed to achieve single innervation. The justification given by Bennett and Robinson (1989) for including this positive feedback is that electrical activity in the nerve terminal could produce electromigration of molecules B in the endplate, so that larger terminals will attract more molecules.

The total amount A_0 of presynaptic substance in each motor neuron is fixed:

$$A_0 = A_n + \sum_{j=1}^{M} A_{nj} + \sum_{j=1}^{M} C_{nj}, \tag{10.5}$$

where N and M are the total numbers of neurons and muscle fibers, respectively. The amount A_{nm} is assumed to be proportional to C_{nm} (thus incorporating a second positive feedback) and A_n:

$$A_{nm} = KC_{nm}A_n, \tag{10.6}$$

where K is a constant.

The total amount B_0 of postsynaptic substance in each muscle fiber is also fixed:

$$B_0 = B_m + \sum_{i=1}^{N} C_{im}. \tag{10.7}$$

Introducing Eqs. (10.5), (10.6), and (10.7) into Eq. (10.4) gives a set of differential equations for how C_{nm} changes over time.

Single innervation is a stable state of the model, and there is an upper limit, proportional to A_0/B_0, on the number of terminals that can be supported by each motor neuron (Rasmussen and Willshaw, 1993). So if the initial amount of polyneuronal innervation is larger than this limit, then terminals will withdraw, even in the absence of competition (intrinsic withdrawal; see section 10.2.2).

Polyneuronal states can also be stable and can co-exist with single innervation states (Van Ooyen and Willshaw, 1999a). This offers an explanation for partial denervation experiments that show that persistent polyneuronal innervation occurs after reinnervation and recovery from prolonged nerve conduction block (see section 10.2.2), while under unblocked conditions single innervation develops (see also section 10.4).

Weak points of the dual constraint model are that (1) it does not make clear the identity of the pre- and postsynaptic resources; (2) a stronger biological justification for the positive feedback loops is needed; and (3) without electrical activity [$\mu = 0$ in Eq. (10.4)], no competitive elimination of connections takes place, which is not in agreement with recent experimental findings (see section 10.2.2).

Joseph and Willshaw (1996) and Joseph et al. (1997) gave a more specific interpretation of the dual constraint model in which A represents the protein agrin, B the acetylcholine receptor (AChR), and C aggregated AChRs. They were able to explain the results produced by focal blockade of postsynaptic AChRs (Balice-Gordon and Lichtman, 1994; see also section 10.2.2).

Harris et al. (1997, 2000)

This model of the development of ocular dominance columns incorporates a combination of Hebbian synaptic modification and activity driven competition for neurotrophins. In the model, each cortical cell has a fixed pool of neurotrophin to distribute over its input connections. The higher the connection strength, the faster the uptake of neurotrophin. Connection strength increases owing to Hebbian LTP at a rate that depends on the amount of neurotrophin taken up (together with the previous assumption, this creates a positive feedback loop). Connection strength decreases owing to heterosynaptic LTD.

The model shows that (1) ocular dominance columns develop normally—even with positive intereye correlations in activity (compare section 10.3.1)—when the available neurotrophin is below a critical amount and (2) column development is prevented when excess neurotrophin is added. A criticism of the model is that it incorporates only physiological plasticity, while anatomical plasticity is (mainly) involved in the formation of ocular dominance columns.

Variable-Resource Models

Elliott and Shadbolt (1998a,b)

This model of the development of the visual system explicitly describes anatomical plasticity and incorporates a role for electrical activity, both in the release and in the uptake of neurotrophin. For the case of a single target (e.g., a cortical cell) with a number of innervating axons (e.g., from the LGN), the rate of change in the number s_i of synapses that axon i has on the target is given by

$$\frac{ds_i}{dt} = \varepsilon s_i \left[\left(T_0 + T_1 \frac{\sum_j s_j a_j}{\sum_j s_j} \right) \frac{(a + a_i)\rho_i}{\sum_j s_j (a + a_j)\rho_j} - 1 \right],$$

(10.8)

where ε is a rate constant, T_0 is a constant representing the activity-independent component of release of neurotrophin by the target; T_1 is a constant for the activity-dependent component; $\sum_j s_j a_j / \sum_j s_j$ is the mean activity of a synapse, where a_j is the level of activity of axon j; $(a + a_i)\rho_i$ represents the capacity of an axon to take up neurotrophin, where a is a constant for activity-independent uptake and ρ_i is the number of neurotrophin receptors per synapse. Equation (10.8) incorporates a positive feedback: neurotrophin increases the number of synapses, while more synapses mean a higher uptake of neurotrophin. The model permits the formation of ocular dominance columns, even in the presence of positively correlated inter-ocular images (compare section 10.3.1). A high level of neurotrophin released in an activity-independent manner prevents the formation of ocular dominance columns.

A criticism of the model is that it is not clear why the activity-dependent release of neurotrophin is taken to depend on the mean activity of the synapse, rather than on the level of activity of the target. Also, in the model, electrical activity directly increases the uptake of neurotrophin, rather than by increasing the number of neurotrophin receptors (Salin et al., 1995; Birren et al., 1992) or the number of synapses (Ramakers et al., 1998).

Jeanprêtre et al. (1996)

Jeanprêtre et al. (1996) were the first to model neuro-trophic signaling in a fully dynamical way, implementing production, degradation, and binding of neurotrophin. They considered a single target that releases neurotrophin and at which there are a number of innervating axons. In the model, each axon has a variable called axonal vigor, which represents its ability to take up neurotrophin and which is proportional to its total number of neurotrophin receptors. The rate of change in vigor depends on the vigor itself (i.e., positive feedback) and increases with the fraction of receptors occupied by neurotrophin, over and above some threshold (the threshold is a constant that represents the value of the axonal vigor that yields zero growth). The system will approach a stable equilibrium point in which a single axon—the one with the lowest threshold—survives.

Criticisms of the model are that (1) the rate of change in axonal vigor (including the positive feedback) is postulated but not explicitly derived from underlying biological mechanisms; and (2) the thresholds do not emerge from the underlying dynamics but need to be assumed.

Van Ooyen and Willshaw (1999b)

Independently from Jeanprêtre et al. (1996), Van Ooyen and Willshaw (1999b) proposed a model of competition that implements neurotrophic signaling in a fully dynamical way and that does not have the above-mentioned drawbacks. For the description of this model, see section 10.4.

10.3.3   Interference Competition

Competition for Space

Competition for space occurs if some essential resource can be obtained only by monopolyzing some portion of space. The resource may be space itself or it may be some immobile resource.

Van Essen et al. (1990)

This model incorporates competition for space together with the idea that the increase in size of a motor neuron terminal depends on how much "scaffold" is incorporated in the underlying basal lamina at the endplate. In the model, a terminal occupies a certain amount of space on the endplate and grows (as a stochastic process) by occupying more space at the expense of the size of other terminals. It is not clear whether the model can account for single innervation, because even after many iterations, a high percentage of muscle fibers remained polyneuronally innervated.

Induced-Fit Model (Ribchester and Barry, 1994)

In the induced-fit model, which was not given in mathematical terms, nerve terminals from different axons have different isoforms of an adhesion molecule, and each endplate may express a number of different complementary isoforms. Nerve terminals induce a conformational change in (or increase the expression of) the complementary adhesion molecules in the endplate so that goodness-of-fit increases. Electrical activity in a terminal accelerates the conformational change. The model was proposed to explain that a block in nerve conduction delays or inhibits elimination of polyneuronal innervation in partially denervated and reinnervated muscle (Taxt, 1983; Barry and Ribchester, 1994, 1995; see also section 10.2.2).

Direct Negative Interactions

In the following models, all of which describe the neuromuscular system, interference competition involves direct negative interactions. Nerve terminals are destroyed or disconnected by the punitive effects of other axons.

Willshaw (1981)

This is the first published formal model of the elimination of polyneuronal innervation in the neuromuscular system. Based on a proposal by O'Brien et al. (1978), Willshaw (1981) assumed that each terminal injects into its endplate a degrading signal, at a rate proportional to its own "survival strength" (the size of the terminal is thought to be proportional to this strength), that reduces the survival strength of all the terminals (including itself) at that endplate. The survival strength of each terminal also increases, at a rate proportional to that strength (positive feedback). Furthermore, the total amount of survival strength supported by each motor neuron is kept constant, i.e., synaptic normalization of the total strength of the output connections.

The model can account for (1) the elimination of polyneuronal innervation, (2) the decrease in spread of motor unit size, (3) the competitive advantage of the terminals of smaller motor units over those of larger ones (Brown and Ironton, 1978), and (4) the increase in motor unit size after neonatal partial denervation (Fladby and Jansen, 1987).

Criticisms of the model are that (1) the positive feedback is not accounted for biologically; and (2)

it uses synaptic normalization of output connections, which implies that not all fibers will show an increase in their total input survival strength during development (see section 10.2).

Nguyen and Lichtman (1996)

This model, which was not given in mathematical terms, has many similarities with Willshaw's (1981) model except that there is an explicit role for electrical activity. In the model, each active synapse, by activating its underlying acetylcholine receptors in the endplate, generates two postsynaptic signals: (1) a punishment signal that spreads over short distances and eliminates the AChRs of neighboring synaptic sites, which instigates the removal of the overlying nerve terminal; and (2) a more locally confined protection signal that neutralizes the punishment signal. The strength of both signals is proportional to the level of activity. Thus, when postsynaptic sites at the same endplate have a different level of activity, the less active ones will generate a weaker protection signal (and a weaker punishment signal) than the more active ones, so that the less active ones lose more AChRs. The loss of AChRs further reduces local postsynaptic activity, leading to an even weaker protection signal, more loss of AChRs, and eventually the removal of the overlying nerve terminal. This positive feedback loop can bring about the removal of all nerve terminals except the most active one. When all the postsynaptic sites are equally active or when they are all inactive, all nerve terminals will be maintained.

The model can account for the observation that when the AChRs of a portion of an endplate are blocked, the blocked AChRs and their directly overlying nerve terminals are eliminated only when a substantial portion remains unblocked (Balice-Gordon and Lichtman, 1994). A criticism of this model (and the next one) is that it relies heavily on electrical activity, while recent experimental results suggest that activity might not play such a decisive role (see section 10.2.2).

Barber and Lichtman (1999)

Barber and Lichtman (1999) put the ideas of Nguyen and Lichtman (1996) into mathematical terms, although the punishment and protection signals are not explicitly modeled. In their model, each synaptic area, A_{mn} for the area that neuron n makes on muscle fiber m, is subjected to two effects: (1) loss of synaptic area, in an amount E_{mn}, through the punishing effect of other axons; and (2) gain or loss of synaptic area, in an amount U_{mn}, through utilization of neuronal resources. Thus,

$$\frac{dA_{mn}}{dt} = -\alpha E_{mn} + \beta U_{mn}, \tag{10.9}$$

where α and β are rate constants.

It is assumed that axons are able to compete effectively only during asynchronous activity and that the punishing effect of an axon is proportional to the amount of neurotransmitter it releases (which in turn is proportional to the axon's terminal size at the endplate and to its mean firing rate), so that

$$E_{mn} = \sum_{i \neq n} f_i A_{mi} (1 - \tau^2 f_n f_i), \tag{10.10}$$

where f_i and f_n are the firing rates of neurons i and n, respectively; the neurons are asynchronously active during a fraction $(1 - \tau^2 f_n f_i)$ of the time, where τ is a constant.

The total amount R of presynaptic resource in each motor neuron is kept constant, so that

$$R = R_{a,n} + f_n \sum_j A_{jn}^\gamma, \tag{10.11}$$

where $R_{a,n}$ is the amount of free resource left in motor

neuron n and $\gamma < 1$ represents the assumption that large synaptic areas are disproportionally less taxing on the resources of the neuron. This total amount of presynaptic resource in each neuron is divided among all its connections, with large synaptic areas receiving a greater share, so that

$$U_{mn} = R_{a,n} \frac{A_{mn}}{\sum_j A_{jn}} = \left(R - f_n \sum_j A_{jn}^\gamma \right) \frac{A_{mn}}{\sum_j A_{jn}}. \tag{10.12}$$

In addition to accounting for the elimination of polyneuronal innervation, the model is able to reproduce the size principle (see section 10.2.2) because the presynaptic resource is utilized more heavily with increased activity of the neuron. The competitive advantage of higher frequency axons early in development is overcome at later stages by the greater synaptic efficacy of axons firing at a lower rate.

10.4   One Model in More Detail

Van Ooyen and Willshaw (1999b) proposed a model of (consumptive) competition that implements neurotrophic signaling in a fully dynamical way. Unlike Jeanprêtre et al. (1996) (see section 10.3.2), they did not need to assume a priori thresholds. Important variables in the model are the total number of neurotrophin receptors that each axon has and the concentration of neurotrophin in the extracellular space. In this model, there is a positive feedback loop between the axon's number of receptors and the amount of neurotrophin bound. Unlike the model of Jeanprêtre et al. (1996), this positive feedback, which enables one or more axons to outcompete the others, was derived directly from underlying biological mechanisms. Following binding to their receptors, neurotrophins can increase the terminal arborization of an axon (see section 10.2.1) and therefore the axon's number of synapses. Because neurotrophin receptors are located on synapses, increasing the number of synapses means increasing the axon's total number of receptors. Thus the more receptors an axon has, the more neurotrophin it will bind, which further increases its number of receptors, so that it can bind even more neurotrophin—at the expense of the other axons.

Neurotrophins might increase the axon's total number of receptors not only by enhancing the terminal arborization of an axon but also by increasing the size of synapses (e.g., Garofalo et al., 1992) or by upregulating the density of receptors (e.g., Holtzman et al., 1992).

10.4.1   Description of the Model

A single target cell is considered at which there are n innervating axons, each from a different neuron (figure 10.4a). Neurotrophin is released by the target into the extracellular space at a (constant) rate σ and is removed by degradation with a rate constant δ. In addition, at each axon i, neurotrophin is bound to receptors with association and dissociation constants $k_{a,i}$ and $k_{d,i}$, respectively. Bound neurotrophin (the neurotrophin–receptor complex) is also degraded, with a rate constant ρ_i. Finally, unoccupied receptors are inserted into each axon at a rate ϕ_i and are degraded with a rate constant γ_i. Thus, the rates of change in the total number R_i of unoccupied receptors on axon i, the total number C_i of neurotrophin–receptor complexes on axon i, and the extracellular concentration L of neurotrophin are

$$\frac{dC_i}{dt} = (k_{a,i}LR_i - k_{d,i}C_i) - \rho_i C_i \qquad (10.13)$$

$$\frac{dR_i}{dt} = \phi_i - \gamma_i R_i - (k_{a,i}LR_i - k_{d,i}C_i) \qquad (10.14)$$

$$\frac{dL}{dt} = \sigma - \delta L - \sum_{i=1}^{n}(k_{a,i}LR_i - k_{d,i}C_i)/v, \qquad (10.15)$$

where v is the volume of the extracellular space. Axons that will end up with no neurotrophin ($C_i = 0$) are assumed to have withdrawn.

The biological effects of neurotrophins—all of which, as explained earlier, can lead to an axon obtaining a higher total number of receptors—are triggered by a signaling cascade that is activated upon binding of neurotrophin to its receptors (Bothwell, 1995). In order for the total number of receptors to increase in response to neurotrophin, the rate of insertion of receptors, ϕ_i, must be an increasing function, f_i (called the growth function), of C_i. To take into account the fact that axonal growth is relatively slow, ϕ_i lags behind $f_i(C_i)$, with a lag given by

$$\tau \frac{d\phi_i}{dt} = f_i(C_i) - \phi_i, \qquad (10.16)$$

where the time constant τ for growth is on the order of days. Immediately setting $\phi_i = f_i(C_i)$ does not change the main results. Van Ooyen and Willshaw (1999b) studied different classes of growth functions, all derived from the general growth function

$$f_i(C_i) = \frac{\alpha_i C_i^m}{K_i^m + C_i^m}. \qquad (10.17)$$

Depending on the values of m and K, the growth function is a linear function (class I: $m = 1$ and K_i much greater than C_i) or a saturating function, which can be either a Michaelis-Menten function (class II: $m = 1$ and K_i not much greater than C_i) or a Hill function (class III: $m = 2$). Within each class, the specific values of the parameters α_i and K_i, as well as those of the other parameters, will typically differ among the innervating axons as a result, for example, of differences in activity or other differences. For example, increased presynaptic electrical activity can increase the axon's total number of receptors (by upregulation: Birren et al., 1992 and Salin et al., 1995; or by stimulating axonal branching: Ramakers et al., 1998),

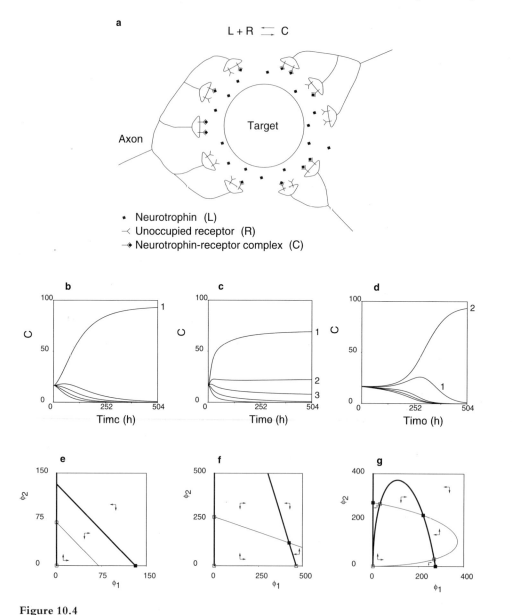

Figure 10.4

The model of Van Ooyen and Willshaw (see section 10.4). (*a*) Target cell with three innervating axons. The target releases neurotrophin, which binds to neurotrophin receptors at the axon terminals. For three different classes of growth functions, (*b–d*) show the development of innervation for a system of five innervating axons, where each axon has a different competitive strength β_i. (*e–g*) The nullcline pictures for a system of two innervating axons [the variables R_i, C_i, $i = 1, 2$ and L are set at which implies that, for example, α_i is increased or γ_i is decreased.

10.4.2 Results of the Model

For class I, starting with any number of axons, elimination of axons takes place until a single axon remains (single innervation), regardless of the rate σ of neurotrophin release (see figure 10.4). For class I, the number of surviving axons cannot be increased by increasing σ because an increased amount of neurotrophin will again become limiting as a consequence of the resulting increase in the size of the winning axon, which shows that the widely held belief that competition is the result of resources being produced in limited amounts is too simplistic. The axon that survives is the one with the highest value of the quantity $\beta_i \equiv [k_{a,i}(\alpha_i/K_i - \rho_i)]/[\gamma_i(k_{d,i} + \rho_i)]$, which is interpreted as the axon's competitive strength. If the growth function is a saturating function (classes II and III) more than one axon may survive (multiple innervation) and then the higher the rate σ of release of neurotrophin, the more axons survive. For class III, stable equilibria of single and multiple innervation can coexist, and which of these will be reached in any specific situation depends on the initial conditions.

For classes I and II, there is just one stable equilibrium point for any set of parameter values and therefore no dependence on initial conditions. For all classes, axons with a high competitive strength β_i survive, and the activity dependence of β_i (e.g., via α_i) means that these are the most active ones, provided that the variation due to other factors does not predominate.

The model can account for the following:

• The development of both single and multiple innervation.

• The coexistence of stable states of single and multiple innervation (class III) in skeletal muscle. Persistent multiple innervation is found in denervation experiments after reinnervation and recovery from prolonged nerve conduction block (Barry and Ribchester, 1995; see section 10.2.2 and figure 10.5).

• Increasing the amount of target-derived neurotrophin delays the development of single innervation (class I) (see section 10.2.2) or increases the number of surviving axons (classes II and III) (e.g., in epidermis; Albers et al., 1994).

• Decreasing the difference in competitive strengths between the different axons (which could be brought about by blocking their activity) delays the develop-

quasi-steady state; in (*e*) and (*f*), $\beta_1 > \beta_2$; in (*g*), $\beta_1 = \beta_2$]. The variable C is expressed in number of molecules and ϕ in number of molecules hr^{-1}. Axons that at the end of the competitive process have no neurotrophin ($C_i = 0$; equivalent to $\phi_i = 0$) are assumed to have withdrawn. In (*e–g*), the bold lines are the nullclines of ϕ_1 and the light lines are the nullclines of ϕ_2 (the *x*- and *y*-axes are also nullclines of ϕ_2 and ϕ_1, respectively). The intersection points of these lines are the equilibrium points. A filled square indicates a stable equilibrium point, an open square an unstable equilibrium point. Vectors indicate direction of change. (*b*) Class I. Elimination of axons takes place until the axon with the highest value of the competitive strength β_i survives. (*c*) Class II. For the parameter settings used, several axons survive. (*d*) Class III. Dependence on initial conditions. Although axon 1 has the highest value of the competitive strength, axon 2 survives because its initial value of ϕ_i is sufficiently higher than that of axon 1. (*e*) Class I. The nullclines do not intersect at a point where both axons coexist. (*f*) Class II. The nullclines intersect at a point where both axons coexist. For a sufficiently lower rate of neurotrophin release, for example, the nullclines would not intersect and only one axon would survive. (*g*) Class III. There is a stable equilibrium point where both axons coexist, as well as stable equilibrium points where either axon is present. For a sufficiently higher value of K_i, for example, the stable equilibrium point where both axons coexist would disappear. (From Van Ooyen, 2001.)

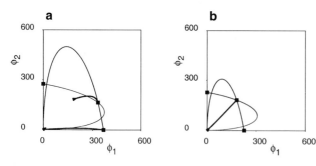

Figure 10.5
For class III, persistent multiple innervation can arise after recovery from nerve conduction block. Shown are the phase-space plots for a system of two innervating axons; for notations, see figure 10.4. The triangles mark the starting points of the trajectories (bold lines). As shown in (*a*), under normal conditions—with electrically active axons that have a different level of activity (values of α_i high and different) and a low initial number of receptors—single innervation develop. When activity is blocked (values of α_i lower and the same), as in (*b*), the same initial conditions lead to multiple innervation. Subsequent restoration of activity means that the nullclines are again as in (*a*), but now the starting values of ϕ_i are those reached as in (*b*), i.e., in the basin of attraction of the polyneuronal equilibrium point. The system goes to this equilibrium and will remain there; i.e., there is persistent polyneuronal innervation. (From Van Ooyen, 2001.)

ment of single innervation or increases the number of surviving axons (the latter only for classes II and III).

• Both presynaptic and postsynaptic activity may be influential, but are not decisive (Ribchester, 1988; Costanzo et al., 2000; see section 10.2.2). For competition to occur, it is not necessary that there be presynaptic activity; differences in the axons' competitive strengths β_i can also arise as a result of differences in other factors than activity. It is also not necessary that there be postsynaptic activity or activity-dependent release of neurotrophin (compare Snider and Lichtman, 1996).

An interesting observation is that the coexistence of several stable equilibria for class III implies that an axon that is removed from a multiply innervated target may not necessarily be able to reinnervate the target ("regenerate") when it is replaced with a low number of neurotrophin receptors (figure 10.6). To stimulate reinnervation, the model suggests that it is

more efficient to increase the number of receptors on the regenerating axons than to increase the amount of neurotrophin, because the latter treatment also makes the existing axons stronger.

10.4.3   *Influence of the Spatial Dimension of the Extracellular Space*

Van Ooyen and Willshaw (1999b) assumed that the concentration of neurotrophin is uniform across the extracellular space, so that all axons "sense" the same concentration. This is a valid assumption if all the axons are close together on the target structure, as, for example, at the endplate on muscle fibers (Balice-Gordon et al., 1993). However, if the target structure is large (e.g., a large dendritic tree), the spatial dimension of the extracelluar space should be taken into account. Modeling local release of neurotrophin along the target and diffusion of neurotrophin in the extracellular space, Van Ooyen and Willshaw (2000) showed

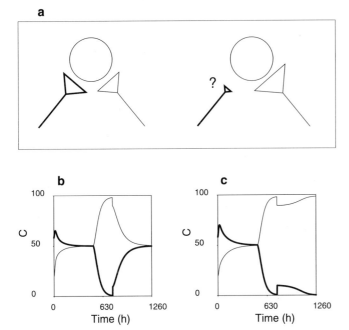

Figure 10.6

(*a*) Removal of an axon from a multiply innervated target and subsequent replacement, for class II (*b*) and class III (*c*). At $t = 504$ hr, axon 1 (bold line) is removed. At $t = 756$ hr, axon 1 is replaced (with initial conditions $\phi_1 = 30$, $R_1 = \phi_1/\gamma$, and $C_1 = 0$). Only for class II can the replaced axon survive. For class III, in order for the replaced axon to survive, a much higher initial value of ϕ_1 would be required. For notations, see figure 10.4. (Modified from Van Ooyen, 2001.)

that the distance between axons mitigates competition, so that if the axons are sufficiently far apart on the target, they can coexist (even under conditions, e.g., a class I growth function, where they cannot coexist with a uniform extracellular space; see figure 10.7). This can explain why (1) when coexisting axons are found on mature muscle cells they are physically separated (Kuffer et al., 1977; Lo and Poo, 1991) and (2) in adulthood a positive correlation exists between the size of the dendritic tree and the number of innervating axons, while in newborn animals neurons of all sizes are innervated by approximately the same number of axons (e.g., in the ciliary ganglion of rabbits; Hume and Purves, 1981; Purves, 1994).

10.4.4 Axons Responding to More than One Type of Neurotrophin

Van Ooyen and Willshaw (2000) considered a single target that releases two types of neurotrophin and at which there are two types of innervating axons (see figure 10.8). Each axon type can respond to both neurotrophin types. The following situations were examined: (1) Individual axons have only a single type of neurotrophin receptor, but this can bind to more than one type of neurotrophin. Different types of axons have different receptor types. (2) Individual axons have more than one type of neurotrophin receptor, and each receptor type binds exclusively to

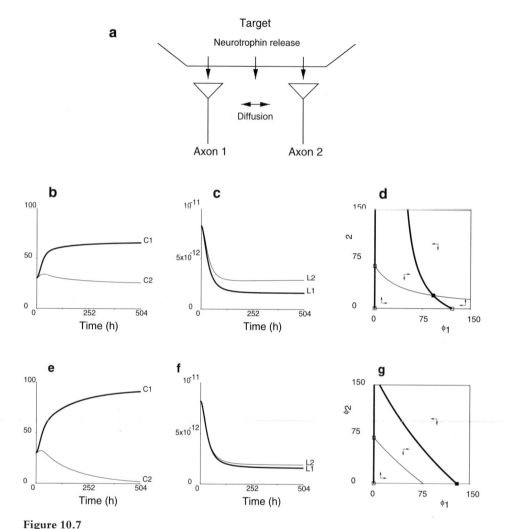

Figure 10.7

Influence of distance between axons on competition. (*a*) There is release of neurotrophin along the target and diffusion of neurotrophin in the extracellular space. Both axons have a class I growth function. (*b–d*) If the axons are relatively far apart, both survive. (*e–g*) If the two axons are close to each other, only one will survive. (*c, f*) The neurotrophin concentrations L_i (in mol l^{-1}) near the axons. (*d, g*) The null-isoclines, in which the bold lines are the null-isoclines of ϕ_1 and the thin lines those of ϕ_2. For other notations, see figure 10.4. (Modified from Van Ooyen and Willshaw, 2000.)

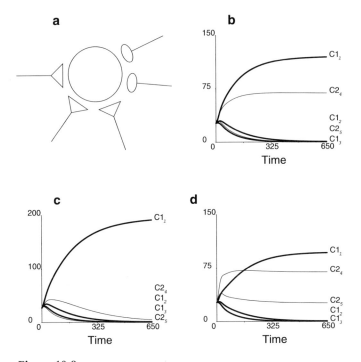

Figure 10.8

(*a*) System of five innervating axons where the target releases two types of neurotrophin, L1 and L2. Axons 1, 2, and 3 have receptor type R1 (which binds preferentially, but not exclusively, to L1), and axons 4 and 5 have receptor type R2 (which binds preferentially, but not exclusively, to L2). In (*b–d*), $C1_i$ ($C2_i$) is the total number of R1-L1 and R1-L2 (R2-L1 and R2-L2) complexes for axon *i*. Except in (*d*), all axons have a class I growth function. Time is in hours. For other notations, see figure 10.4. (*b*) When the receptor specificity is high, there is competitive exclusion within each group, but coexistence among groups. (*c*) When the receptor specificity is low, only one axon overall survives. In (*d*), the second group of axons (axons 4 and 5) has a class II growth function, the first group (axons 1, 2, and 3) class I. Axons 1, 4, and 5 survive. (Modified from Van Ooyen and Willshaw, 2000.)

one type of neurotrophin. Different types of axons have these receptor types in different proportions. The results show that for both (1) and (2), different types of axons can coexist (even under conditions, e.g., a class I growth function, where they cannot coexist with a single type of neurotrophin) if they respond to the neurotrophins with sufficiently different "affinities." For (1), this means that each type of receptor should bind preferentially, but not necessarily exclusively, to one type of neurotrophin. For (2), this means that the receptor content among different types of axons should be sufficiently different. By having axons respond with different affinities to more than one type of neurotrophin, the model can account for competitive exclusion among axons of one type while at the same time there is coexistence with axons of another type innervating the same target (figure 10.8b,d). This occurs, for example, on Purkinje cells, where climbing fibers compete with each other during development until only a single one remains, which coexists with parallel fibers innervating the same Purkinje cell (Crepel, 1982).

10.4.5   Parallels with Population Biology

In population biology, competition has been studied in many formal models (e.g., Yodzis, 1989; Grover, 1997). Parallels with axonal competition would allow results from population biology to be applied to neurobiology. Van Ooyen and Willshaw (2000) showed that the equations describing axonal competition are of the same form as those describing consumer–resource systems (Yodzis, 1989). By making quasi-steady-state approximations—on the slow time scale of ϕ_i—for R_i and C_i (i.e., $dR_i/dt = dC_i/dt = 0$), they showed that Eqs. (10.13)–(10.16) can be rewritten as

$$\frac{d\phi_i}{dt} = \phi_i[g_i(L,\phi_i) - \lambda_3] \qquad (10.18)$$

$$\frac{dL}{dt} = \sigma - \delta L - \sum_{i=1}^{n} \phi_i h_i(L), \qquad (10.19)$$

where function $g_i(L,\phi_i)$ encompasses the growth function, and function $h_i(L) \equiv \lambda_1 L/(\lambda_{2,i} + L)$ includes the kinetics of binding neurotrophin to receptors. (All λs are constants.) Note that under the quasi-steady-state approximations, $\phi_i = \rho_i C_i + \gamma_i R_i$. Thus ϕ_i is a measure of the total number of neurotrophin receptors (unoccupied plus bound to neurotrophin) on axon i. In population biological terms, ϕ_i is the size of the population of consumer species i; L is the size of the resource population; $h_i(L)$ is the functional response of the consumer, which describes how much resource is consumed per individual consumer per unit of time; and $g_i(L,\phi_i) - \lambda_3$ is the numerical response of the consumer, which describes the change in the consumer population expressed per individual per unit of time in response to (in general) both re-

source and consumer. For class I of the general growth function, $g_i(L,\phi_i) = g_i(L) = \lambda_{4,i}L/(\lambda_{2,i} + L)$.

The form in which the classical Lotka-Volterra competition equations are given, i.e., without direct reference to what the consumer species are competing for, is obtained from Eqs. (10.18) and (10.19) by making a quasi-steady-state approximation for the resource, i.e., $dL/dt = 0$. This gives an expression for L in terms of ϕ_i, which can then be inserted into Eq. (10.18). For example, for class I, if we assume for simplicity that all $\lambda_{2,i}$ are the same and δ can be neglected, we obtain

$$\frac{d\phi_i}{dt} = \phi_i\left[\frac{\lambda_{4,i}}{(\lambda_1/\sigma)\sum_{i=1}^{n}\phi_i} - \lambda_3\right]. \qquad (10.20)$$

10.5   Discussion

The model by Van Ooyen and Willshaw (1999b) links competition in the development of nerve connections with the underlying actions and biochemistry of neurotrophins. It can account for the development of single and multiple innervation, as well as for several other experimental findings, including the observation that activity is influential but not decisive in competition.

The model suggests that the regulation of axonal growth by neurotrophins is crucial to the competitive process in the development, maintenance, and regeneration of nerve connections. Among the many axonal features that can change during growth in response to neurotrophin [the degree of arborization (and consequently the number of synapses), the size of synapses, and the density of neurotrophin receptors], the consequent change in the axon's total number of neurotrophin receptors, which changes its capacity for removing neurotrophin, is what drives the competition.

Being a variable resource model (see section 10.3.2), this model has the advantage that its variables and parameters are directly interpretable in terms of the underlying biology (e.g., release, degradation, and binding of neurotrophin; insertion and turnover of receptor). This makes it also more straightforward to extend the model.

Future Modeling Studies

Axons in the model by Van Ooyen and Willshaw (1999b) have only a single target, whereas in the neuromuscular system, for example, each axon innervates a number of targets, so that there will also be competition among branches of the same axon for neurotrophin receptors (which are produced in the soma). Furthermore, the effects of activity have not yet been studied explicitly (e.g., the activity-dependent release of neurotrophin).

In general, a challenge for future modeling studies is to investigate whether explicitly implementing the putative underlying mechanisms of competition makes a difference in models in which competition is involved. For example, Harris et al. (1997) and Elliott and Shadbolt (1998b) showed that implementing the putative underlying mechanism of activity-dependent competition permits the formation of ocular dominance columns in the presence of positively correlated interocular images. Ocular dominance columns do not occur under these conditions when competition is enforced using multiplicative normalization.

Further challenges for modeling competition include (1) accounting for the observation, in the visual system as well as in the neuromuscular system, that activity is influential but may not be decisive; and (2) combining physiological plasticity (changes in synaptic strength) with anatomical plasticity (changes in axonal arborization) (as in Elliott et al., 2001).

Future Experimental Studies

Further experimental studies are necessary to find out what type(s) of competition is (are) involved in the formation of nerve connections. More types of competition may be involved at the same time, e.g., consumptive competition plus interference competition. Recent findings (see sections 10.2.2 and 10.2.3), both in the neuromuscular and in the visual system, have supported a role for neurotrophic factors in consumptive competition.

The model by van Ooyen and Willshaw (1999b), which implements consumptive competition for neurotrophins, can be tested experimentally. The model predicts that axons that are being eliminated will have a low number of neurotrophin receptors. The shape of the growth function [i.e., the dose-response curve between neurotrophin and axonal growth; see Eq. (10.17)], which determines what type of innervation can develop, can be determined experimentally in vitro by measuring, for different concentrations of neurotrophin, the axon's total number of neurotrophin receptors over all its synapses.

In assessing the role of electrical activity in competition, it is important to know exactly how activity has been changed, including postsynaptic activity (and whether decreased levels of activity increase or decrease the release of neurotrophin; see Snider and Lichtman, 1996), the absolute level of presynaptic activity, and the relative differences in activity among innervating axons. The models suggest that all these could in principle have different effects.

Finally, synapse elimination is thought to be a process distinct from "Wallerian" degeneration—a synchronous, obliterative response to nerve injury in which nerve terminals are degraded and undergo phagocytosis (e.g., Winlow and Usherwood, 1975). However, an interesting alternative paradigm with the potential to offer insights into mechanisms of synapse elimination is provided by the Wld^S mutant mouse and its transgenic derivatives (Ribchester et al., 1995; Gillingwater and Ribchester, 2001). These mice have slow Wallerian degeneration or none. Sciatic nerve axotomy in Wld^S mice induces synaptic boutons to withdraw from motor endplates in a fashion that strongly resembles synapse elimination. Very recently, the Wld^S genotype has been used to form the genetic background for thy1-CFP transgenic mice, in which motor axons and synaptic terminals endogenously express cyan fluorescent protein (Gillingwater et al., 2002). These mice, which have endogenously fluorescent synapses that are protected from Wallerian degeneration, offer many advantages that should facilitate further descriptive, experimental, and computational analyses of synapse elimination and its molecular mechanisms.

References

Albers, K. M., Wright, D. E., and Davies, B. M. (1994). Overexpression of nerve growth factor in epidermis of transgenic mice causes hypertrophy of the peripheral nervous system. *J. Neurosci.* 14: 1422–1432.

Alsina, B., Vu, T., and Cohen-Cory, S. (2001). Visualizing synapse formation in arborizing optic axons in vivo: Dynamics and modulation by BDNF. *Nat. Neurosci.* 4: 1093–1101.

Balice-Gordon, R. J., and Lichtman, J. W. (1993). In vivo observations of pre- and postsynaptic changes during the transition from multiple to single innervation at developing neuromuscular junctions. *J. Neurosci.* 13: 834–855.

Balice-Gordon, R. J., and Lichtman, J. W. (1994). Long-term synapse loss induced by focal blockade of postsynaptic receptors. *Nature* 372: 519–524.

Balice-Gordon, R. J., and Thompson, W. J. (1988). Synaptic rearrangements and alterations in motor unit properties in neonatal rat extensor digitorum longus muscle. *J. Physiol.* 398: 191–210.

Balice-Gordon, R. J., Chua, C. K., Nelson, C. C., and Lichtman, J. W. (1993). Gradual loss of synaptic cartels precedes axon withdrawal at developing neuromuscular junctions. *Neuron* 11: 801–815.

Barber, M. J., and Lichtman, J. W. (1999). Activity-driven synapse elimination leads paradoxically to domination by inactive neurons. *J. Neurosci.* 19: 9975–9985.

Barry, J. A., and Ribchester, R. R. (1994). Effects of recovery from nerve conduction block on elimination of polyneuronal innervation in partially denervated and reinnervated rat muscle. *J. Physiol* 476: 62–63.

Barry, J. A., and Ribchester, R. R. (1995). Persistent polyneuronal innervation in partially denervated rat muscle after reinnervation and recovery from prolonged nerve conduction block. *J. Neurosci.* 15: 6327–6339.

Bennett, M. R., and Robinson, J. (1989). Growth and elimination of nerve terminals at synaptic sites during polyneuronal innervation of muscle cells: Atrophic hypothesis. *Proc. Roy. Soc. London B* 235: 299–320.

Betz, W. J., Caldwell, J. H., and Ribchester, R. R. (1979). The size of motor units during post-natal development of rat lumbrical muscle. *J. Physiol.* 297: 463–478.

Beumer, K. J., Rohrbough, J., Prokop, A., and Broadie, K. (1999). A role for PS integrins in morphological growth and synaptic function at the postembryonic neuromuscular junction of *Drosophila. Development* 126: 5833–5846.

Bi, G.-Q., and Poo, M.-M. (2001). Synaptic modification by correlated activity: Hebb's postulate revisited. *Annu. Rev. Neurosci.* 24: 139–166.

Bienenstock, E. L., Cooper, L. N., and Munro, P. W. (1982). Theory for the development of neuron selectivity: Orientation specificity and binocular interaction in visual cortex. *J. Neurosci.* 2: 32–48.

Birren, S. J., Verdi, J. M., and Anderson, D. J. (1992). Membrane depolarization induces p140[trk] and NGF responsiveness, but not p75[LNGFR], in MAH cell. *Science* 257: 395–397.

Bothwell, M. (1995). Functional interactions of neurotrophins and neurotrophin receptors. *Annu. Rev. Neurosci.* 18: 223–253.

Boulanger, L., and Poo, M. (1999). Gating of BDNF-induced synaptic potentiation by cAMP. *Science* 284: 1982–1984.

Brown, M. C., and Ironton, M. R. (1978). Sprouting and regression of neuromuscular synapses in partially denervated mammalian muscles. *J. Physiol. (London)* 278: 325–348.

Brown, M. C., Jansen, J. K. S., and Van Essen, D. (1976). Polyneuronal innervation of skeletal muscle in newborn rats and its elimination during maturation. *J. Physiol. (London)* 261: 387–422.

Brown, M. C., Hopkins, W. G., and Keynes, R. J. (1982). Short- and long-term effects of paralysis on the motor innervation of two different neonatal mouse muscles. *J. Physiol.* 329: 439–450.

Cabelli, R. J., Hohn, A., and Shatz, C. J. (1995). Inhibition of ocular dominance column formation by infusion of NT-4/5 or BDNF. *Science* 267: 1662–1666.

Callaway, and Van Essen, D. C. (1989). Slowing of synapse elimination by alpha-bungarotoxin superfusion of the neonatal rabbit soleus muscle. *Dev. Biol.* 131: 356–365.

Callaway, E. M., Soha, J. M., and Van Essen, D. C. (1987). Competition favouring inactive over active motor neurons during synapse elimination. *Nature* 328: 422–426.

Cohen-Cory, S., and Fraser, S. E. (1995). Effects of brain-derived neurotrophic factor on optic axon branching and remodelling in vivo. *Nature* 378: 192–196.

Colman, H., and Lichtman, J. W. (1992). "Cartellian" competition at the neuromuscular junction. *Trends Neurosci.* 15: 197–199.

Costanzo, E. M., Barry, J. A., and Ribchester, R. R. (1999). Co-regulation of synaptic efficacy at stable polyneuronally innervated neuromuscular junctions in reinnervated rat muscle. *J. Physiol.* 521: 365–374.

Costanzo, E. M., Barry, J. A., and Ribchester, R. R. (2000). Competition at silent synapses in reinnervated skeletal muscle. *Nat. Neurosci.* 3: 694–700.

Crair, M. C., Horton, J. C., Antonini, A., and Stryker, M. P. (2001). Emergence of ocular dominance columns in cat visual cortex by 2 weeks of age. *J. Comp. Neurol.* 430: 235–249.

Crepel, F. (1982). Regression of functional synapses in the immature mammalian cerebellum. *Trends Neurosci.* 5: 266–269.

Crowley, J. C., and Katz, L. C. (1999). Development of ocular dominance columns in the absence of retinal input. *Nat. Neurosci.* 2: 1125–1130.

Crowley, J. C., and Katz, L. C. (2000). Early development of ocular dominance columns. *Science* 290: 1321–1324.

Elliott, T., and Shadbolt, N. R. (1998a). Competition for neurotrophic factors: Mathematical analysis. *Neur. Comput.* 10: 1939–1981.

Elliott, T., and Shadbolt, N. R. (1998b). Competition for neurotrophic factors: Ocular dominance columns. *J. Neurosci.* 18: 5850–5858.

Elliott, T., Maddison, A. C., and Shadbolt, N. R. (2001). Competitive anatomical and physiological plasticity: A neurotrophic bridge. *Biol. Cybern.* 84: 13–22.

English, A. W., and Schwartz, G. (1995). Both basic fibroblast growth factor and ciliary neurotrophic factor promote the retention of polyneuronal innervation of developing skeletal muscle fibers. *Dev. Biol.* 169: 57–64.

Feng, G., Mellor, R. H., Bernstein, M., Keller-Peck, C., Nguyen, Q. T., Wallace, M., Nerbonne, J. M., Lichtman, J. W., and Sanes, J. R. (2000). Imaging neuronal subsets in transgenic mice expressing multiple spectral variants of GFP. *Neuron* 28: 41–51.

Fladby, T., and Jansen, J. K. S. (1987). Postnatal loss of synaptic terminals in the partially denervated mouse soleus muscle. *Acta Physiol. Scand.* 129: 239–246.

Funakoshi, H., Belluardo, N., Arenas, E., Yamamoto, Y., Casabona, A., Persson, H., and Ibáñez, C. F. (1995). Muscle-derived neurotrophin-4 as an activity-dependent trophic signal for adult motor neurons. *Science* 268: 1495–1499.

Garofalo, L., Ribeiro-da-Silva, A., and Cuello, C. (1992). Nerve growth factor-induced synaptogenesis and hypertrophy of cortical cholinergic terminals. *Proc. Natl. Acad. Sci. U.S.A.* 89: 2639–2643.

Gillingwater, T. H., and Ribchester, R. R. (2001). Compartmental neurodegeneration and synaptic plasticity in the Wld(s) mutant mouse. *J. Physiol.* 534: 627–639.

Gillingwater, T. H., Thomson, D., Mack, T. G., Soffin, E. M., Mattison, R. J., Coleman, M. P., and Ribchester, R. R. (2002). Age-dependent synapse withdrawal at axotomised neuromuscular junctions in Wld(s) mutant and Ube4b/ Nmnat transgenic mice. *J. Physiol.* 543: 739–755.

Grover, J. P. (1997). *Resource Competition.* London: Chapman and Hall.

Hantai, D., Rao, J. S., and Festoff, B. W. (1988). Serine proteases and serpins: Their possible roles in the motor system. *Rev. Neurol. (Paris)* 144: 680–687.

Harris, A. E., Ermentrout, G. B., and Small, S. L. (1997). A model of ocular dominance column development by competition for trophic factor. *Proc. Natl. Acad. Sci. U.S.A.* 94: 9944–9949.

Harris, A. E., Ermentrout, G. B., and Small, S. L. (2000). A model of ocular dominance column development by competition for trophic factor: Effects of excess trophic factor with monocular deprivation and effects of antagonist of trophic factor. *J. Comput. Neurosci.* 8: 227–250.

Henneman, E. (1985). The size-principle: A deterministic output emerges from a set of probabilistic connections. *J. Exp. Biol.* 115: 105–112.

Holtzman, D. M., Li, Y., Parada, L. F., Kinsman, S., Chen, C.-K., Valletta, J. S., Zhou, J., Long, J. B., and Mobley, W. C. (1992). p140[trk] mRNA marks NGF-responsive forebrain neurons: Evidence that *trk* gene expression is induced by NGF. *Neuron* 9: 465–478.

Horton, J. C., and Hocking, D. R. (1996). An adult-like pattern of ocular dominance columns in striate cortex of newborn monkeys prior to visual experience. *J. Neurosci.* 16: 1791–1807.

Huisman, J. (1997). "The Struggle for Light." PhD thesis, University of Groningen, Netherlands.

Hume, R. I., and Purves, D. (1981). Geometry of neonatal neurones and the regulation of synapse elimination. *Nature* 293: 469–471.

Jansen, J. K. S., and Fladby, T. (1990). The perinatal reorganization of the innervation of skeletal muscle in mammals. *Prog. Neurobiol.* 34: 39–90.

Jeanprêtre, N., Clarke, P. G. H., and Gabriel, J.-P. (1996). Competitive exclusion between axons dependent on a single trophic substance: A mathematical analysis. *Math. Biosci.* 133: 23–54.

Jordan, C. L. (1996). Ciliary neurotrophic factor may act in target musculature to regulate developmental synapse elimination. *Dev. Neurosci.* 18: 185–198.

Joseph, S. R. H., and Willshaw, D. J. (1996). The role of activity in synaptic competition at the neuromuscular junction. *Adv Neural Infor. Proc. Syst.* 8: 96–102.

Joseph, S. R. H., Steuber, V., and Willshaw, D. J. (1997). The dual role of calcium in synaptic plasticity of the motor endplate. In *Computational Neuroscience—Trends in Research 1997,* J. Bower, ed. pp. 7–12. New York: Plenum.

Kashani, A. H., Chen, B. M., and Grinnell, A. D. (2001). Hypertonic enhancement of transmitter release from frog motor nerve terminals: Ca^{2+} independence and role of integrins. *J. Physiol.* 530: 243–252.

Keller-Peck, C. R., Feng, G., Sanes, J. R., Yan, Q., Lichtman, J. W., and Snider, W. D. (2001a). Glial cell line-derived neurotrophic factor administration in postnatal life results in motor unit enlargement and continuous synaptic remodeling at the neuromuscular junction. *J. Neurosci.* 21: 6136–6146.

Keller-Peck, C. R., Walsh, M. K., Gan, W. B., Feng, G., Sanes, J. R., and Lichtman, J. W. (2001b). Asynchronous synapse elimination in neonatal motor units: Studies using GFP transgenic mice. *Neuron* 31: 381–394.

Kuffer, D., Thompson, W., and Jansen, J. K. S. (1977). The elimination of synapses in multiply-innervated skeletal muscle fibers of the rat: Dependence on distance between endplates. *Brain Res.* 138: 353–358.

Kwon, Y. W., and Gurney, M. E. (1996). Brain-derived neurotrophic factor transiently stabilizes silent synapses on developing neuromuscular junctions. *J. Neurobiol.* 29: 503–516.

LeVay, S., Wiesel, T. N., and Hubel, D. H. (1980). The development of ocular dominance columns in normal and visually deprived monkeys. *J. Comput. Neurol.* 191: 1–51.

Lewin, G. R., and Barde, Y.-A. (1996). Physiology of the neurotrophins. *Annu. Rev. Neurosci.* 19: 289–317.

Lohof, A. M., Ip, N. Y., and Poo, M. M. (1993). Potentiation of developing neuromuscular synapses by the neurotrophins NT-3 and BDNF. *Nature* 363: 350–353.

Lohof, A. M., Delhaye-Bouchaud, N., and Mariani, J. (1996). Synapse elimination in the central nervous system: Functional significance and cellular mechanisms. *Rev. Neurosci.* 7: 85–101.

Lo, Y.-J., and Poo, M.-M. (1991). Activity-dependent synaptic competition in vitro: Heterosynaptic suppression of developing synapses. *Science* 254: 1019–1022.

Magchielse, T., and Meeter, E. (1986). The effect of neuronal activity on the competitive elimination of neuromuscular junctions in tissue culture. *Dev. Brain Res.* 25: 211–220.

Miller, K. D. (1996). Synaptic economics: Competition and cooperation in correlation-based synaptic competition. *Neuron* 17: 371–374.

Miller, K. D., and MacKay, D. J. C. (1994). The role of constraints in Hebbian learning. *Neur. Comput.* 6: 100–126.

Miller, K. D., Keller, J. B., and Stryker, M. P. (1989). Ocular dominance column development: Analysis and simulation. *Science* 245: 605–615.

Nelson, P. G., Fields, R. D., Yu, C., and Liu, Y. (1993). Synapse elimination from the mouse neuromuscular junction in vitro: A non-Hebbian activity-dependent process. *J. Neurobiol.* 24: 1517–1530.

Nguyen, Q. T., and Lichtman, J. W. (1996). Mechanism of synapse disassembly at the developing neuromuscular junction. *Curr. Opin. Neurobiol.* 6: 104–112.

Nguyen, Q. T., Parsadanian, A. S., Snider, W. D., and Lichtman, J. W. (1998). Hyperinnervation of neuromuscular junctions caused by GDNF overexpression in muscle. *Science* 279: 1725–1729.

O'Brien, R. A. D., Östberg, A. J. C., and Vrbová, G. (1978). Observations on the elimination of polyneuronal innervation in developing mammalian skeletal muscles. *J. Physiol. (London)* 282: 571–582.

Poo, M. M. (2001). Neurotrophins as synaptic modulators. *Nat. Rev. Neurosci.* 2: 24–32.

Purves, D. (1988). *Body and Brain: A Trophic Theory of Neural Connections.* Cambridge, Mass.: Harvard University Press.

Purves, D. (1994). *Neural Activity and the Growth of the Brain.* Cambridge: Cambridge University Press.

Purves, D., and Lichtman, J. W. (1980). Elimination of synapses in the developing nervous system. *Science* 210: 153–157.

Purves, D., and Lichtman, J. W. (1985). *Principles of Neural Development.* Sunderland, Mass.: Sinauer.

Ramakers, G. J. A., Winter, J., Hoogland, T. M., Lequin, M. B., Van Hulten, P., Van Pelt, J., and Pool, C. W. (1998). Depolarization stimulates lamellipodia formation and axonal but not dendritic branching in cultured rat cerebral cortex neurons. *Dev. Brain Res.* 108: 205–216.

Rasmussen, C. E., and Willshaw, D. J. (1993). Presynaptic and postsynaptic competition in models for the development of neuromuscular connections. *Biol. Cybern.* 68: 409–419.

Ribchester, R. R. (1988). Activity-dependent and -independent synaptic interactions during reinnervation of partially denervated rat muscle. *J. Physiol. (London)* 401: 53–75.

Ribchester, R. R. (1992). Cartels, competition and activity-dependent synapse elimination. *Trends Neurosci.* 15: 389.

Ribchester, R. R. (1993). Co-existence and elimination of convergent motor nerve terminals in reinnervated and paralysed adult rat skeletal muscle. *J. Physiol.* 466: 421–441.

Ribchester, R. R. (2001). Development and plasticity of neuromuscular connections. In *Brain and Behaviour in Human Neural Development*, A. F. Kalverboer and A. Gramsbergen, eds. Dordrecht: Kluwer Academic Press, pp. 261–341.

Ribchester, R. R., and Barry, J. A. (1994). Spatial versus consumptive competition at polyneuronally innervated neuromuscular junctions. *Exp. Physiol.* 79: 465–494.

Ribchester, R. R., and Taxt, T. (1984). Repression of inactive motor-nerve terminals in partially denervated rat muscle after regeneration of active motor axons. *J. Physiol. (London)* 347: 497–511.

Ribchester, R. R., Tsao, J. W., Barry, J. A., Asgari-Jirhandeh, N., Perry, V. H., and Brown, M. C. (1995). Persistence of neuromuscular junctions after axotomy in mice with slow Wallerian degeneration (C57BL/WldS). *Eur. J. Neurosci.* 7: 1641–1650.

Ribchester, R. R., Thomson, D., Haddow, L. J., and Ush-karyov, Y. A. (1998). Enhancement of spontaneous transmitter release at neonatal mouse neuromuscular junctions by the glial cell line-derived neurotrophic factor (GDNF). *J. Physiol.* 512: 635–641.

Ridge, R. M. A. P., and Betz, W. J. (1984). The effect of selective, chronic stimulation on motor unit size in developing rat muscle. *J. Neurosci.* 4: 2614–2620.

Salin, T., Mudo, G., Jiang, X. H., Timmusk, T., Metsis, M., and Belluardo, N. (1995). Up-regulation of *trkB* mRNA expression in the rat striatum after seizures. *Neurosci. Lett.* 194: 181–184.

Sanes, J. R., and Lichtman, J. W. (1999). Development of the vertebrate neuromuscular junction. *Annu. Rev. Neurosci.* 22: 389–442.

Sanes, J. R., Apel, E. D., Burgess, R. W., Emerson, R. B., Feng, G., Gautam, M., Glass, D., Grady, R. M., Krejci, E., Lichtman, J. W., Lu, J. T., Massoulie, J., Miner, J. H., Moscoso, L. M., Nguyen, Q., Nichol, M., Noakes, P. G., Patton, B. L., Son, Y. J., Yancopoulos, G. D., and Zhou, H. (1998). Development of the neuromuscular junction: Genetic analysis in mice. *J. Physiol. (Paris)* 92: 167–172.

Shatz, C. J., and Stryker, M. P. (1978). Ocular dominance columns in layer IV of the cat's visual cortex and the effects of monocular deprivation. *J. Physiol. (London)* 281: 267–283.

Snider, W. D., and Lichtman, J. W. (1996). Are neurotrophins synaptotrophins? *Mol. Cell. Neurosci.* 7: 433–442.

Song, S., Miller, K. D., and Abbott, L. F. (2000). Competitive Hebbian learning through spike-timing dependent synaptic plasticity. *Nat. Neurosci.* 3: 919–926.

Stoop, R., and Poo, M. M. (1996). Synaptic modulation by neurotrophic factors: Differential and synergistic effects of brain-derived neurotrophic factor and ciliary neurotrophic factor. *J. Neurosci.* 16: 3256–3264.

Stryker, M. P., and Harris, W. A. (1986). Binocular impulse blockade prevents the formation of ocular dominance columns in cat visual cortex. *J. Neurosci.* 6: 2117–2133.

Taxt, T. (1983). Local and systemic effects of tetrodotoxin on the formation and elimination of synapses in reinnervated adult rat muscle. *J. Physiol.* 340: 175–194.

Thompson, W. J. (1983). Synapse elimination in neonatal rat muscle is sensitive to pattern of muscle use. *Nature* 302: 614–616.

Thompson, W. J., and Jansen, J. K. S. (1977). The extent of sprouting of remaining motor units in partly denervated immature and adult rat soleus muscle. *Neuroscience* 4: 523–535.

Thompson, W. J., Kuffler, D. P., and Jansen, J. K. S. (1979). The effect of prolonged reversible block of nerve impulses on the elimination of polyneuronal innervation of newborn rat skeletal muscle fibers. *Neuroscience* 4: 271–281.

Turrigiano, G. G., Leslie, K. R., Desai, N. S., Rutherford, L. C., and Nelson, S. B. (1998). Activity-dependent scaling of quantal amplitude in neocortical neurons. *Nature* 391: 892–896.

Van Essen, D. C. (1997). A tension-based theory of morphogenesis and compact wiring in the central nervous system. *Nature* 385: 313–318.

Van Essen, D. C., Gordon, H., Soha, J. M., and Fraser, S. E. (1990). Synaptic dynamics at the neuromuscular junction: Mechanisms and models. *J. Neurobiol.* 21: 223–249.

Van Ooyen, A. (2001). Competition in the development of nerve connections: A review of models. *Network: Comput. Neural Syst.* 12: R1–R47.

Van Ooyen, A., and Willshaw, D. J. (1999a). Poly- and mononeuronal innervation in a model for the development of neuromuscular connections. *J. Theor. Biol.* 196: 495–511.

Van Ooyen, A., and Willshaw, D. J. (1999b). Competition for neurotrophic factor in the development of nerve connections. *Proc. Roy. Soc. London B.* 266: 883–892.

Van Ooyen, A., and Willshaw, D. J. (2000). Development of nerve connections under the control of neurotrophic factors: Parallels with consumer-resource systems in population biology. *J. Theor. Biol.* 206: 195–210.

Vyskocil, F., and Vrbova, G. (1993). Nonquantal release of acetylcholine affects polyneuronal innervation on developing rat muscle-fibers. *Eur. J. Neurosci.* 5: 1677–1683.

Walsh, M. K., and Lichtman, J. W. (2003). In vivo timelapse imaging of synaptic takeover associated with naturally occurring synapse elimination. *Neuron* 37: 67–73.

Wickelgren, I. (2000). Neurobiology. Heretical view of visual development. *Science* 290: 1271–1273.

Willshaw, D. J. (1981). The establishment and the subsequent elimination of polyneuronal innervation of developing muscle: Theoretical considerations. *Proc. Roy. Soc. London B* 212: 233–252.

Willshaw, D. J., and Von der Malsburg, Ch. (1976). How patterned neural connections can be set up by self-organisation. *Proc. Roy. Soc. London B* 194: 431–445.

Winlow, W., and Usherwood, P. N. (1975). Ultrastructural studies of normal and degenerating mouse neuromuscular junctions. *J. Neurocytol.* 4: 377–394.

Yan, H. Q., Mazow, M. L., and Dafny, N. (1996). NGF prevents the changes induced by monocular deprivation during the critical period in rats. *Brain Res.* 706: 318–322.

Yodzis, P. (1989). *Introduction to Theoretical Ecology.* New York: Harper & Row.

Zhu, P. H., and Vrbova, G. (1992). The role of Ca^{2+} in the elimination of polyneuronal innervation of rat soleus muscle-fibers. *Eur. J. Neurosci.* 4: 433–437.

Zoubine, M. N., Ma, J. Y., Smirnova, I. V., Citron, B. A., and Festoff, B. W. (1996). A molecular mechanism for synapse elimination: Novel inhibition of locally generated thrombin delays synapse loss in neonatal mouse muscle. *Dev. Biol.* 179: 447–457.

Models for Topographic Map Formation

David Willshaw and David Price

11

The successful functioning of the nervous system relies to a large degree on the existence of precise patterns of connectivity within and between populations of nerve cells. One fundamental question for neurobiology is how such patterns of connections are formed. In particular, how are connections made so as to form a geographic map of one structure onto another, as found in all vertebrate visual systems? This chapter has three parts. (1) We review experimental and theoretical methodologies that address the question of how maps of connections are formed in the nervous system. (2) We review the main hypotheses for neural map making that have been put forward. (3) We then present the major experimental evidence for map making and the computational models that have emerged. We outline the challenges posed by new experimental evidence relating to the underlying molecular biology of the formation of nerve connections emerging from the recent results of genetic manipulations.

11.1 Neurobiological Maps

In neurobiology, the existence of a "map" is taken to refer to the fact that the response properties of a population of nerve cells become distributed across a second population of nerve cells, by virtue of the connections made with these cells by the first population. One example of a neural map is the projection of the vertebrate retina onto the contralateral optic tectum in nonmammalian vertebrates (Gaze, 1958), to which the retina projects directly. The projection is of

such precision that stimulation of each small region of the retina produces activity in a small area of the tectum, the projections of these areas onto the tectum being arranged to form a geographic map of the retina across the tectal surface. Another example is that individual cells in the binocularly innervated mammalian visual cortex, which receives projections from both eyes via the lateral geniculate nucleus, respond to activation of the ganglion cells in just one eye; the pattern of eye preference thus formed over the cortex resembles a pattern of zebra stripes (LeVay et al., 1975; see also chapter 12).

The use of the term "map" in the neurobiological context should be compared with its use in the more general mathematical sense where it describes the association (mapping) of elements of one set with those in a second set; ordered topographic mappings are just one particular type of mathematical map, where the sets of elements have a geometric structure.

Both anatomical and physiological means have been used to investigate the nature of the map between two neural structures. To investigate point-to-point mappings, the oldest method involves the production of a local lesion in one set of cells. The localized structural changes that result within its target population are then taken as evidence for a link between the two populations of cells. More recently, the application of axonal tracer molecules has enabled the path of axons to be followed from cell body to axonal destination (Mesulam, 1982; Cook and Rankin, 1984). The electrophysiological approach is to stimulate cells in one small part of a structure and to record the effects of the stimulation in the target structure. The stimulation can be direct, via local injection of current, or indirect. For example, presentation of a local stimulus within the visual field of an animal will cause stimulation of that part of the retina onto which the visual field is projected according to the principles of geometrical optics. By repeating this procedure and stimulating different parts of the retina, a map of the retina onto its target structure can be built up (Gaze, 1958). The electrophysiological approach is potentially more versatile because it allows the construction of maps of features that have more complex properties than simple area-to-area relationships.

In this chapter we restrict ourselves to the problem of how the geometric relationships within one set of cells are transferred to a second set of cells so that a geographic or topographic map of the first structure is formed over the second structure. There are several well-characterized examples. In the somatosensory system in rodents, the two-dimensional array of whiskers on the snout projects through midbrain nuclei to form a somatotopic map in the somatosensory cortex (Rose and Mountcastle, 1959), called the barrel field (because the cellular arrangement in this part of the cortex resembles barrels). This ordered projection is indirect, as is the ordered projection of the retina onto the visual cortex in mammals.

In contrast, in amphibians and fish, the retinotopic projection of the retina onto the optic tectum, the main visual center, is direct. Early in development, the axons of the retinal ganglion cells in each retina travel along the optic nerve, cross the nerve from the other eye at the optic chiasma, travel up the wall of the diencephalon, and arrive at the optic tectum, where they are distributed across the surface of the tectum so as to give rise to a map of the entire retinal surface over the entire surface of the optic tectum (Gaze, 1958).

The problem that the axons have to "solve" can be formulated as: What are the signals that guide them to their site of termination? The relative simplicity of the retinotectal system, its accessibility during development, and the relative ease with which experiments can be carried out to test particular hypotheses of map making have made it a favorite system for computational neuroscientists. Most of the mathematical and computer models that have been developed for the formation of ordered nerve connections relate to the retinotectal system in lower vertebrates. As a result, this review is focused on this particular neurobiological system.

11.2 Strategies for Model Building in Developmental Neurobiology

In this chapter, most of the discussion is about particular models of specific developmental processes. The term "model" should be distinguished from a theory or hypothesis, which in this context describes a general idea, or collection of ideas, about a developmental process. A model is a specific realization of a theory or hypothesis. Models can be informal or formal. Informal models are those described, for example, in pictorial or verbal language (see also the introduction to this volume). We concentrate here on formal models, which are those described in a mathematical formalism, the properties of which in most cases are examined by establishing the underlying mathematical relationships analytically or, more usually, numerically, by using a computer.

The use of formal models to test hypotheses relating to specific biological questions is an activity that is now common to many areas of biology, including developmental biology and neurobiology. The models considered here are formulated as sets of equations representing the actions of the cellular or subcellular elements and their interactions in the biological system under consideration. Solution of the equations, either analytically or by computer simulation, specifies how, according to the model, the systems under consideration will behave under the given conditions. The modeling work that we describe should be distinguished from modeling of a purely mathematical nature, where it is attempted to describe in quantitative form how the key measurable attributes of a system depend on other parameters but without requiring the proposed mathematical relationships to have any interpretation in the underlying biology.

Models are usually constructed in one of two ways (Sejnowski et al., 1988): top-down modeling is concerned with constructing a model system containing the machinery that enables it to carry out a specific computation or have specific properties; bottom-up modeling is concerned with investigating the properties of a model system that arise from the assumed interactions among its elements. Most models for the development of connections are top-down because the developmental problem to be solved is well specified.

11.2.1 Pattern Formation

The functioning of the nervous system depends critically on the roles adopted by specific cells and their relationships with other cells. The basic question addressed in all modeling enterprises in developmental neurobiology focuses on how the individual members of a set of cells acquire differences from one another that enable them (1) early in development to adopt different developmental paths and differentiate into different structures (see chapters 1–4); and (2) later in development, to make different patterns of connections. Both stages of development can be said to involve pattern formation (Wolpert, 1969; Slack, 1991). Similar theoretical concepts have been used for the two stages, although they involve different types of patterns. In the first case the pattern is a property of the cells themselves and in the second case the pattern is in the relation between nerve cells.

As applied to the development of nerve connections in the retinotectal system, all models concentrate on the source of the information acquired by individual cells that enables different cells to act differently. The key questions concern:

- Acquisition of the information
  - From the genes or from the environment?
  - Acquired by retinal cells, tectal cells, or both?
- Signaling of the information to the cells
  - Molecular or electrical signaling?
  - Diffusion or active transport?

There are many levels at which computational neuroscience models can be formulated, such as the system, cellular, synaptic, or subsynaptic levels (Sejnowski et al., 1988). Given that we are discussing models for the formation of nerve connections, it is natural to concentrate on the synaptic level. At this level, different models may be expressed in different amounts of detail. Two examples of models that are expressed in different amounts of detail are (1) a set of instructions (often expressed anthropomorphically) for how axons find their tectal partners, such as "find the tectal cell with the label that is identical to your label"; and (2) a set of mathematical equations calculating the net force that acts on an axonal growth cone as calculated from the repulsive influences of its neighbors in the optic pathway.

The first, crude, retinotectal maps were constructed from the results of axon degeneration studies (Attardi and Sperry, 1963), but the first maps created with any precision were constructed by extracellular recording from the optic tectum of goldfish and of the anura *Rana pipiens* and *Xenopus laevis* (Gaze, 1970; Sharma, 1972). In *Xenopus*, at least fifty distinguishable recording positions are arranged in topographic order (Gaze,

1970). The other important attribute of such maps is that they always have a specific orientation; all retinotectal maps are arranged so that nasal field (and therefore, by camera inversion, temporal retina) projects to rostral tectum and dorsal field (ventral retina) to medial tectum (figure 11.1).

Unlike the situation in the mammalian visual system, the optic nerve in amphibians and fish can regenerate following surgical damage. Given that originally experiments were more easily carried out on adults than on neonates, most of the early data were obtained from studies on the regeneration of connections. The first ideas about the formation of nerve connections were derived from Langley (1895), whose data were obtained from information on the regeneration of his own peripheral nerves. As a consequence, most of the theories and most of the models for the development of ordered nerve connections are focused on regeneration studies, meaning that in almost all models, features related to the changing morphology of the system during development are not included.

11.3   Overview of Hypotheses for Map Formation

We now describe the main classes of hypotheses that have been advanced for the formation of ordered maps of connections. Despite the fact that most of the main hypotheses were developed between the 1940s and the 1970s, they are still relevant today and no new type of hypothesis has been advanced since then; most "new" hypotheses are combinations of older ones. These hypotheses can be divided into three groups involving (1) specific cellular properties, (2) specific relationships between cells, or (3) external factors.

As a prelude to the discussion of the three major classes of hypothesis, the *retrograde modulation hypothesis*

formulated by Weiss (1937a,b) is mentioned. In this hypothesis, it is assumed that growing nerve fibers make contact with their target at random. Different retinal cells send out electrical signals of different types, with each tectal cell tuned to respond to the signal that is characteristic of a different retinal location. In this way, specificity between individual retinal and tectal cells is established. However, since connections are assumed to be made at random, there is no anatomical map, and so this hypothesis does not address the fundamental question of how such maps are formed. This hypothesis is mentioned mainly for its historical relevance because it had an important influence on the first major class of hypothesis.

1. *Cellular properties—chemoaffinity*. The notion of chemoaffinity is associated with the name of Sperry (1943, 1963), a student of Weiss, who formulated the concept as a reaction to the retrograde modulation hypothesis. On the basis of degeneration studies carried out on normal adult goldfish, he proposed that there are preexisting sets of biochemical markers that label both retinal and tectal cells and that the ordered pattern of connections observed during development is generated by the matching together of cells with the matching labels (figure 11.2A).

2. *Intercellular relationships—neighbor matching*. The idea of this hypothesis is that there is a mechanism that forces the retinal axons from cells of neighboring origins to innervate neighboring tectal cells (Lettvin, cited in Chung, 1974). This mechanism has to be supplemented by a separate mechanism specifying the orientation of the map, which cannot be supplied by a mechanism working on intercellular relations only. Two ways of supplying neighborhood information have been suggested.

• *Electrical activity*. This derives from the idea that cells that are close together in the retina have correlated firing patterns. It is assumed that neighboring tectal

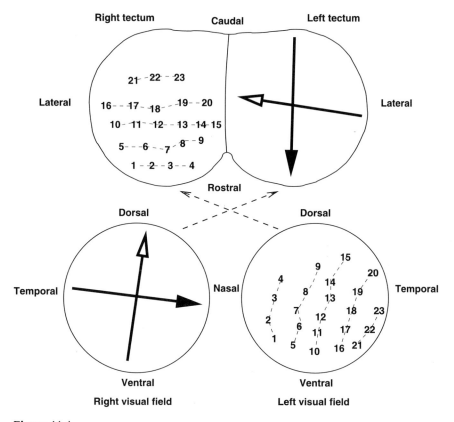

Figure 11.1
The ordered projection of a visual field onto the contralateral tectum in an adult *Xenopus laevis*. Each of the twenty-three numbered regions in the left visual field denotes the field position where a small light spot elicits maximal response at the correspondingly labeled recording electrode on the right optic tectum. As an aid to demonstrating the orderliness of the projection, electrode recording positions in the same straight line running from lateral and medial tectum have been joined with dotted lines and so have the corresponding visual field positions. The map of the left visual field (and hence retina) onto the right optic tectum is ordered and in a specific orientation, with the nasal field projecting to rostral tectum and the ventral field to lateral tectum. The similar direct projection from the right visual field to the left optic tectum is shown schematically, with the two arrows indicating the polarity of the projection.

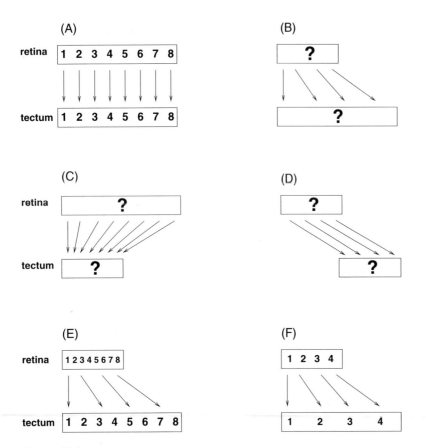

Figure 11.2
A schematic view of the projection from a one-dimensional retina to a one-dimensional tectum in normal and surgically altered conditions displaying systems matching. (*A*) The normal case of a full-sized retina projecting to a full-sized tectum. According to Sperry's hypothesis of chemoaffinity, this is achieved by each retinal ganglion cell carrying a unique identifying label (in this case, one of the numbers from 1 to 8), which enables it to contact the tectal cell with the corresponding label. (*B, C, D*) The projections formed in adult goldfish after transection of the optic nerve and surgical removal of half of the retina (*B*), half of the tectum (*C*), or half of the retina and half of the tectum (*D*). In all cases, the remaining retina reestablishes an ordered projection onto the remaining tectum. In these cases, the distribution of Sperry's postulated sets of labels across the retina and the tectum is unknown. (*E, F*) The distribution of labels across a surgically constructed half-retina and a whole tectum under two different hypotheses. (*E*) According to the hypothesis of regulation, the half-retina regenerates the entire set of labels, which allows it to innervate the entire tectum in an ordered fashion. (*F*) According to the hypothesis of induction, the retinal labels remain unchanged, but the tectal labels are altered to match those in the retina, again enabling the axons to make the appropriate connections by recognizing their tectal partner carrying the matching label.

cells have strongly correlated firing patterns and that connections between cells are reinforced according to their level of activity ("cells that fire together wire together"). If all retinal cells that are neighbors connect to tectal cells that are neighbors, a topographically ordered map of the retina onto the tectum will result (Chung, 1974; Willshaw and Von der Malsburg, 1976).

• *Molecular mechanisms—marker induction.* Properties other than those of electrical activity which have spatial distributions, such as freely moving molecules, can be exploited. It has been suggested that the tectal cells acquire the ability to make contact with selected retinal cells through signals passed to them from the retinal cells which cause retinal labels, or markers, to be induced into the tectum. This hypothesis could also be regarded as an extension of the chemoaffinity hypothesis (Von der Malsburg and Willshaw, 1977; Willshaw and Von der Malsburg, 1979).

3. *Mapping involving external properties.* Two members of this class can be identified:

• *Fiber ordering.* The idea of this hypothesis is that the fibers attain their order before contact is made with the target structure. This could be done, for example, by axons matching to labels that are placed in the pathway rather than on the tectum itself (Attardi and Sperry, 1963) (which would be a variant of the chemoaffinity hypothesis). Another suggestion is that since axons are already ordered as they leave the retina, they could simply maintain that order all the way to the optic tectum (Horder and Martin, 1979; Rager and Von Oeynhausen, 1979).

• *The timing hypothesis.* In this hypothesis, the earliest growing fibers to reach their target make connections with the earliest differentiating postsynaptic cells. This would assume a mechanism that converts positional information into temporal information (Jacobson, 1960; Gaze, 1960).

11.4 Cellular Properties—Chemoaffinity

Based on his observations that after the optic nerve in the adult newt was cut and the eye rotated, the animal's response to visual stimulation was not adapted to the eye's rotation, Sperry (1943, 1944, 1945) formulated his doctrine of chemoaffinity. According to this hypothesis, both the retinal and the tectal cells carry prespecified, distinguishing labels of a chemical origin. The making of connections involves matching each retinal cell with the tectal cell carrying the matching label. Originally the matching process was regarded as analogous to fitting keys to locks (figure 11.2A). In later papers, Sperry (1963) went on to propose an informal model for this process. Following related ideas about specification of the developing limb, he proposed that each axis of the retina is labeled by a different label in a graded fashion, so that there are as many labels as there are axes of variation. In this way, a different collection of labels is assigned to each retinal location. He assumed that similar gradients were present in the tectum (see also section 11.4.4). Each incoming axon would then find the tectal cell with the matching collections of labels to which it would make contact.

There are several general points of interest about this hypothesis. First, the process of matching retinal and tectal elements is assumed to take place entirely at the optic tectum; second, as applied to development, this hypothesis assumes a fully grown retina connecting with a fully grown tectum. We now know that both structures and the pathway between them develop as connections are made, but there was no provision for these changes in the theory. At a more conceptual level, this hypothesis places a significant burden on the genome to provide enough information to specify every label used by the developing nervous system. Finally, it is assumed that the sets of matching labels are developed independently yet perfectly in step; a slight deviation in the ideal program of growth will lead to errors in the connections made.

Much experimental work has been carried out on the development and regeneration of the ordered retinotectal projection of the retina onto the optic tectum in amphibians and fish to test Sperry's hypothesis. The main experimental issues have been (1) whether the labels that are proposed to exist are fixed or are plastic, which would be signaled by a plasticity of connections under certain circumstances; and (2) the mode of acquisition of these labels by the retina and tectum.

11.4.1   Plasticity of Labels and Connections

Sperry's original idea was that the labels are the property of the tissue itself and so could not be changed by external influences, such as those arising when his goldfish "sees the world upside down" after one of its eyes has been rotated. This idea was supported by his subsequent work showing that the optic fibers from a surgically diminished half-retina in goldfish regenerated to the appropriate half of the tectum (Attardi and Sperry, 1963). However, later experiments showed that the half-retinal projection expanded in order to cover the whole tectum (Schmidt et al., 1978). This result is an example of those obtained from the extensive series of the so-called mismatch experiments. After surgery to reduce the size of the retina or of the tectum, coupled with transection of the optic nerve, optic nerve fibers are challenged to reinnervate the optic tectum that remains (Horder, 1971; Yoon, 1972; Gaze and Keating, 1972; Schmidt et al., 1978). The fundamental result is that whatever the size of the two participating structures, an ordered map of the retina will be formed across the entire tectal surface, in the normal orientation (figures 11.2B, 11.2C, and 11.2D).

The phrase "systems matching" was coined to describe this type of connectivity pattern (Gaze and Keating, 1972). This result is found not only in the regeneration of connections as seen during the development of the visual system. In *Xenopus*, the retina and tectum are still developing as the projection between them develops. They grow in different ways, and from the very earliest stages of development at which a map can be recorded, there is an ordered projection of the retina onto the tectum (Gaze et al., 1974). This implies a continual adjustment of the pattern of connections during development.

Consider, for example, cells from near the middle of the retina, which are among the earliest retinal cells generated. Eventually they come to project to cells at the center of the optic tectum, but they cannot project there initially because the appropriate tectal cells have not yet been born. Electrophysiological mapping shows that these cells initially project to the front of the tectum and gradually move in a caudal direction to reach their final position (figure 11.3). The nonlinearities in the projection pattern seen at developmental stages reflect this changing pattern of connections during development (Gaze et al., 1974). The conclusion that retinal axons move their site of termination during development is supported by electron microscope studies demonstrating the degeneration of synapses during development. In experiments on *Rana pipiens* tadpoles, Reh and Constantine-Paton (1983) investigated how the termination sites of retinal ganglion cell terminals labeled with horseradish peroxidase (HRP) change during development. They showed that the fibers could travel for up to 1.4 mm in the rostrocaudal direction, which represents a substantial fraction of the entire extent of the adult tectum. Together with related electrophysiological

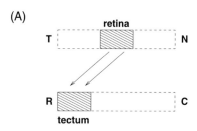

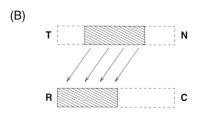

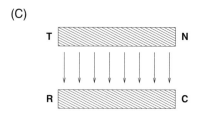

Figure 11.3
A diagram of how connections are made during the development of the retinotectal projection in *Xenopus laevis*. The diagram shows how the cells arranged along the temperonasal axis of the retina project in order across the rostrocaudal axis of the tectum at three different stages of development. (*A*) Initially, the first-born retinal cells, at the center of the retina, project to the first-born tectal cells, in rostral tectum. (*B, C*) Two stages of development, during which new retinal cells are added at the periphery, whereas new tectal cells are added to caudal tectum only. As a result, individual cells change their partners until the adult configuration is reached. The retinal and tectal cells existing at each of the three development stages are shown by the shaded regions.

and electron microscope studies (Gaze et al., 1979b), this established that the retinal ganglion cell terminals continually change their tectal partners during development.

A second set of studies on development exploits the properties of the experimentally induced compound eye projection in *Xenopus* (Gaze et al., 1963). A compound eye is made early in development by replacing a half-eye rudiment by another half-eye rudiment of different embryonic origin. The eye that develops is called a compound eye. Many different combinations of compound eye are possible, the most common ones being double nasal eyes (NN) made from two nasal half-rudiments, double temporal (TT), and double ventral (VV) compound eyes. In the adult, the eyes are of normal size and appearance (apart from some abnormalities in pigmentation). A single optic nerve develops and innervates the contralateral tectum in the normal fashion. However, the projection made by these compound eyes as assessed by extracellular recording is grossly abnormal. Each half-eye corresponding to the two half-eye rudiments that were brought together to make the compound eye projects in order across the entire optic tectum (Gaze et al., 1963) instead of being localized to just one half-tectum (figure 11.4).

If connections are formed through the matching of labels carried by the retinal axons and the tectal cells, these two sets of experimental results provide evidence that the labels are plastic. The question then arises as to the mechanisms underlying this plasticity. Some people have argued that the labels do not change but that the mapping function between them does; alternatively, that there are other mechanisms at play that can override the effects of the fixed labels. Whichever of these interpretations is correct, it is clear that the version of chemoaffinity as first presented by Sperry is not compatible with the experimental findings.

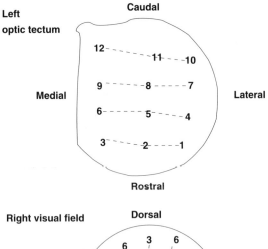

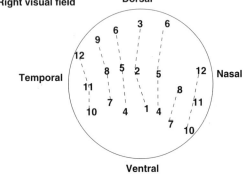

Figure 11.4
The ordered projection of the visual field onto the contralateral tectum recorded from an adult *Xenopus laevis* with one double nasal compound eye. Conventions are as in figure 11.1. This double nasal compound eye was made by replacing the temporal half of a right eye rudiment by a nasal half from a right eye rudiment. In this map each position on the tectum from which recordings were made can be stimulated by two distinct regions in the visual field, arranged symmetrically about the dorsoventral axis.

11.4.2 Specification of the Retina

If cells from different parts of the retina and tectum are labeled according to their place of origin, at some stage in development different parts of the retina (and the tectum) must acquire their own identity. The issue of how and when this occurs has been discussed in terms of how the different axes of the retina and the tectum are specified. When do the axes of the retina become polarized to enable, for example, nasal retina to develop connections with caudal tectum, temporal retina with rostral tectum, dorsal retina with lateral tectum, and so on?

Following rotation of a *Xenopus* eye rudiment prior to embryonic stage 28, the retinotectal map developed in the adult was found to be indistinguishable from a normal map. In contrast, following rotation of the eye rudiment at stage 32, the adult map was found to be ordered and the difference in map orientation compared with that in the normal case was equal to the amount by which the eye rudiment had been rotated (Jacobson, 1967; Gaze et al., 1979b). The interpretation was that the retinal axes were laid down between these two stages. Further experiments suggested that (1) the nasotemporal and dorsoventral axes of the eye are specified independently, one after another (Jacobson, 1967) (which mirrored similar inferences about the specification of the axis of the limb); and (2) the axis of one of two half-eye rudiments when put together in the same orbit to form a compound eye could undergo respecification after the time of axial polarization by the presence of the other half-eye (Hunt and Frank, 1975).

These findings were controversial. However, some of the issues have been resolved by the use of albino tissue as an indicator of retinal origin. This suggested that in some of the controversial cases the retina had not been respecified; rather, the eye rudiment that had been rotated had died and was replaced by regenerating tissue, which had the normal orientation (Gaze et al., 1979a).

A large series of experiments was carried out to assess the maps developed from *Xenopus* compound eyes made in many different combinations (Straznicky and Gaze, 1980). Examples are V_rV_r (the compound eye is assembled from two ventral halves from right-eye rudiments) and NV_rNV_l (nasoventral right half with nasoventral left half). In all cases, the orientation of each part-map was predictable from knowledge of the origin of the corresponding half making up the compound eye maps. In a series of experiments using "pie-slice" compound eyes (Willshaw et al., 1983), the pie-slice part of the map was either normal (indicating loss of the pie slice) or had an orientation predictable from knowledge of the place of origin of the pie slice.

11.4.3 Specification of the Optic Tectum

A limited amount of work has been done on assessing how the optic tectum becomes specified, using a similar paradigm of comparing the effects of rotating tectal precursors at very early stages. Chung and Cooke (1975, 1978) rotated portions of *Xenopus* pretectal tissue. In the cases where the tissue that gives rise to the diencephalon (normally immediately anterior to the tectum) had been rotated as well, the adult retinotectal map was found to be rotated. In cases where there had been no rotation of diencephalic tissue, a normal map resulted. Based on a relatively small number of experiments, the inference was that the presumptive diencephalon contains an organizer (a "beacon") that determines the polarity of tectal tissue.

This type of experiment attempting to manipulate the tectal axis has been repeated in a different form at the genetic level. Itasaki and Nakamura (1992) transplanted mesencepalic alar plate into the diencephalon of chick embryos. They found that insertion of this tissue reverses the normal rostrocaudal gradient of expression of the gene *engrailed* in the optic tectum. *Engrailed* is a homeobox gene, originally discovered in *Drosophila* (Morata and Lawrence, 1975; Kornberg, 1981) and now known to have homologs in many vertebrates (Patel et al., 1989). It has a gradient of expression along the rostrocaudal axis of the chick optic tectum, with a high level in caudal tectum, to which nasal fibers normally project, and a low level in rostral tectum, which is innervated by temporal fibers (Itasaki and Nakamura, 1996). It may control the ligands of the Eph class of receptors (see later discussion).

The effect of reversing the rostrocaudal axis of expression of *engrailed* is that nasal fibers come to innervate rostral tectum; conversely, temporal fibers innervate caudal tectum rather than rostral tectum. Using a retroviral gene transfer technique to produce animals with high levels of *engrailed* expression at specific regions throughout chick tectum, Friedman and O'Leary (1996) and Itasaki and Nakamura (1996) found that nasal fibers preferentially innervate the areas of high expression of *engrailed*, whereas temporal fibers do not. This suggests that there is a causal link between expression of *engrailed* and the establishment of retinotectal maps.

The use of large-scale screening of zebrafish mutants has found a number of genes affecting the mapping of the retina onto the contralateral tectum. Mutations in some genes are related to abnormal mapping of connections along the dorsoventral axis and in other genes to similar abnormalities along the anterior-posterior (rostrocaudal) axis. In some cases the arrangement of optic fibers within the optic pathway is normal; in other cases it is abnormal. These results may provide evidence for genes that control the establishment of gradients of guidance cues arranged along these axes (Trowe et al., 1996). However, the precision of the retinotectal projection in these studies is too low to allow any firm conclusions.

11.4.4 *Evidence for Gradients of Molecules*

At the time when the idea of chemospecificity was proposed, the molecular labels required by the theory were hypothetical objects and they were justified on the basis of necessity rather than on direct experimental evidence. Since the 1970s, many molecules have been considered as candidates for Sperry's postulated chemical labels. Initial work was carried out in vitro. It was investigated whether specificity involves preferential adhesions between retinal cells and tectal cells. Cells prepared from central chick neural retina were found to adhere preferentially to medial tectum, and cells from dorsal retina to lateral tectum (Roth and Marchase, 1976; Gottlieb et al., 1976). Similar results were obtained using axonal tips instead of entire neurons (Halfter et al., 1981). More recent experiments have tested the growth responses of retinal axons.

Walter et al. (1987) allowed retinal cells to extend axons onto a membrane made up of tectal strips derived alternatively from rostral and caudal tectum. Ganglion cells from nasal retina were found to innervate each type of strip equally. In contrast, cells from temporal retina extended axons onto strips of rostral origin only. The effect seems to be one involving repulsion of cells of temporal origin by caudal cells. This was shown directly in experiments where after tectal cell membranes were contacted by a growth cone of temporal origin, the growth cone filopodia withdrew, leading to the collapse and retraction of the growth cone (Cox and Bonhoeffer, 1990). This notion of repulsion has become central to modern ideas of how molecules interact to form ordered patterns of nerve connections.

The evidence cited here is suggestive of the existence of molecular labels, but does not indicate the identity of the labels. A number of candidate molecules have been proposed, such as toponymic molecule (TOP; Trisler and Schneider, 1981) and repulsive guidance molecule (RGM; Stahl et al., 1990). Recent interest has focused on the Eph receptors, the largest known subfamily of receptor tyrosine kinases, and their associated ligands, now known as ephrins (Flanagan and Vanderhaeghen, 1998; Nakamura, 2001; Wilkinson, 2001). The molecular compositions of Eph receptors have been known for some years. They have been found to be expressed in the developing and adult nervous system (Tuzi and Gullick, 1994), and the family of associated ligands, the ephrins, has recently been cloned (Flanagan and Vanderhaeghen, 1998). The ephrins, discovered in mouse superior colliculus (Cheng and Flanagan, 1994), bind to the Eph receptors found in retinal ganglion cells. These pairs of molecules seem to fulfill the requirements for the labels needed for chemospecificity. In the mouse, there is a gradient of the receptor EphA5 across the nasotemporal axis of the retina, whereas there is a complementary gradient of ephrinA2 from anterior to posterior superior colliculus. In chicks, there is a similar set of gradients in the retina and tectum. EphA3 is expressed in a decreasing gradient from the temporal to the rostral pole, and in the optic tectum ephrinA5 and ephrinA2 are arranged in a gradient increasing from rostral to caudal tectum. In all cases, the gradients in the two structures do not match, but are complementary. For example, temporal retina projects to rostral tectum, but in temporal retina the level of Eph receptor is at its highest value, whereas in rostral tectum, the level of the ephrins is lowest (figure 11.5B).

The following results from the chick suggest the involvement of the Eph receptors and ephrins in map formation:

1. These molecules are produced at the time when retinal axons travel into the tectum to make their connections.

(A) **(B)**

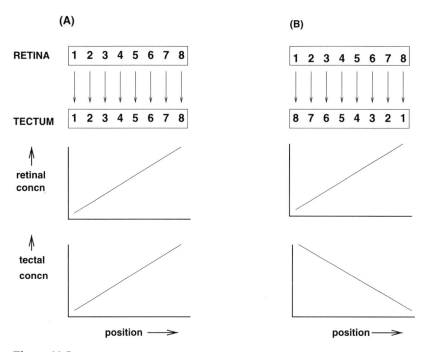

Figure 11.5
The distribution of the molecular gradients underlying the formation of nerve connections in a schematic one-dimensional retinotectal projection. (*A*) According to the hypothesis of chemospecificity, the retinal and tectal gradients are matching; i.e., axons from the high end of the retinal gradient project to the region of the tectum that is at the high end of the tectal gradient and similarly, axons from the low end of the retina project to the low end of the tectal gradient. (*B*) Conversely, in chick, mouse, and goldfish, retinal axons with a high density of Eph receptors project to the part of the tectum with a low density of ephrins and vice versa.

2. In experiments where ephrinA2 was overexpressed at the low end of its gradient:
• Incoming axons destined for the low end of the ephrinA2 gradient avoided this region.
• Axons that normally map to high ephrinA2 regions of the tectum were unperturbed by the overexpression.

From these experiments, it was inferred that connections are made in the following way (Nakamoto et al., 1996): The ligand ephrinA2 is distributed over the optic tectum, a gradient being built up from ros-

tral (low end) to caudal (high end) tectum. Axons of temporal origin normally project to rostral tectum. They have large numbers of EphA3 receptors on their surfaces, making them highly sensitive to ephrinA2. These axons enter rostrally and grow until the ever-increasing amount of ephrinA2 that they encounter forces them to stop. By contrast, axons of nasal origin normally map to the back of the tectum, where ephrinA2 levels are high. This is possible because these axons have few EphA3 receptors, rendering them nearly oblivious to the ligand, which allows them to pass through increasing concentrations of it as they cross the tectum.

11.4.5 Models Based on Chemoaffinity

Sperry's model of neuronal specificity (Sperry, 1943, 1945, 1963) has already been described, as applied to the retinotectal system of lower vertebrates. Several issues about the labels that were proposed to exist were also discussed, together with various informal models.

Gierer (1983) proposed that axons grow in the direction of the maximal slope of a growth parameter, or potential. Each axon has its own potential assigned to it, which is assumed to be a function of the coordinates of its retinal cell of origin and its current location on the tectum. The direction of growth of an axon is down the steepest slope of its potential, thereby decreasing its value. The axon will continue to change its position until its potential cannot be diminished any further. The form of the potential function is such that at this point the axon will have found its correct location in the tectum. This model can be thought of as a computationally more plausible model than that of Nakamoto et al. (1996) for how molecular gradients give rise to maps of connections (which Gierer's work predates considerably). As such, Gierer's model is a straightforward demonstration of how gradients can provide guidance signals. However, there is no provision for plasticity of connections under conditions of retinotectal mismatch.

Models Providing for Plasticity of Connections

If two sets of nerve cells interconnect to form an ordered map of connections on the basis of the labels carried by the participating cells, and if after surgical diminution of the retina or tectum the retinal cells project to tectal positions that they would not have occupied in the absence of surgery (Gaze and Keating,

1972), then surgery must have changed the labels. Various proposals have been made to explain this.

Regulation

To account for the map seen in the mismatch experiments, it was suggested that removal of cells triggers a reorganization of labels in the surgically affected structure (Meyer and Sperry, 1973). Following retinal hemiablation, for example, the set of labels initially deployed over the remaining half-retina would become rescaled so that the set of labels possessed by a normal retina would now be spread across the reduced retina (figure 11.2E). This would allow the entire reduced retina to project in order across the entire tectum. The mechanism was called regulation by analogy with similar findings in morphogenesis, where a complete structure can regenerate from a partial structure (Weiss, 1939). However, this is a post hoc explanation and lacks predictive power (Gaze and Keating, 1972).

Plasticity through Competition?

Gaze and Keating (1972) suggested that it may be possible for systems matching to occur without the necessity for changes in labels. Retinal axons could compete for space, and if each retinal axon has the same amount of "synaptic strength," the set of axons can expand or contract to fill the amount of tectal space available. Prestige and Willshaw (1975) formalized the notion of chemospecificity by distinguishing between two types of chemical matching schemes. In schemes of type I, each retinal cell has affinity for a small group of tectal cells and less for other cells (i.e., chemoaffinity). Cells that develop connections according to this scheme will make specific connections with no scope for flexibility. In schemes of type II, all axons have high affinity for making connections at one end of the tectum and progressively less for tectal cells elsewhere. Conversely, tectal cells have high affinity for axons from one pole of the retina and less from others; there is graded affinity between the two sets of cells.

Prestige and Willshaw (1975) explored models of type II where the affinities were fixed. Simulations showed that ordered maps (albeit in one dimension only) can be formed if competition is introduced by limiting the number of contacts that each cell can make, which ensures an even spread of connections; without competition, the majority of the connections would be between the retinal and tectal cells of highest affinity. In order to produce plasticity when the two systems are of different sizes, the additional assumption had to be made that the number of connections made by each cell can be altered. This is equivalent to introducing a form of regulation, even though the labels as such are not changed.

Eph/ephrin–Based Models

Nakamoto et al. (1996) proposed an informal model for topographic mapping to account for the following findings:

1. Ephrins and their receptors are arranged in countergradients; i.e., axons from the high end of the retinal gradient normally project to the cells at the low end of the tectal gradient and vice versa.

2. The interactions between Eph receptors and ephrins cause repulsion.

According to this model, all axons have an equal tendency to grow toward posterior tectum. In the final, stable state, once the mapping has been set up, this tendency is counterbalanced by an equivalent amount of negative signal. The amount of negative signal is determined by the number of receptors bound by ligand. In the simplest case, if the axon has an amount R of receptor that binds to an amount L of ligand, the strength of the negative signal would be the product $R \times L$. The same amount of signal would result from either high receptor and low ligand, low receptor and high ligand, or a medium amount of both quantities. This would account for the necessity of having countergradients.

There are several problems with this type of mechanism. First, the model specifies how order can be arranged along one dimension only and cannot be generalized to two dimensions. The fact that optic fibers grow in along the rostrocaudal axis of the tectum is exploited in the model when accounting for the ordering of connections along this axis; this fact cannot also be used for the ordering of connections along the mediolateral axis. It could be that ordering of axons along the second axis is controlled by a separate set of gradients; or that there is an entirely different mechanism for exploiting the information supplied from molecular gradients. The second problem is that this model is concerned only with how to generate matching gradients in the retina and tectum. Like all models employing fixed labels, it will not account for the variety of results from both experimental and normal situations that demonstrate plasticity of connections in the system (Gaze and Keating, 1972). Finally, the model has one of the fundamental weaknesses of Sperry's proposal itself, which is that there is a lack of information concerning how, in early development, two precise gradients could be set up and kept in step with each other during the growth of two independently developing structures.

11.5 Intercellular Relationships—Neighbor Matching

A very different hypothesis from that of chemoaffinity is that the nature of the contacts made by any given axon depends on the contacts formed by its neighbors. This suggests a different type of relation between retinal and tectal cells, in particular that there is no fixed relation between them.

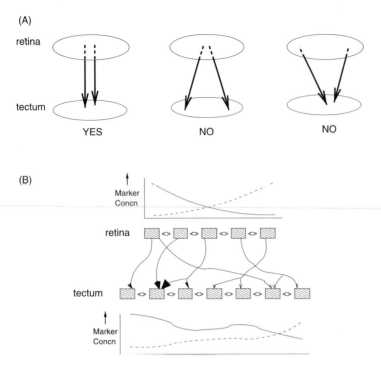

Figure 11.6

Neighbor-matching hypotheses for the formation of the retinotectal map. (*A*) A mechanism that operates (1) to connect each pair of neighboring retinal cells to neighboring tectal cells (left-hand diagram) and (2) to prevent non-neighboring retinal cells from connecting to neighboring tectal cells (center diagram) and (3) vice versa (right-hand diagram), will generate an ordered retinotectal map. (*B*) Illustration of the marker induction scheme. Retinal cells acquire markers of two distinct types in concentrations shown in the top graph. Tectal cells acquire markers from the retinal cells through the retinotectal synapses, in amounts in proportion to their synaptic strength, and through intratectal exchange. The distribution of the two markers over the tectum is shown in the bottom graph and is continually refined according to the similarity between retinal and tectal markers. This leads to a gradual refinement in the tectal marker profile and the pattern of connections until a copy of the retinal marker distribution has been reproduced in the tectum, with an ordered mapping of the retina onto the tectum.

The underlying idea is that there is a mechanism that ensures that axons with nearby cells of origin in the retina tend to make connections on nearby tectal cells (figure 11.6A). This could be the result of spontaneous activity in the retina, which is formed into small islands of highly correlated activity because of the pattern of intraretinal connectivity. Connections between these active cells and similar islands of activity in the tectum would be reinforced by an activity-dependent Hebbian mechanism (Hebb, 1949; Chung, 1974; Willshaw and Von der Malsburg, 1976). Alternatively, there could be a set of labels assigned to the retinal cells with a graded spatial distribution; cells of nearby retinal origin carry similar labels and therefore establish contact on the same part of the tectum on the basis of their carrying similar labels (Von der

Malsburg and Willshaw, 1977; Willshaw and Von der Malsburg, 1979). This distinction gives rise to the issue of the evidence for activity-dependent versus activity-independent mechanisms for map formation.

11.5.1 Activity Dependence versus Activity Independence

Various experimental results suggest that both activity-independent and activity-dependent influences are essential for the establishment of ordered nerve connections (see also chapters 10 and 12). It could be that activity-independent effects direct the nerve fibers onto the tectum and are responsible for producing an initial coarsely grained map and that the refinement of connections into the precision seen in the adult is driven by activity. One possible type of activity is correlated activity. The existence of strongly correlated spontaneous activity has been demonstrated among ganglion cells, in the adult at least (Arnett, 1978; Rodieck, 1967). This makes spontaneous activity a plausible basis for the formation of ordered maps.

Evidence for Activity-Dependent Mechanisms
Most investigations have been on the effect of neural activity on map formation in the regeneration of adult projections, and some have examined the effect of tetrodotoxin (TTX), which blocks voltage-sensitive sodium channels and therefore prevents the propagation of nerve impulses. Meyer (1983) and Schmidt and Edwards (1983) made repeated injections of TTX into the goldfish eye after the optic nerve was crushed. They found that the maps regenerated in the presence of TTX were ordered, but not with the same amount of detail expected after regeneration without application of TTX. Cook and Rankin (1986) investigated the refinement of the regenerated goldfish optic projection under conditions of abnor-

mal electrical stimulation produced by keeping the animals under stroboscopic illumination. They made a careful quantitative study in which retrograde transport of HRP from a standard tectal injection site was employed to measure the precision of the mapping. For all the retinal cells labeled, they calculated the mean-squared distance of each labeled retinal cell to its nearest neighbor and expressed this as a proportion of retinal area. In normal animals, the precision calculated in this way has a value of 1 percent. Early in regeneration, without exposure to stroboscopic illumination, labeled retinal cells are very widely dispersed, and over the next 2 or 3 months, precision is improved until the normal figure of 1 percent is reached. After 3 months of exposure to strobe light, the precision of the mapping was still very diffuse and corresponded to that seen 1 month after regeneration under normal lighting conditions.

Evidence for Activity-Independent Mechanisms
In addition to the evidence relating to chemoaffinity discussed earlier, evidence for the role of activity-independent mechanisms comes from work showing that maps of connections can develop in the absence of neural activity or where the normal activity patterns have been disturbed.

The Californian newt *Taricha torosa* manufactures TTX, to which it is insensitive. Harris (1980) grafted eye rudiments on this newt from a newt that is tetrodotoxin sensitive. This "silent" eye developed as normal and innervated the optic tectum apparently normally. In this eye, all action potentials were blocked, yet neuroanatomical studies revealed a retinotopic map. However, only a low degree of order could be demonstrated owing to the particular anatomical techniques employed. The transplanted silent eye could maintain an optic projection even in the presence of competition from the electrically active normal eye.

It may be that activity-independent mechanisms act to arrange fibers in the optic tract as they approach the tectum. Stuermer (1990) showed that the initial growth of optic fibers into the tectum of the zebra fish is not affected by blocking activity with TTX. Cook and Becker (1988) investigated the regeneration of an orderly projection from the retina to the tectum in goldfish after the optic nerve was cut. In normal animals, axons from ventral retina travel in the medial brachium of the optic tract and axons from dorsal retina in the lateral brachium. After regeneration, ventral and dorsal retina were represented in both brachia. Optic nerve fibers reached the optic tectum some 3 weeks after the optic nerve was cut. Three weeks later, the number of inappropriately directed fibers had decreased, owing probably to elimination of external collaterals rather than cell death. Cook and Becker (1988) found that this refinement was not affected by intraocular injection of TTX or by subjecting the animals to continuous stroboscopic illumination. In previous experiments this had been found to impair the refinement of the map itself, as mentioned earlier.

11.5.2   Models for Map Making Employing Neural Activity

Activity-based models originate in the model for the development of orientation specificity in the visual cortex that was proposed by Von der Malsburg (1973). He showed that when connections between two sets of cells are modified according to the amount of simultaneous pre- and postsynaptic activity present for each synapse (Hebb, 1949), retinal activity patterns in the form of straight lines of activated cells will cause the cortex to acquire orientation preferences. The pattern of preferences in the orientation map produced corresponds approximately to what is measured experimentally.

The first activity-based model for the development of retinotopy, called the neural activity model, was that of Willshaw and Von der Malsburg (1976). This was the first of two proposals that these authors made for how neighborhood mechanisms for making connections could be realized in the nervous system, the second being the marker induction model (Von der Malsburg and Willshaw, 1977).

Both models are based on the idea that ordered maps of connections will form if all pairs of neighboring retinal cells connect with tectal cells that also are neighbors (figure 11.6). The models differ in that in one model molecules are used to implement the neighborhood mechanism and in the other model neural activity is used.

In the neural activity version, Willshaw and Von der Malsburg (1976) assumed that the retina is spontaneously active and that through the action of short-range lateral interconnections, cells that are closer together are more likely to be spontaneously active concurrently than cells that are farther apart. If initially connections are made between the retina and tectum at random, spontaneously active retinal cells will come to excite cells at random positions on the tectal surface and the tectal cells that are closer together would be also more likely to fire in synchrony than cells farther apart (owing to lateral interconnections in the tectum). The synapses between retinal and tectal cells that are simultaneously active would then be strengthened, in a Hebbian fashion (Hebb, 1949). It was demonstrated by computer simulation how such a mechanism could lead to the formation of ordered neural mappings. To specify the polarity of the map, some of the retinal cells were assumed to be connected initially to the tectal cells in the right orientation (but not necessarily in the right position). The simulation results illustrated the plasticity of connections that is allowed under such conditions, which will account for the results of the size-disparity experiments.

Takeuchi and Amari (1979) and Amari (1980) carried out a one-dimensional analysis of a continuous version of the neural activity model, as applied to retinocortical connections. They showed that when the width of the input stimuli is smaller than the extent of the lateral interactions in the cortex, an ordered map results; when the width is greater, the map is ordered on a large scale, but on a small scale it breaks up into blocks. Later work showed that the area of the cortical sheet occupied by the parts of the retinal sheet that are stimulated relatively frequently during development occupy proportionally more of the cortex sheet (Amari, 1983). More recently, this analysis has been extended to the case where the cortical sheet is made up of excitatory and inhibitory cells (Da Silva Filho, 1992).

The predictions and assumptions made from the neural activity neighborhood type of model require detailed verification, much of which is lacking. There is some evidence for the assumption that the retina is spontaneously active and that the amount of correlated activity between two points is positively correlated with the distance between them (Rodieck, 1967; Arnett, 1978; Galli and Maffei, 1988; see also chapter 12). At a more general level, much careful work has been done to demonstrate the activity-dependent sharpening up of retinotectal projections in experimental paradigms (such as adult goldfish) where activity can be manipulated (see, for example, Schmidt and Edwards, 1983; Cook and Rankin, 1986; Rankin and Cook, 1986; Cook, 1987, 1988). Many researchers talk of activity-dependent "refinement" of connections, implying that this may be a secondary mechanism that refines a map that is already formed. It is possible that the primary mechanism is the one that forms the initial crude map, i.e., the mechanism for specifying the polarity of the map.

11.5.3 *Neighbor Matching Not Involving Electrical Activity*

Marker Induction

Another mechanism by which axons from cells of neighboring origin can be distinguished from axons from other cells is one that uses sets of labels (or markers) assigned to retinal cells. These markers are then induced, through the initial connections formed, into the tectal cells (figure 11.6B). Markers spread between tectal cells by a mechanism such as diffusion. Nearby tectal cells will then acquire similar markers. A mechanism that strengthens connections according to the similarity between retinal and tectal markers will cause the tectal cells to acquire markers and at the same time an ordered retinotectal map will develop (Willshaw and Von der Malsburg, 1979). The experimental evidence that directly relates to this idea is now described. In this case, the theoretical work preceded the experimental findings.

A surgically constructed half-retina regenerates an ordered projection to the entire optic tectum, rather than innervating the half of the tectum that it would innervate if it were part of a normal eye. This could be because the retinal labels had been changed by the surgery, the tectal labels had been changed, or both sets of labels had changed. Schmidt et al. (1978) carried out a set of experiments in adult goldfish. By forcing a surgically reduced half-retina to innervate the optic tectum together with a normal retina, he was able to use the projection made by the normal retina to calibrate the putative labels carried by the tectum and the experimental retina.

He removed one half of a retina from an adult goldfish and allowed the remaining half-retina to regenerate its connections to the entire contralateral tectum. He then diverted this projection to the ipsilateral tectum, which carried a projection from a normal retina, and showed that it innervated, in order, only its "appropriate" half. This established that the half-retina was still a half-retina in terms of the labels that it was assumed to carry.

In a complementary set of experiments, he diverted the projection from a whole retina onto an ipsilateral tectum that carried an expanded projection from a half-retina. In this case, only the half of the retina that had a matching origin with the experimental half-retina established an ordered projection on the ipsilateral tectum, showing that the tectum carried a half-set of labels. These results demonstrated that in a surgically constructed half-eye that made an expanded projection onto the optic tectum, the retinal labels did not change, but the tectum acquired a half-set of labels; i.e., the labels were induced from the retina.

Formal Models—Marker Induction

Von der Malsburg and Willshaw (1977) suggested that in all cases the retina retains its labels and the tectum's labels are modifiable. The idea is that the retina is labeled by a set of markers continually generated at fixed locations within the retinal surface and subject to lateral transport and degradation. At steady state, this sets up a fixed set of markers assigned to each retinal location, nearby retinal cells carrying similar sets of markers. As connections are formed, the markers are induced through the synapses already made into the tectum, which holds no markers initially. The rate of transfer of markers over a synapse is in proportion to the strength of that synapse. At each tectal site, markers from the various retinal cells innervating it become blended together with markers from adjacent tectal regions. By this means, each tectal cell acquires a characteristic set of markers. Synapses are progressively strengthened, in proportion to the similarity between the markers carried by the corresponding retinal and tectal cells. Each tectal cell becomes specific to the retinal cells carrying the markers most similar to its own set. This sets off a positive feedback

mechanism, resulting in each tectal cell becoming more and more specific to particular retinal cells and thereby attracting more and more markers of this type. Because nearby retinal cells and nearby tectal cells carry similar sets of markers, retinal neighbors tend to project to tectal neighbors (figure 11.6B). Provided that the initial pattern of innervation is biased to favor the desired orientation of the map, the result is that the set of retinal markers and a retinotopic map is induced onto the tectum in the desired orientation.

According to the marker induction model, the way in which the retina plays a role in establishing connections is somewhat similar to the way the periphery is involved in the development of barrel fields in the somatosensory cortex. The same basic idea has been applied to the problem of the elimination of superinnervation in developing muscle (Ribchester and Barry, 1994; see also chapter 10).

This model solves the problem of how a set of markers (or labels) in one structure can be reproduced in a second set in a way that is resistant to variations in the developmental program for the individual structures. It is able to account for the systems-matching sets of results (Gaze and Keating, 1972) as well as those on the reinnervation of the optic tectum following graft translocation and rotation, which suggest that in some but not all cases different parts of the optic tectum have acquired specificities for individual retinal fibers (Jacobson and Levine, 1975; Levine and Jacobson, 1974; Hope et al., 1976; Yoon, 1971, 1980; Gaze and Hope, 1983).

This model is consistent with the conclusions reached by Schmidt et al. (1978), as reviewed in the previous subsection. Furthermore, according to the marker induction model, each half-eye of a *Xenopus* compound eye contains a half-set of labels even though it projects across the entire tectum (Willshaw and Von der Malsburg, 1979). Evidence for this is discernible in the results from experiments by Straznicky and Tay (1982) on diverting compound eye projections in a manner analogous to that used by Schmidt et al. (1978) in their regeneration experiments and from the fact that the position of retinal axons growing from compound eyes in the optic tract is characteristic of the position occupied by fibers from the appropriate half of a normal eye (Straznicky et al., 1979). In addition, the nonlinearities in the extracellularly recorded maps of compound eye projections during development (Straznicky et al., 1981) are predicted directly from the marker induction model.

What counts against the marker induction model is that Schmidt's experimental results (Schmidt et al., 1978) have never been confirmed. In addition, a problem shared by all molecular-based mechanisms of this type (until very recently) is that there is little direct evidence for sets of labels in retina and tectum.

Recent experiments on mice pose intriguing new challenges for the marker induction type of model. Using a knock-in assay, Brown et al. (2000) introduced the EphA3 receptor into the mouse retina, which normally has EphA5 receptors only. In the knock-in mouse, some retinal ganglion cells had EphA5 receptors and some had both EphA3 and EphA5 receptors, both receptors being distributed across the nasotemporal axis of the retina. The retinocollicular maps developed were wholly abnormal, with one half of the colliculus receiving an ordered projection from the cells containing EphA5 and the other half an ordered projection from the cells containing both EphA5 and EphA3 receptors. These results strongly suggest an inductive effect, driven by the abnormal retinal distribution of Eph receptors acting as retinal labels (figure 11.7). One challenge to theorists is whether a retinal gradient can induce and maintain the tectal gradient even though the two gradients run in completely opposite directions.

The Arrow Model

The arrow model (Hope et al., 1976), which is contemporary with the neural activity model (Willshaw and Von der Malsburg, 1976), sought to explain the results of systems matching, mentioned earlier. The assumptions made in the model are that each retinal fiber has to "know" whether it is in the correct relative position with respect to the neighboring fibers on the tectum. In the model, fibers that are in the incorrect relative positions are able to swap their positions, and the process of comparison of fiber positions is repeated. Starting from any initial pattern of connections, an ordered map of connections in the correct orientation will result. Provided that each axon is in addition given a certain degree of random exploratory behavior, the set of retinal fibers will come to occupy, in order and in the correct orientation, the appropriate area of tectum available (systems matching). This model has no fixed labels of the type suggested by Sperry (1943, 1944, 1963); all that each fiber is required to know is information about the desired polarity of the map.

An advantage (and disadvantage) of this type of model is that it is immediately falsifiable. It predicts that the maps produced by allowing optic nerve fibers to reinnervate the adult tectum after a portion of the tectum has been removed, rotated, and then replaced will be a normal map, with the small portion of the map identified with the rotated part of the tectum being rotated by a corresponding amount. However, since there is no information about absolute position on the tectum, if two parts of the tectum are interchanged without rotation (translocation), the arrow model will predict a normal map. A variety of experimental results have been obtained, but the interpretation of the authors of the arrow model [who themselves carried out translocation experiments, which are mentioned in the original paper (Hope et al., 1976)] is that the maps obtained after the

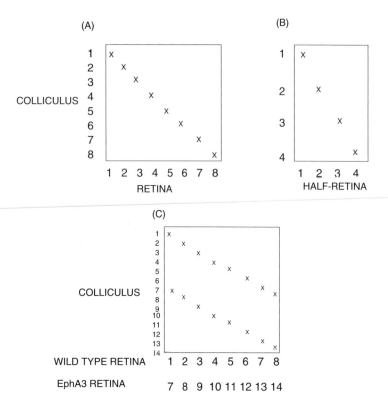

Figure 11.7

Proposal for how the retinocollicular maps found in EphA3 knock-in mice (Brown et al., 2000) can be accounted for by marker induction. According to this hypothesis, the full set of retinal markers is induced in order over the entire target structure. (*A*) The markers (here represented by the numbers $1, 2, \ldots, 8$) distributed over the full retina are induced onto the colliculus in numerical order to produce an ordered retinocollicular projection. (*B*) Similarly, the markers $1, 2, 3, 4$ from a half-retina are induced over the colliculus to give an expanded half-retina projection. (*C*) Following Brown et al. (2000), the retinal cells containing the extra Eph3 receptors have their markers augmented, giving rise to two populations of cell markers, namely $1, 2, \ldots, 8$ (wild type) and $7, 8, \ldots, 14$. The full population of markers now ranging from 1 to 14 is induced onto the colliculus in numerical order, giving rise to the double projection observed; compare with figure 6 of Brown et al. (2000).

translocation experiments are not normal, but contain matching translocated portions, which falsifies their own model.

11.6 Mapping Involving External Properties

Two members of the class of mappings that involve external properties can be identified: fiber ordering and the timing hypothesis.

11.6.1 Fiber Ordering—The Retinotectal Projection in Fish

It could be argued that the topographic mapping problem will have been solved if the nerve fibers arrive at their target structure in order. Fibers are ordered initially, insofar that their retinal cells of origin are arranged in order across the retina. If this order is preserved to the point of synaptogenesis, a map of the first structure will be imposed on the second structure without there being any need for the action of any complex mechanism of the types already discussed.

In the simplest possible case, fibers could grow out alongside those from their neighboring cells of origin and maintain this ordering all the way to the target. This can be tested by looking for order in the mature pathway. One vertebrate with a remarkably precise ordering of axons within the optic nerve is the cichlid fish (Anders and Hibbard, 1974; Scholes, 1979, 1981; Rusoff, 1984; Presson et al., 1985). The axons of the retinal ganglion cells pass into the optic nerve in order. The nerve cross-section just behind the eye resembles a long, folded-up ribbon. There is a polar coordinate representation of retinal position in the nerve cross-section, the radial dimension represented along the long axis and the circumferential dimension along the short axis. The retina grows in rings. The axons from each annulus of newly formed cells add on

as a band of unmyelinated fibers at one end of the ribbon. It is inferred that newly growing fibers travel over the substrate laid down by the older fibers. This order is maintained toward the optic chiasma as gradually the ribbon becomes shorter and fatter. The folds coagulate until finally the cross-section becomes roughly circular (figure 11.8).

Despite this evidence of a high degree of order within the optic pathway, as an explanation of how the ordered retinotectal map is formed, it is unsatisfactory:

1. After the optic chiasma, the fibers change their relative positions very abruptly, each group of fibers of the same age seeming to stay together. The fibers then form the lateral and medial brachia and are led onto the tectum. It has not been possible to follow the course of these fibers as they change their position.

2. It is not known how optic fibers leave the two brachia to innervate the tectum. The situation may be complicated by the possibility that the fibers are destined for more than one area of termination; i.e., two or more sets of innervating fibers are intermingled.

3. The polarity of the retinotectal map formed means that, in principle, fibers cannot be led onto the optic tectum without grossly violating their relations with their neighbors. Let us make the simplifications that all optic nerve fibers leave the optic nerve head together, and in order, and travel to the tectum together. Follow the planar cross-section of the growth cones as they travel to the tectum (this can be thought of as moving a disk representing the retina along the optic pathway, which eventually projects onto the surface of the optic tectum). Maps in some polarities can be produced without fibers losing contact with one another on the way to the tectum; to produce the polarity of the map that is observed, the disk has to be flipped over, which corresponds to a gross violation of neighborhood relations between fibers (figure 11.9).

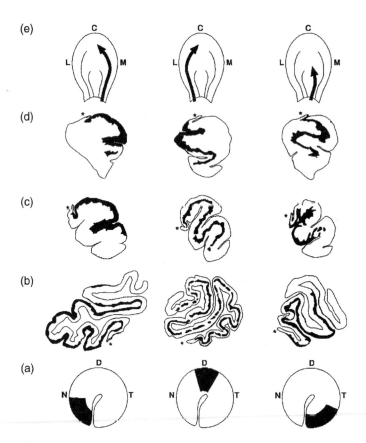

Figure 11.8

The arrangement of fibers in the optic nerve of a cichlid fish. The series shows the pattern of osmiophilic degeneration of fibers in the optic pathway after selective labeling of nasal retina (left-hand column), dorsal retina (middle column), and temporal retina (right-hand column) with horseradish peroxidase. (*a*) The distribution of label in the retina. (*b*) Immediately behind the optic nerve head, the optic nerve in cross-section is a very long and thin ribbon. There is an ordered representation of retina within the ribbon, the radius being represented along the long axis and the circumference along the short axis, i.e., a representation in polar coordinates. Unmyelinated, presumably newly arriving axons are grouped at one end of the ribbon, shown by a star. (*c*) A more posterior section of the optic nerve, showing that the ribbon has become shorter and wider, yet the polar coordinate representation has remained. (*d*) A little in front of the optic tectum, there has been a rearrangement of fiber positions within the cross-sections, over a very small distance, which disturbs their neighborhood relationships. (*e*) The fibers being led onto the optic tectum. N, nasal; T, temporal; D, dorsal; V, ventral; R, rostral; C, caudal; M, medial; L, lateral. (Reproduced from Scholes, 1981, with permission.)

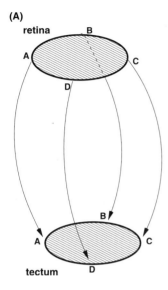
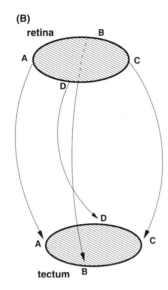

Figure 11.9

(*A*) Illustration of how an ordered map between the retina and tectum could be attained by axons traveling directly from one structure to the other while maintaining their relative positions. (*B*) Maps with certain polarities cannot be achieved in this way; in the case shown, the required pattern of interconnections is attained only at the expense of a massive crossing over of fibers.

The biological case is more complicated than this. Fibers do not grow out or arrive together and they undergo specific transformations that occur within the three-dimensional optic pathway, but the argument remains the same.

4. It may be that the ordering of fibers in the adult is an unfaithful historical record of developmental events; direct experimental evidence is lacking.

The channel catfish (*Ictalurus punctatus*) (Dunn-Meynell and Sharma, 1986) has a different ordering of fibers in the optic nerve. In the first part of the optic nerve, the ganglion cell axons are arranged in thirteen separate optic papillae. In the first part of the nerve cross-section, the papillae are arranged in a U-shaped formation. Fibers of dorsal origin are at one tip of the U and ventral fibers at the other tip, with fibers of both nasal and temporal origin at the base of the U. At the level of the optic chiasma, the U shape has flattened out while retaining the relative ordering of the papillae. In the optic tract, a substantial reordering of fibers takes place.

In other species, there is more evidence about the behavior of fibers throughout the entire visual pathway, but at a less detailed level. Along the pathway between the retina and colliculus of the quokka wallaby (*Setonix brachyurus*) (Chelvanayagam et al., 1998), fibers of nasal and temporal origin remain in the medial and lateral regions of the pathway. But dorsal and ventral fibers exchange positions, thereby achieving the required uniaxial inversion of the retinocollicular map.

These studies illustrate that there are cases where fibers maintain contact with their neighbors for much of the distance along the pathway, which could contribute to the formation of a map but without being the whole story. The cichlid fish seems to be an extreme case where there is a high degree of ordering in the optic pathway. There are signs that similar types of ordering exist in many other species of fish, but to a lesser extent or obscured by such factors as the pattern of fasciculation (Bunt, 1982).

11.6.2 *The Timing Hypothesis*

In invertebrates, experiments involving the surgical ablation of target cells during development show that connections are made on the basis of the time of arrival of axons at their target structure (Lopresti et al., 1973). In chicks, Rager and Von Oeynhausen (1979) proposed that axons arriving at one area of the tectum (such as the rostral edge) make contacts there and occupy this space, preventing later-arriving fibers from innervating this region and forcing them to contact regions nearby. In this way a temporal pattern is converted into a spatial pattern. It is fairly simple to imagine how such a mechanism could specify the ordering of optic fiber terminals along the direction of ingrowth of optic fibers, but not along the direction perpendicular to it. The basic problem is that timing provides for variation along one dimension only, whereas most maps are at least two-dimensional. Moreover, timing is inherently unstable, and on this hypothesis, a disruption of timing relations would result in a disruption in the map. This is contrary to the experimental findings. Maps are not affected by a delay in the arrival of optic fibers made by diverting optic fibers to take a different route. In surgically constructed *Xenopus* compound eyes made up of halves of different ages, the normal temporal pattern of axonal outgrowth is reversed but the initial pattern of innervation is normal (Holt, 1984).

11.7 Other Models

A wide variety of models for map making exists, and it is impossible to include them all here. We have concentrated on the ones that are either historically or fundamentally significant or about which we have specialist knowledge. Here we mention several other models to illustrate the variety of different approaches that have been adopted.

11.7.1 *Hybrid Models*

Whitelaw and Cowan (1981) combined the idea of a gradient of adhesive selectivity proposed by Prestige and Willshaw (1975) with synaptic updating as in the neural activity model (Willshaw and Von der Malsburg, 1976). The amount by which individual synaptic strengths are assumed to change follows the prescription of the neural activity model, but then the raw changes in strength are multiplied by the degree of adhesion between the corresponding retinal and tectal cells. In a modification to the model, a postulated random depolarization that is due to spontaneous release of transmitter is added to account for the finding that a map will form in the absence of externally applied electrical activity (Cowan and Friedman, 1990).

Overton and Arbib (1982) developed a pair of models called the branch arrow model (BAM) and the extended branch arrow model (XBAM), as extensions of the arrow model (Hope et al., 1976). In BAM, each axon terminates at a number of different sites on the tectum, and the overall movement of each retinal fiber is determined by the sum total of the influences from each terminating region. In addition, retinal axons can interact only if their cells of origin are within a certain specified distance of one another. It was found that ordered maps form only when this distance is less than two-thirds of the size of the retina. In XBAM, a mechanism of retinotectal interaction according to chemospecific markers is added.

Both of these classes of models incorporate a large amount of biological realism and account for a wide variety of experimental results. However, this is at the cost of surrendering predictive power.

11.7.2   *Optimization Models*

In models of the optimization type, the developmental process involves minimization of the amount of some physical quantity, e.g., the length of connections between cells, which can be calculated for each configuration of connections and which has an extreme value when the configuration is in its required state. If a procedure to systematically modify connections can be devised that will make the configuration more optimal at each stage, then continued application of this procedure will result in the required configuration being reached. In physics, such a physical quantity is called an energy.

Fraser's Energy Model
Fraser (1981) introduced the idea of an "adhesive free energy." This is a number that can be calculated for every state on a system and that describes how well the various constraints postulated to control the development of the mapping have been satisfied. In order of increasing importance, the constraints relate to:

1. Gradients of adhesive specificity along the dorsoventral and the nasotemporal axis of the retina together with a matching gradient on the tectum

2. The degree of adhesion between retinal and tectal cells, which varies according to relative position

3. The degree of competition among retinal axons for tectal space, which is also position independent

4. A tendency for axons that occupy nearby places on the tectum to stabilize their connections if their cells of origin are also neighbors

This model accounts for a variety of phenomena, but the advantages of Fraser's hypothesized energy function are mathematical rather than biological.

Dimension Reduction
The idea used in dimension reduction is that each point on the two-dimensional surface of the visual cortex that codes, for example, the three coordinates of space, ocularity preference, and direction of orientation represents a position in a high-dimension "stimulus space" (see also chapter 12). Models for how such mappings can be produced have been devised. The elastic net (Durbin and Willshaw, 1987) is one such model. It was devised originally to solve the Traveling Salesman Problem, the classic computer science problem of finding the shortest circuit around a given set of points (Lawler et al., 1986). This problem can be thought of as constructing a mapping from a two-dimensional space onto a one-dimensional space while maintaining neighborhood relationships. When applied to neural mapping problems, a model based on the elastic net can account for the production of retinotopy, ocular dominance, and orientation maps simultaneously (Erwin et al., 1995). The elastic net was not designed for biological plausibility, and this is its drawback. It may be interesting to note that it was in fact developed from the marker induction model for retinotopy (Willshaw and Von der Malsburg, 1979), and many of the biological features of that model were removed. In light of this fact, it is curious that the marker induction model could never be applied successfully to the development of all three types of map; Von der Malsburg (1979) showed how it could account for retinotopy and ocularity by assuming that the development of retinotopy precedes that of ocularity.

11.8   Conclusions

The problem of understanding how ordered maps are generated within the nervous system is now more than a century old. As far as the establishment of retinotopic maps is concerned, interest in the underlying theoretical basis of this phenomenon has seen two decades of intense activity. The 1940s saw the first theoretical accounts, in which ideas of chemoaffinity were introduced and developed; from mid-1970 to mid-1980, new concepts were introduced (particularly those concerning the role of competition and neural activity), old concepts were refined, and a plethora of models developed. While many of the models could fit the available facts, the critical experimental findings concerning the nature of cell-to-cell recognition at the synaptic level remained unanswered. In the early years of the twenty-first century, the new data concerning the molecular basis of retinotectal mapping promise to introduce a new decade of modeling activity and perhaps finally the answers that all of us in the field are seeking.

Acknowledgments

Our thanks to Fiona Jamieson for helping us to write this paper and to Arjen van Ooyen for his very helpful comments and for his extraordinary patience in waiting for the chapter. David Willshaw acknowledges the support of the United Kingdom Medical Research Council under Programme Grant PG911632.

References

Amari, S. (1980). Topographic organization of nerve fields. *Bull. Math. Biol.* 42: 339–364.

Amari, S. (1983). Field theory of self-organizing nets. *IEEE Trans. Syst., Man, Cybernetics* 13: 741–748.

Anders, J. J., and Hibbard, E. (1974). The optic system of the teleost *Cichlasoma meeki. J. Comp. Neurol.* 158: 145–154.

Arnett, D. W. (1978). Statistical dependence between neighboring retinal ganglion cells in goldfish. *Exp. Brain Res.* 32: 49–53.

Attardi, D. G., and Sperry, R. W. (1963). Preferential selection of central pathways by regenerating optic fibers. *Exp. Neurol.* 7: 46–64.

Brown, A., Yates, P. A., Burrola, P., Ortuno, D., Vaidya, A., Jessell, T. M., Pfaff, S. L., O'Leary, D. D., and Lemke, G. (2000). Topographic mapping from the retina to the midbrain is controlled by relative but not absolute levels of EphA receptor signalling. *Cell* 102: 77–88.

Bunt, S. M. (1982). Retinotopic and temporal organization of the optic nerve and tracts in the adult goldfish. *J. Comp. Neurol.* 206: 209–226.

Chelvanayagam, D. K., Dunlop, S. A., and Beazley, L. D. (1998). Axon order in the visual pathway of the quokka wallaby. *J. Comp. Neurol.* 390: 333–341.

Cheng, H.-J., and Flanagan, J. G. (1994). Identification and cloning of ELF-1, a developmentally expressed ligand for the Mck4 and Sek receptor tyrosine kinases. *Cell* 79: 157–168.

Chung, S.-H. (1974). In search of the rules for nerve connections. *Cell* 3: 201–205.

Chung, S.-H., and Cooke, J. (1975). Polarity of structure and of ordered nerve connections in the developing amphibian brain. *Nature* 258: 126–132.

Chung, S.-H., and Cooke, J. (1978). Observations on the formation of the brain and nerve connections following embryonic manipulation of the amphibian neural tube. *Proc. Roy. Soc. London B* 201: 335–373.

Cook, J. E. (1987). A sharp retinal image increases the topographic precision of the goldfish retinotectal projection during optic nerve regeneration in stroboscopic light. *Exp. Brain Res.* 68: 319–328.

Cook, J. E. (1988). Topographic refinement of the goldfish retinotectal projection: Sensitivity to stroboscopic light at different periods during optic nerve regeneration. *Exp. Brain Res.* 70: 109–116.

Cook, J. E., and Becker, D. L. (1988). Retinotopical refinement of the regenerating goldfish optic tract is not linked to activity-dependent refinement of the retinotectal map. *Development* 104: 321–329.

Cook, J. E., and Rankin, E. C. C. (1984). Use of a lectin-peroxidase conjugate (WGA-HRP) to assess the retinotopic precision of goldfish optic terminals. *Neurosci. Lett.* 48: 61–66.

Cook, J. E., and Rankin, E. C. C. (1986). Impaired refinement of the regenerated retinotectal projection of the goldfish in stroboscopic light: A quantitative WGA-HRP study. *Exp. Brain Res.* 63: 421–430.

Cowan, J. D., and Friedman, A. E. (1990). Development and regeneration of eye-brain maps: A computational model. In *Advances in Neural Information Processing Systems*, Vol. 2, D. Touretzky, ed. pp. 3–10. Morgan Kaufman Publishers, San Mateo, California.

Cox, E. C., Muller, B., and Bonhoeffer, F. (1990). Axonal guidance in the chick visual system: Posterior tectal membranes induce collapse of growth cones from temporal retina. *Neuron* 2: 31–37.

Da Silva Filho, A. R. C. (1992). "Investigation of a Generalised Version of Amari's Continuous Model for Neural Networks." PhD thesis, University of Sussex, UK.

Dunn-Meynell, A. A., and Sharma, S. C. (1986). The visual system of channel catfish (*Icatlurus punctatus*) 1. Retinal ganglion cell morphology. *J. Comp. Neurol.* 247: 32–55.

Durbin, R., and Willshaw, D. J. (1987). An analogue approach to the Travelling Salesman Problem using an elastic net method. *Nature* 126: 689–691.

Erwin, E., Obermayer, K., and Schulten, K. (1995). Models of orientation and ocular dominance columns in the visual cortex: A critical comparison. *Neur. Comput.* 7: 425–468.

Flanagan, J. G., and Vanderhaeghen, P. (1998). The ephrins and Eph receptors in neural development. *Annu. Rev. Neurosci.* 21: 309–345.

Fraser, S. E. (1981). A different adhesion approach to the patterning of neural connections. *Dev. Biol.* 79: 453–464.

Friedman, G. C., and O'Leary, D. D. M. (1996). Retroviral misexpression of *engrailed* genes in the chick optic tectum perturbs the topographic targeting of retinal axons. *J. Neurosci.* 16: 5498–5509.

Galli, L., and Maffei, L. (1988). Spontaneous impulse activity of rat retinal ganglion cells in prenatal life. *Science* 242: 90–91.

Gaze, R. M. (1958). The representation of the retina on the optic lobe of the frog. *Quart. J. Exp. Physiol.* 43: 209–224.

Gaze, R. M. (1960). Regeneration of the optic nerve in amphibia. *Int. Rev. Neurobiol.* 2: 1–40.

Gaze, R. M. (1970). *The Formation of Nerve Connections.* London: Academic Press.

Gaze, R. M., and Hope, R. A. (1983). The visuotectal projection following translocation of grafts within an optic tectum in the goldfish. *J. Physiol. (London)* 344: 257–275.

Gaze, R. M., and Keating, M. J. (1972). The visual system and "neuronal specificity." *Nature* 237: 375–378.

Gaze, R. M., Jacobson, M., and Szekely, G. (1963). The retinotectal projection in *Xenopus* with compound eyes. *J. Physiol. (London)* 165: 384–499.

Gaze, R. M., Keating, M. J., and Chung, S. H. (1974). The evolution of the retinotectal map during development in *Xenopus. Proc. Roy. Soc. London B* 185: 301–330.

Gaze, R. M., Feldman, J. D., Cooke, J., and Chung, S.-H. (1979a). The orientation of the visuo-tectal map in *Xenopus*: Development aspects. *J. Embryol. Exp. Morphol.* 53: 39–66.

Gaze, R. M., Keating, M. J., Ostberg, A., and Chung, S. H. (1979b). The relationship between retinal and tectal growth in larval *Xenopus*: Implications for the development of the retinotectal projection. *J. Embryol. Exp. Morphol.* 53: 103–143.

Gierer, A. (1983). Model for the retinotectal projection. *Proc. Roy. Soc. London B* 218: 77–93.

Gottlieb, D., Rock, K., and Glaser, L. (1976). A gradient of adhesive specificity in developing avian retina. *Proc. Natl. Acad. Sci. U.S.A.* 73: 410–414.

Halfter, W., Claviez, M., and Schwarz, U. (1981). Preferential adhesion of tectal membranes to anterior embryonic chick retina neurites. *Nature* 292: 67–70.

Harris, W. A. (1980). The effect of eliminating impulse activity on the development of the retinotectal projection in salamanders. *J. Comp. Neurol.* 194: 303–317.

Hebb, D. (1949). *The Organization of Behavior.* New York: Wiley.

Holt, C. E. (1984). Does timing of axon outgrowth influence retinotectal topography in *Xenopus? J. Neurosci.* 4: 1130–1152.

Hope, R. A., Hammond, B. J., and Gaze, R. M. (1976). The arrow model: Retinotectal specificity and map formation in the goldfish visual system. *Proc. Roy. Soc. London B* 194: 447–466.

Horder, T. J. (1971). Retention, by fish optic nerve fibres regenerating to new terminal sites in the tectum, of "chemospecific" affinity for their original sites. *J. Physiol. (London)* 216: 53–55.

Horder, T. J., and Martin, K. C. (1979). Morphogenetics as an alternative to chemospecificity in the formation of nerve connections. *Symp. Soc. Exp. Biol.* 32: 275–539.

Hunt, R. K., and Frank, E. D. (1975). Neuronal locus specificity: Trans-repolarization of *Xenopus* embryonic retina after the time of axial specification. *Science* 189: 563–565.

Itasaki, N., and Nakamura, H. (1992). Rostro-caudal polarity of the optic tectum in birds—correlation of en gradient and topographic order in retinotectal projection. *Neuron* 8: 787–798.

Itasaki, N., and Nakamura, H. (1996). A role for gradient *en* expression in positional specification on the optic tectum. *Neuron* 16: 55–62.

Jacobson, M. (1960). "Studies in the Organisation of Visual Mechanisms in Amphibians." PhD thesis, Edinburgh University, Scotland.

Jacobson, M. (1967). Retinal ganglion cells: Specification of central connections in larval *Xenopus laevis. Science* 155: 1106–1108.

Jacobson, M., and Levine, R. L. (1975). Plasticity in the adult frog brain: Filling in the visual scotoma after excision or translocation of parts of the optic tectum. *Brain Res.* 88: 339–345.

Kornberg, T. (1981). *Engrailed*: A gene controlling compartment and segment formation in *Drosophila. Proc. Natl. Acad. Sci. U.S.A.* 78: 1095–1099.

Langley, J. N. (1895). Note on regeneration of preganglionic fibers of the sympathetic ganglion. *J. Physiol. (London)* 18: 280–284.

Lawler, E. L., Lenstra, J. K., Rinnooy Kan, A. H. G., and Shmoys, D. B. (1986). *The Traveling Salesman Problem.* Chichester, UK: Wiley.

LeVay, S., Hubel, D. H., and Wiesel, T. N. (1975). The pattern of ocular dominance columns in macaque monkey visual cortex revealed by a reduced silver stain *J. Comp. Neurol.* 159: 559–576.

Levine, R. L., and Jacobson, M. (1974). Deployment of optic nerve fibers is determined by positional markers in the frog's tectum. *Exp. Neurol.* 43: 527–538.

Lopresti, V., Macagno, E. R., and Levinthal, C. (1973). Structure and development of neuronal connections in isogenic organisms: Cellular interactions in the development of the optic lamina of *Daphnia. Proc. Natl. Acad. Sci. U.S.A.* 70: 433–437.

Mesulam, M.-M. (1982). Principles of horseradish peroxidase neurohistochemistry and their applications for tracing neural pathways. In *Tracing Neural Connections with Horseradish Peroxidase,* M.-M. Mesulam, ed. pp. 1–155. Chichester, UK: Wiley.

Meyer, R. L. (1983). Tetrodotoxin inhibits the formation of refined retinotopography in goldfish. *Dev. Brain. Res.* 6: 293–298.

Meyer, R. L., and Sperry, R. W. (1973). Tests for neuroplasticity in the anuran retinotectal system. *Exp. Neurol.* 40: 525–539.

Morata, G., and Lawrence, P. A. (1975). Control of compartment development in the *engrailed* gene in *Drosophila. Nature* 255: 614–617.

Nakamoto, M., Cheng, H.-J., Friedman, G. C., McLaughlin, T., Hansen, M. J., Yoon, C. H., O'Leary, D. D. M., and Flanagan, J. G. (1996). Topographically specific effects of ELF-1 on retinal axon guidance in vitro and retinal axon mapping in vivo. *Cell* 86: 755–766.

Nakamura, H. (2001). Regionalization and acquisition of polarity in the optic tectum. *Prog. Neurobiol.* 65: 473–488.

Overton, K. J., and Arbib, M. A. (1982). The extended branch-arrow model of the formation of retino-tectal connections. *Biol. Cybern.* 45: 157–175.

Patel, N. H., Martin-Blanco, E., Coleman, K. G., Poole, S. J., Ellis, M. C., Kornberg, T. B., and Goodman, C. S. (1989). Expression of engrailed proteins in arthropods, annelids and chordates. *Cell* 58: 955–968.

Presson, J., Fernald, R. D., and Max, M. (1985). The organization of retinal projections to the diencephalon and pretectum in the cichlid fish *Haplochromis burtoni. J. Neurosci.* 235: 360–374.

Prestige, M. C., and Willshaw, D. J. (1975). On a role for competition in the formation of patterned neural connexions. *Proc. Roy. Soc. London B* 190: 77–98.

Rager, G., and Von Oeynhausen, B. (1979). Ingrowth and ramification of retinal fibers in the developing optic tectum of the chick embryo. *Exp. Brain Res.* 33: 65–78.

Rankin, E. C. C., and Cook, J. E. (1986). Topographic refinement of the regenerating retinotectal projection of the goldfish in standard laboratory conditions: A quantitative WGA-HRP study. *Exp. Brain Res.* 63: 409–420.

Reh, T. A., and Constantine-Paton, M. (1983). Retinal ganglion cell terminals change their projection sites during larval development of *Rana pipiens. J. Neurosci.* 4: 442–457.

Ribchester, R., and Barry, J. (1994). Spatial versus consumptive competition at polyneuronally innervated neuromuscular junctions. *Exp. Physiol.* 79: 465–494.

Rodieck, R. W. (1967). Maintained activity of cat retinal ganglion cells. *J. Neurophysiol.* 30: 1043–1071.

Rose, J. E., and Mountcastle, V. B. (1959). Touch and kinesthesis. In *Handbook of Physiology*, Vol. 1, W. H. Field and V. Hall, eds. Washington D.C.: American Physiological Society, pp. 387–430.

Roth, S., and Marchase, R. B. (1976). An *in vitro* assay for retinotectal specificity. In *Neuronal Recognition*, S. H. Barondes, ed. pp. 227–248. New York: Plenum.

Rusoff, A. (1984). Paths of axons in the visual system of perciform fish and implications of these paths for rules governing axonal growth. *J. Neurosci.* 4: 141–1428.

Schmidt, J. T., and Edwards, D. L. (1983). Activity sharpens the map during the regeneration of the retinotectal projection in goldfish. *Brain Res.* 269: 29–39.

Schmidt, J. T., Cicerone, C. M., and Easter, S. S. (1978). Expansion of the half retinal projection to the tectum in goldfish: An electrophysiological and anatomical study. *J. Comp. Neurol.* 177: 257–278.

Scholes, J. (1979). Nerve fibre topography in the retinal projection to the tectum. *Nature* 278: 620–624.

Scholes, J. (1981). Ribbon optic nerves and axonal growth patterns in the retinal projection to the tectum. In *Development of the Nervous System*, D. R. Garrod and J. D. Feldman, eds. Cambridge: Cambridge University Press, pp. 181–214.

Sejnowski, T. J., Koch, C., and Churchland, P. S. (1988). Computational neuroscience. *Science* 241: 1299–1306.

Sharma, S. C. (1972). The retinal projection in adult goldfish: An experimental study. *Brain Res.* 39: 213–223.

Slack, J. M. W. (1991). *From Egg to Embryo: Determinative Events in Early Development*. Cambridge: Cambridge University Press.

Sperry, R. W. (1943). Visuomotor co-ordination in the newt (*Triturus viridescens*) after regeneration of the optic nerve. *J. Comp. Neurol.* 79: 33–55.

Sperry, R. W. (1944). Optic nerve regeneration with return of vision in anurans. *J. Neurophysiol.* 7: 57–69.

Sperry, R. W. (1945). Restoration of vision after crossing of optic nerves and after contralateral transplantation of the eye. *J. Neurophysiol.* 8: 15–28.

Sperry, R. W. (1963). Chemoaffinity in the orderly growth of nerve fiber patterns and connections. *Proc. Natl. Acad. Sci. U.S.A.* 50: 703–710.

Stahl, B., Muller, B., Boxberg, Y., Cox, E. C., and Bonhoeffer, F. (1990). Biochemical characterization of a putative axonal guidance molecule of the chick visual system. *Neuron* 5: 735–743.

Straznicky, K., and Gaze, R. M. (1980). Stable programming for map orientation in fused eye fragments in *Xenopus. J. Embryol. Exp. Morph.* 55: 123–142.

Straznicky, K., and Tay, D. (1982). Retinotectal map formation in dually innervated tecta: A regeneration study in *Xenopus* with one compound eye following bilateral optic nerve section. *J. Comp. Neurol.* 206: 119–130.

Straznicky, K., Gaze, R. M., and Horder, T. J. (1979). Selection of appropriate medial branch of the optic tract by fibers of ventral retinal origin during development and in regeneration: An autoradiographic study in *Xenopus*. *J. Embryol. Exp. Morph.* 50: 253–267.

Straznicky, K., Gaze, R. M., and Keating, M. J. (1981). The development of the retinotectal projections from compound eyes in *Xenopus*. *J. Embryol. Exp. Morph.* 58: 79–91.

Stuermer, C. A. O. (1990). Retinotopic organization of the developing retinotectal projection in the zebrafish embryo under TTX-induced neural-impulse blockade. *J. Neurosci.* 10: 3615–3626.

Takeuchi, A., and Amari, S. (1979). Formation of topographic maps and columnar microstructures in nerve fields. *Biol. Cybern.* 35: 63–72.

Trisler, G. D., and Schneider, M. D. (1981). A topographic gradient of molecules in retina can be used to identify neuron position. *Proc. Natl. Acad. Sci. U.S.A.* 78: 2145–2149.

Trowe, T., Klostermann, S., Baier, H., Granato, M., Crawford, A. D., Grunewald, B., Hoffmann, H., Karlstrom, R. O., Meyer, S. U., Muller, B., Richter, S., Nusslein-Volhard, C., and Bonhoeffer, F. (1996). Mutations disrupting the ordering and topographic mapping of axons in the retinotectal projection of the zebrafish, *Danio rerio*. *Development* 123: 439–450.

Tuzi, N. L., and Gullick, W. J. (1994). Eph, the largest known family of putative growth factor receptors. *Br. J. Cancer* 69: 417–421.

Von der Malsburg, C. (1973). Self-organization of orientation sensitive cells in the striate cortex. *Kybernetik* 14: 85–100.

Von der Malsburg, C. (1979). Development of ocularity domains and growth behavior of axon terminals. *Biol. Cybern.* 32: 49–62.

Von der Malsburg, C., and Willshaw, D. J. (1977). How to label nerve cells so that they can interconnect in an ordered fashion. *Proc. Natl. Acad. Sci. U.S.A.* 74: 5176–5178.

Walter, J., Henke-Fahle, S., and Bonheoffer, F. (1987). Avoidance of posterior tectal membranes by temporal retinal axons. *Development* 101: 909–913.

Weiss, P. (1937a). Further experimental investigations on the phenomenon of homologous response in transplanted amphibian limbs. I. Functional observations. *J. Comp. Neurosci.* 66: 181–209.

Weiss, P. (1937b). Further experimental investigations on the phenomenon of homologous response in transplanted amphibian limbs. II. Nerve regeneration and the innervation of transplanted limbs. *J. Comp. Neurosci.* 66: 481–536.

Weiss, P. (1939). *Principles of Development*. New York: Holt.

Whitelaw, V. A., and Cowan, J. D. (1981). Specificity and plasticity of retinotectal connections: A computational model. *J. Neurosci.* 1: 1369–1387.

Wilkinson, D. G. (2001). Multiple roles of Eph receptors and ephrins in neural development. *Nat. Neurosci. Rev.* 2: 155–164.

Willshaw, D. J., and Von der Malsburg, C. (1976). How patterned neural connexions can be set up by self-organisation. *Proc. Roy. Soc. London B* 194: 431–445.

Willshaw, D. J., and Von der Malsburg, C. (1979). A marker induction mechanism for the establishment of ordered neural mappings: Its application to the retinotectal problem. *Phil. Trans. Roy. Soc. London B* 287: 203–243.

Willshaw, D. J., Fawcett, J. W., and Gaze, R. M. (1983). The visuotectal projections made by *Xenopus* "pie slice" compound eyes. *J. Embryol. Exp. Morph.* 74: 29–45.

Wolpert, L. (1969). Positional information and the spatial pattern of cellular differentiation. *J. Theor. Biol.* 25: 1–47.

Yoon, M. (1971). Reorganization of retinotectal projection following surgical operations on the optic tectum in goldfish. *Exp. Neurol.* 33: 395–411.

Yoon, M. (1972). Transposition of the visual projection from the nasal hemiretina onto the foreign rostral zone of the optic tectum in goldfish. *Exp. Neurol.* 37: 451–462.

Yoon, M. G. (1980). Retention of the topographic addresses by reciprocally translated tectal re-implant in adult goldfish. *J. Physiol.* 308: 197–215.

Development of Ocular Dominance Stripes, Orientation Selectivity, and Orientation Columns

12

N. V. Swindale

Neurons in the somatosensory and visual cortices respond to spatially localized and specific kinds of stimuli. For example, many visual cortex neurons have a preference for stimulation through one of the two eyes (ocular dominance) and for stimuli of a particular orientation (orientation selectivity). This chapter reviews the variety of models that have been proposed to explain the development of ocular dominance and orientation selectivity in the mammalian visual cortex.

12.1 Introduction

12.1.1 Columnar Organization and Maps in the Adult Visual Cortex

The response properties of neurons in the somatosensory and visual cortices tend to remain unchanged with position perpendicular to the cortical surface (Mountcastle, 1957; Hubel and Wiesel, 1962). This property, termed columnar organization, means that for many purposes the cortex can be treated as a two-dimensional sheet. The term "column" is generally used (somewhat loosely) to define a set of cells having boundaries perpendicular to the cortical surface and spanning all the cortical layers, with some property in common, such as eye preference or selectivity for a particular orientation.

In contrast to columnar invariance, response properties often change systematically with sideways (tangential) movement through the cortex; these forms of variation are called mappings (see also chapter 11).

The properties that have ordered mappings include (1) the spatial receptive field, i.e., the specific region of visual space within which a stimulus must be present in order for a cell to respond; (2) ocular dominance, i.e., a preference for stimulation through one of the two eyes; and (3) orientation selectivity, i.e., a preference for a bar or an edge of a particular orientation. Ordered maps of other receptive field properties, such as preference for spatial frequency or for a particular direction of stimulus motion, have also been demonstrated. Additional maps may exist, although most of the possibilities are speculative (Swindale, 2000).

12.1.2 The Cortex as a Dimension-Reducing Map

In this section, a general framework for thinking about maps is presented. Each cell in the visual cortex can be thought of as representing (when maximally active) the presence in the image of a specific combination of stimulus features, e.g., the presence of an edge of a particular orientation moving in a particular direction across a specific small region of visual space in a particular eye. This set of properties defines a point $\mathbf{w}$ in an N-dimensional stimulus space $\mathbf{S}^N$ whose axes are the stimulus parameters of interest (figure 12.1). There is thus a mapping between an N-dimensional stimulus space and the two-dimensional sheet of the cortex. This type of mapping has been termed a dimension-reducing mapping (Durbin and Mitchison, 1990).

While one typically thinks of a map in terms of the way in which some property varies across a surface, it

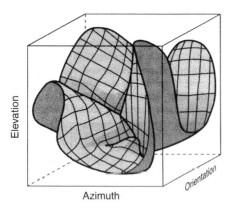

Figure 12.1
A stimulus space, **S**. The folded sheet represents the visual cortex; points on the grid represent position in cortical coordinates. The receptive field properties of each point in the cortex determine the position of the corresponding point of the sheet in **S**. A 3-D space is shown here, with the dimensions representing the receptive field position (elevation and azimuth) and preferred orientation.

is often useful to consider the mapping in the inverse direction, i.e., to consider the map as a projection of a 2-D surface into **S**. The complex foldings of the 2-D cortical sheet within **S** are thought to be subject to two constraints: a continuity constraint and a completeness constraint, which act in opposition. The continuity constraint means that the mapping should be locally smooth—neighboring points in the cortex have similar receptive fields (i.e., map to neighboring points in **S**)—while the completeness constraint means that the cortex should fill functionally important regions of **S** as completely as possible. Since for $N > 2$, not all points in **S** can be mapped to the cortex in a neighborhood-preserving way, the mapping must necessarily be incomplete. One might require, however, that some part of cortex come within some minimum distance of every functionally important point in **S**.

Alternatively, because receptive fields are not infinitely narrow, one can think of them as occupying small regions of stimulus space. More generally, one can think of the cortex as filling **S** with neural activity, and one interpretation of completeness is that the density of activity should be as uniform as possible (Swindale, 1991). The implications of this particular approach will be returned to later on. First, some specific mappings are described in more detail.

12.1.3   The Retinotopic Map

Each visual cortex (i.e., on the left and the right side of the brain) contains a topographic map of the contralateral visual field (see also chapter 11). This mapping exists because there is a topographically precise mapping from the retina to the layers of the lateral geniculate nucleus (LGN) and because there is a similarly precise mapping from each of these layers to layer IV of the visual cortex. To a first approximation, equal areas of the visual cortex are innervated, via the LGN, by equal numbers of retinal ganglion cells. Because ganglion cell density is highest in the central region of the retina, the visual world is not linearly scaled onto the cortex, but is distorted so that the magnification factor (square millimeters of cortex per square degree of visual angle) is greatest in the central visual field and least in the peripheral field. Although it is an important detail for modelers, relatively little is known about the local precision of the topography. The most precise mapping possible would have LGN axons connecting to cells in layer IV in a pattern that matched the locations of the ganglion cells driving the LGN axons. It is possible that the retinotopic map approaches this precision in layer IVCβ of the macaque monkey (Blasdel and Lund, 1983; Hubel et al., 1974; Blasdel and Fitzpatrick, 1984), where receptive fields are small and circularly symmetrical and the ret-

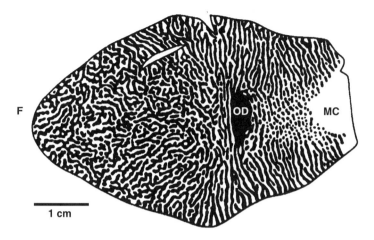

Figure 12.2
The complete pattern of ocular dominance stripes in the flattened visual cortex of a macaque monkey. A complete retinotopic map of the contralateral visual field is represented within the area of cortex shown. F, the region corresponding to the fovea; OD, the region corresponding to the optic disk; MC, the monocular segment. (From Florence and Kaas, 1992.)

inotopic map shows little disorder. In subprimate species, such as cats and ferrets, the degree of precision is probably much less than this (Albus, 1975).

12.1.4 Ocular Dominance Stripes

Eye dominance or ocular dominance columns form a pattern of periodic branching stripes with a width of about 0.5 mm, interdigitated with regions of the same width that prefer stimulation through the other eye (figure 12.2). The pattern reflects the fact that the inputs from the left and right eyes, relayed through separate layers of the LGN, terminate in nonoverlapping stripes in layer IV of the cortex. The similarity between ocular dominance stripes and other patterns such as fingerprints and the striped and spotted patterns found on the body surfaces of fish, frogs, and zebras has often been commented on and suggests the possibility that their development might be understood in terms similar to those postulated to explain

pattern formation in these other systems (Murray, 1989).

12.1.5 Orientation Domains

In addition to a preference for stimulation via one or the other eye, most visual cortex neurons respond to bars or edges flashed or moved across the receptive field at specific orientations. One of the causes of this selectivity in those cells that receive inputs directly from the LGN is a receptive field organization in which one or more regions of ON responsiveness alternate with regions of OFF responsiveness (Hubel and Wiesel, 1962). The preferred orientation of these cells, which are called simple cells, can be predicted from the orientation of the line (or lines) that best separates the ON and OFF regions.

The overall layout of orientation preference is continuous and periodic (Hubel and Wiesel, 1974; Swindale et al., 1987; Bonhoeffer and Grinvald, 1991;

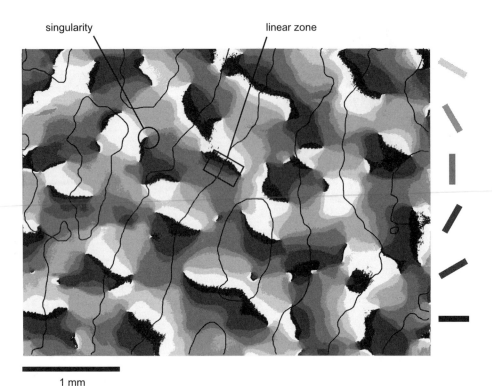

Figure 12.3
A map of orientation preference in the macaque monkey. Each gray level represents a unique orientation preference, as shown by the key on the right. A similar though not identical gray scale is used to represent orientation preference in figures 12.5b and 12.6. A singularity (circle) and a linear zone (rectangle) are indicated. Black lines mark the boundaries between ocular dominance columns. (Reproduced and redrawn from data in Blasdel, 1992, and Obermayer and Blasdel, 1993.)

Blasdel, 1992). Iso-orientation domains—regions of the map where orientation preference lies within a defined range—have a well-defined periodicity of about 1 mm in the cat and about 0.6 mm in the monkey (figure 12.3). The maps also include singular points where a single complete 180 degree set of domains meet. Because of their appearance in color-coded images, the singularities and the regions immediately surrounding them are often called pinwheels. They are characterized as positive if the orientation rotates in a clockwise direction when a clockwise cir-

cuit is made around the singularity, and as negative if it rotates in a counterclockwise direction. Areas of cortex between the pinwheel regions often contain iso-orientation domains that run in a roughly parallel direction; these are called linear zones.

12.1.6 Relationships between Different Visual Maps

Structural relationships between the different maps have been demonstrated in a number of instances, although they are often weak. As mentioned earlier, there is a precise retinotopic map in layer IVC of the monkey, and cells have small, circularly symmetrical receptive fields. The retinotopic map interacts with the map of eye dominance in the following way: As layer IVC is traversed in a tangential direction, receptive field positions in the corresponding eye shift in a constant direction. As the boundary between neighboring ocular dominance stripes is crossed, the receptive field shifts into the other eye, to a location corresponding to that mapped to the center of the adjacent stripe (Hubel et al., 1974; Blasdel and Fitzpatrick, 1984). This type of z-folding is one way of ensuring that the entire visual field of each eye gets represented in the half of the cortex area that constitutes one set of eye dominance stripes.

A relationship between the retinotopic and orientation maps in area 17 of the cat was reported by Das and Gilbert (1997). They found fractures across which there were simultaneous jumps in both preferred orientation and receptive field position; in general, the retinotopic gradient and the orientation gradient were strongly correlated. It remains to be seen whether such correlations are a general feature of visual cortex maps; there is evidence from the tree shrew that they may not be (Bosking et al., 1997).

Combined maps of eye dominance and orientation (Bartfeld and Grinvald, 1992; Hübener et al., 1997) show a tendency for singularities to be located in the centers of eye dominance stripes and for isoorientation domains to run across the boundaries of eye dominance columns at right angles (figure 12.3). A possible explanation for these orthogonal gradient relationships is that they maximize coverage uniformity (Swindale et al., 2000), i.e., the uniform representation of all combinations of the parameters represented in the map.

12.2   Development of Visual Cortex Maps— Neurobiological Background

An important issue is the extent to which environmentally driven patterns of neural activity determine the map structures. One possibility is that many aspects of visual cortex map organization (e.g., ocular dominance and orientation specificity) might be entirely the result of postnatal visual experience. At one point this seemed a realistic possibility, particularly when it was reported that stimulus specificity was absent in the visual cortex of very young kittens (Barlow and Pettigrew, 1971) and that rearing kittens in an environment with lines of a single orientation resulted in a visual cortex containing neurons whose orientation selectivities all matched the orientation experienced (Blakemore and Cooper, 1970; Hirsch and Spinelli, 1970). This view was probably strengthened by the fact that the earliest computational models of visual cortex development were able to show that environmentally driven patterns of neural activity could account for the development of orientation selectivity (Von der Malsburg, 1973), ocular dominance stripes (Von der Malsburg and Willshaw, 1976), and retinotopic maps (Willshaw and Von der Malsburg, 1976; see also chapter 11).

Subsequent work has not confirmed this extreme environmentalist viewpoint. It is now clear that many forms of stimulus selectivity, and their columnar organization, are either present at birth or can be shown to be present in animals reared in the dark from the time of eye opening. For example, in macaque monkeys (whose eyes are open at birth), orientation selectivity (Wiesel and Hubel, 1974), orientation columns (Blasdel et al., 1995), and ocular dominance columns (Horton and Hocking, 1996) are present at birth. In kittens (whose eyes open at 7–10 days after birth), orientation-selective neurons can be recorded at 6–8 days of age (Albus and Wolf, 1984; Braastad and Heggelund, 1985), while orientation and ocular dominance columns are present in rudimentary form in normal or visually inexperienced kittens at 15 days of age (Crair et al., 1998).

Given these observations, attention has been devoted to roles that spontaneously occurring patterns of neural activity might play in the initial formation of ocular dominance and orientation maps (Katz and Shatz, 1996). These experiments generally suggest that spontaneous activity plays a crucial role (however, see Crowley and Katz, 2000, and section 12.5). For example, in kittens, silencing retinal activity by intraocular injections of tetrodotoxin abolishes ocular dominance columns (Stryker and Harris, 1986). Blocking activity in retinal ON-center ganglion cells by intraocular injection of DL-2-amino-4-phosphonobutyric acid (APB) prevents the development of orientation selectivity (Chapman and Gödecke, 2000).

Given the probable importance of spontaneous neural activity in the earliest stages of map development, it is not surprising that a disturbance of visual experience early in life can affect the later stages of map development in many different ways. Closing one eye during the first 2–6 weeks of age in kittens, or before 3 months of age in macaque monkeys, causes the ocular dominance stripes representing the closed eye to shrink in area, while the stripes representing the open eye expand and take over the territory vacated by the closed eye (Hubel et al., 1977; Shatz and Stryker, 1978). The interpretation of the analogous experiment in which animals are reared in an environment where lines of a single orientationpredominate has been less straightforward. However, a recent study in the cat using optical recording, which avoids many of the problems inherent in earlier studies, showed that iso-orientation domains corresponding to the orientation experienced do

increase in size (Sengpiel et al., 1999). This confirms that neurons can change their orientation preference in response to environmentally driven patterns of stimulation.

Overall, these experiments support a restricted environmentalist viewpoint in which visually driven patterns of activity do not play a role in initially setting up the map, but can sculpt and modify a preexisting map by causing local shrinkage or expansion of columns. Whether early visually driven activity can go beyond this and change more global details, such as column periodicity, remains controversial (see section 12.4.2).

12.2.1   Role of Spontaneous Retinal Activity

Since many aspects of visual cortex map formation occur either in utero (as in primates) or postnatally in the absence of visual experience (as in cats and ferrets), visually driven activity cannot be the primary factor that establishes receptive field structure and columnar organization. Attention therefore has to be focused on patterns of spontaneous neural activity that occur before the eyes open. At a very early stage, before many synaptic connections have been made, cortical neurons are coupled by gap junctions, and small domains of cells exhibit coordinated transient elevations in Ca^{2+} levels (Kandler and Katz, 1998). These events could play a role in establishing common feature selectivity in the earliest stages of map formation.

Another form of spontaneous activity occurs in the embryonic retina, where ganglion cells fire in irregular bursts (Galli and Maffei, 1988). These bursts are correlated in neighboring ganglion cells and form waves that spread across the retina (Wong, 1999). Models for this behavior have been presented (Burgi and Grzywacz, 1994; Feller et al., 1997). While retinal waves are a likely candidate for a mechanism to refine topography and enforce laminar segregation of retinal inputs in the LGN, they end a few days before the emergence of ocular dominance and orientation maps in ferrets, so their role in these aspects of map formation is uncertain. It is not known whether retinal waves occur in kitten or primate retinas, so their role in map formation in these species is also uncertain.

Spontaneous bursting has been demonstrated in developing ferret LGN (Weliky and Katz, 1999). This shows positive interocular correlations that are dependent on feedback from the visual cortex. Much more remains to be discovered about this phenomenon, particularly with respect to its spatiotemporal patterning, its persistence beyond the period of eye opening, and the role of cortical feedback and patterning in the cortex itself. These details are likely to be critical for future models.

12.2.2 Factors Affecting the Development of Ocular Dominance Columns

A brief summary of some experimental results pertinent to models of the formation of ocular dominance columns is given here:

• Monocular deprivation during the critical period causes the ocular dominance stripes for the closed eye to shrink and those for the open eye to expand (Hubel et al., 1977; Shatz and Stryker, 1978; LeVay et al., 1980); this can occur after segregation is complete.

• Stripes shrunken by monocular deprivation can re-expand if the deprived eye is opened and the normal eye closed (reverse suturing) (LeVay et al., 1980; Swindale et al., 1981).

• Silencing retinal activity abolishes segregation (Stryker and Harris, 1986).

• The effects of monocular deprivation can be blocked by infusing the cortex with N-methyl-D-aspartate (NMDA) receptor antagonists (Bear and Rittenhouse, 1999).

• Infusion of the γ-aminobutyric acid (GABA) agonist muscimol into the cortex of monocularly deprived kittens (which will cause cortical neurons to hyperpolarize, so that their inputs will fail to evoke action potentials) causes strengthening of the inputs from the deprived eye and a weakening of the inputs from the normal eye (Reiter and Stryker, 1988).

• Infusion of neurotrophins (NT-4/5 or brain-derived neurotrophic factor) into kitten visual cortex blocks the formation of ocular dominance columns (Cabelli et al., 1995).

• Monocular deprivation by lid suture produces a bigger ocular dominance shift than monocular TTX injection (Rittenhouse et al., 1999).

12.2.3 Factors Affecting the Development of Orientation Columns

Any model of orientation column development ought to be able to explain the following observations:

• Orientation preferences should vary smoothly over most parts of the map, except in singularities and (possibly) short fracture regions.

• The power spectrum of the orientation vectors should have a strong nonzero peak.

• The map should contain half-rotation (i.e., 180-degree) singularities of positive and negative sign, with an irregular spacing and a density in the range of 2.0–3.5 per λ^2, where λ is the dominant wavelength as determined by Fourier spectral analysis.

• Singularities should be grouped so that approximately 70–80 percent of nearest-neighbor pairs are of opposite sign (Obermayer and Blasdel, 1997).

• There should be no statistical relationship between orientation and orientation gradient angle.

• Spatial segregation on ON and OFF center inputs within cortical receptive fields should be a primary determinant of orientation selectivity, since blocking activity in the ON pathway during development abolishes orientation selectivity (Chapman and Gödecke, 2000).

12.3   An Overview of Models of Visual Cortex Map Formation

The history of modeling of visual cortex maps begins with the demonstration by Von der Malsburg (1973) that oriented patterns of activity in a retina, connected initially at random by Hebbian synapses to a cortex with lateral short-range excitatory and long-range inhibitory connections, could give rise to neurons that are selective for specific orientations laid out in a spatially organized map. The components and assumptions of most models have changed very little since then (figure 12.4). In the following sections I describe these components and how they have typically been used.

12.3.1   Retinal Inputs

Inputs to the model cortex are typically represented in terms of static activation levels in one or more two-dimensional arrays that are assumed to correspond to sheets of cells in the retina or the LGN. These values are assumed to represent firing rates averaged over a period of time that is brief compared with the time scale of synaptic weight change. Models of ocular dominance column formation of course use two such arrays, one for the left and one for the right eye. Additional layers may be used to represent the activities of ON and OFF ganglion cell types; for example, four

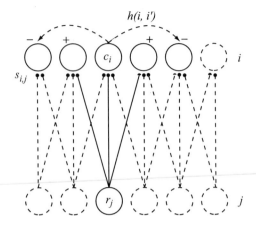

Figure 12.4
Generic structure of visual cortex map models. Units in an input layer (the LGN or retina) with activities r_j make connections with strengths $s_{i,j}$ to units in an output layer (the cortex) that have activities c_i. Local interactions between the cortical units are described by an interaction function $h(i, i')$, which is typically positive for short distances and negative for intermediate distances between points i and i'.

sheets of cells, corresponding to the ON and OFF layers in the two eyes, were used by Erwin and Miller (1998). Most, if not all, models assume that the input layers project directly to the cortex. Thus, either the LGN is assumed to pass on retinal activity without changing it, or activity is assumed to arise directly within the LGN.

Very simple patterns of activity have typically been assumed, including randomly oriented lines, randomly positioned circular or elongated blobs, or noise into which local correlations have been introduced. For models of ocular dominance column formation, one technique is to assign random values to the points in each array, smooth them with a Gaussian kernel, and then add a fraction of the values in each eye to those in the other. This creates a stimulus with defined local and intereye correlations. Very few simulations have used natural images, which is probably justified given the ability of cortical maps to develop in the dark, nor have there been many attempts to use statistics based on the known spontaneous firing patterns of retinal or LGN neurons (Elliott and Shadbolt, 1999, is an exception).

12.3.2 The Cortex

In most, if not all, models, the cortex is represented by a single sheet of cells receiving connections from the input layers. In many cases, a fixed "Mexican hat" pattern of short-range excitatory and intermediate-range inhibitory connections between cortical points has been assumed (although without much in the way of empirical justification). Very few modelers have assumed the presence of more cortical layers, or of feedback from the cortex to the input layer(s).

For a given pattern of activity in the input layers, activity in each cortical unit is generally assumed to be determined by the sum of the input activities times the relevant weights plus a contribution from the activities of surrounding cortical units via the lateral interaction function. A threshold, or other response nonlinearity, may be applied to the outputs of the cortical units. Because the activity pattern in the cortical sheet cannot be determined simply—it may take many iterations before a stable cortical response to a particular input is generated—most models have found ways of sidestepping or simplifying this step. The ways in which this has been done are discussed in section 12.4.

12.3.3 Learning Rules

Many models (e.g., Von der Malsburg, 1973; Miller et al., 1989; Goodhill, 1993) are based on the simplest form of Hebbian learning, in which synaptic weights are increased by an amount proportional to the product of pre- and postsynaptic activation levels. Be-

cause these values cannot be negative, this means that weights cannot decrease. Without an additional regulatory mechanism, competition—the process in which an increase in the strength of some connections leads to the weakening of others—cannot occur. In these models, competition is typically implemented by ensuring that the sum of the synaptic strengths onto individual neurons (i.e., each point in the cortical array) remains a constant at each learning step, a procedure termed normalization. Normalization can be implemented by division or by subtraction. The choice is not trivial because it can significantly affect the way development proceeds (Miller and MacKay, 1994; Goodhill and Barrow, 1994; Wiskott and Sejnowski, 1998; see also chapter 10). In some models, alternative ways of implementing competition and avoiding normalization have been explored (see sections 12.4.5 and 12.4.6).

12.3.4 Initial Conditions

Many models assume some degree of topographic order in the initial set of connections between the input and cortical layers. In some models, connections from the input layer are assumed to be made initially within a small region of the cortical layer, described by an arbor function, which is in exact topographic correspondence with the input layer. The arbor function is fixed in size and position, so that a given afferent can only make or modify its connections with a fixed cortical region. Other models allow for a less rigid retinotopy and consequently permit less rigid initial conditions and more varied retinotopic outcomes. Initial connection strengths are generally assumed to be random. Periodic boundary conditions are often assumed for convenience; i.e., distances over the input and output arrays are calculated as though left and right edges, and the top and bottom edges, are contiguous.

12.4   The Models

While this general framework for modeling is about as simple as it could be, it can still prove unwieldy and slow when implemented on a computer. As a result, many modelers have found ways of further simplifying the calculations. This has had worthwhile consequences because it has resulted in a variety of related simpler models whose behaviors are easier to understand. Some of the ways in which simplification has been achieved are the following:

• *Linear Hebbian models.* In these models it is assumed that learning can be averaged over input patterns. This means that the quantities that are explicitly represented in the calculations are time-averaged spatial correlation functions present in each of the input layers, rather than explicit patterns of neuronal activity.

• *Competitive Hebbian models.* In these models the calculation of cortical activity following a stimulus is simplified by assuming that synaptic modification occurs only in the most active cortical unit and its nearest neighbors.

• *Lateral interaction models.* Here all the interactions are lumped into a single lateral interaction function. This leads to models that are computationally simple and that bring out similarities with more general theories of pattern formation.

• *Feature-based or dimension-reduction models.* In these, the description of the input space is simplified by representing stimuli as points in a feature space (section 12.1.2). This leads to very abstract models whose construction seems far removed from biology, yet which have been remarkably successful in explaining the phenomenology of visual cortex maps.

• *Models that avoid explicit normalization rules.* These include models based on competition for neuro-

trophins (section 12.4.5) and the BCM learning rule (section 12.4.6).

These classes of model are described in more detail in the following sections.

12.4.1   Linear Hebbian Models

Linear Hebbian models avoid the explicit representation of activity patterns in the input and output layers; instead, they assume that changes in synaptic strength are determined by time- and space-averaged patterns of correlation in the input layers. This simplification follows if the responses of the units in a layer are linearly related to the activities in the input layer and if the change in weights following each activity pattern is small. In this case, the learning rule can be expressed simply in terms of the time-averaged spatial correlations in the input patterns.

The Ocular Dominance Column Model of Miller et al.

The model of Miller et al. (1989) assumes two input layers, L and R, with four corresponding correlation functions, C^{LL}, C^{RR}, C^{LR}, and C^{RL}, specifying how the correlations in neural firing rates vary with lateral separation in the LGN layers. Inputs from a location j in the LGN are assumed to make contact with a retinotopically corresponding cortical neuron centered on a location i in the cortex and spread over a surrounding region described by a fixed arborization function $A(i - j)$. (It is assumed that any position j in the LGN maps directly to an equal position i in the cortex, so that LGN and cortical coordinates are interchangeable.) The arborization function is 1 over a small square region and zero elsewhere. The strengths of the connections at time t are given by the functions $s^L(i, j, t)$ and $s^R(i, j, t)$. Lateral cortical interactions are described by a Mexican hat function h, which is a radially symmetrical difference of Gaussians with a fixed width. The contribution of a synapse $s(k,l)$ to the correlation value associated with a second synapse $s(i,j)$ is assumed to be proportional to the product of the correlation value associated with the separation between the cells of origin in the LGN [i.e., $C(j-l)$], the strength of the synapse itself [i.e., $s(k,l,t)$], and the value of the lateral interaction function for separation of the synapses in the cortex [i.e., $h(i-k)$]. This gives the following learning rule:

$$s^L(i,j,t+1) = s^L(i,j,t) + \varepsilon A(i-j)\sum_{k,l} h(i-k)$$

$$\times \, [C^{LL}(j-l)s^L(k,l,t)$$

$$+ C^{LR}(j-l)s^R(k,l,t)], \quad (12.1)$$

where ε is a constant determining the overall growth rate. The corresponding equation for s^R is obtained by interchanging L and R. Initial connection strengths are random within a small range. At each time step, after updating the connection strengths, a subtractive normalization procedure is carried out. Separate limits are also put on the maximum and minimum synaptic strengths.

For the model to work, it is sufficient that the within-eye correlations C^{LL} and C^{RR} are positive Gaussian functions, while the between-eye correlations C^{LR} and C^{RL} are either zero or negative. Under these conditions, cortical receptive fields, which are initially binocular and equal in size to the arbor function, gradually become smaller and monocular, while individual afferent arbors become smaller and often break up into patches confined to neighboring ocular dominance stripes. As a result of these changes, a striped pattern of ocular dominance develops. When the cortical interaction function contains both short-range excitatory and long-range inhibitory components, the spacing of the stripes is determined by the

position of the peak in the Fourier transform of the cortical interaction function $h(x)$. Narrower within-eye correlation functions result in more binocular cells at the borders of the stripes and smaller receptive field sizes. When $h(x)$ is purely excitatory, segregation occurs, provided a constraint maintaining the total strength of individual axonal arbors is applied.

Further Applications of Correlation-Based Models

The conceptual framework offered by Eq. (12.1) can be extended to explain the development of orientation columns (Miyashita and Tanaka, 1992; Miller, 1994) and the joint development of orientation and ocular dominance columns (Erwin and Miller, 1998).

For orientation selectivity, cortical inputs are again represented by two sheets of cells, but in this case they represent ON-center and OFF-center LGN cells. The model now has to produce a segregation of inputs within individual receptive fields, rather than receptive fields that are entirely dominated by one or the other layer. This will happen if (1) a difference of Gaussians is used to describe the correlations, an upright 1 (positive near the origin) for ON-ON and OFF-OFF correlations, and an inverted 1 for ON-OFF and OFF-ON interactions; and (2) the functions change sign in a distance less than the width of the arbor function. This causes the development of receptive fields that are divided into two (or, occasionally, more) regions of ON and OFF responsiveness, from which an orientation preference can be calculated. This changes continuously over the surface of the cortex; singularities are present; and individual iso-orientation domains are morphologically similar to those observed in the monkey and cat. The overall orientation pattern, however, lacks a well-defined periodicity. Periodicity is a prominent characteristic of real orientation maps, and this suggests that the model needs modification.

However, the model's basic premise—that interactions between ON- and OFF-center afferents establish the initial map of orientation preference—is supported by experimental results showing that inactivation of ON-center pathways during early development blocks the emergence of orientation selectivity (Chapman and Gödecke, 2000).

To explain the joint development of orientation selectivity and ocular dominance, the model has four input layers: L-ON, L-OFF, R ON, and R-OFF, with a corresponding 4 × 4 matrix of correlation functions. Erwin and Miller (1998) discuss the conditions that must be satisfied in order for the model to produce ocular dominance segregation as well as correlated orientation maps in the two eyes. Although these conditions can be satisfied, the model, like the simpler ON-OFF model, fails to generate periodic orientation columns. It is also unable to reproduce the tendency toward orthogonal intersection of ocular dominance column boundaries and iso-orientation domains, as observed in cats and monkeys. However, a weak tendency for orientation singularities to lie in the centers of the ocular dominance stripes can be produced by a two-stage model in which the correlation functions change over time in such a way that ON-OFF segregation occurs in advance of L-R segregation.

Linear correlation-based models would appear to be limited in terms of how well they can describe the appearance of cortical feature maps. As implemented by Miller and his colleagues, the arbor function has the undesirable effect of imposing a fixed retinotopy, while the normalization rules that are integral to the way the models work are complex and not derivable from known cellular mechanisms. In spite of this, the models are important because of the simplicity of the underlying assumption, namely, that Hebbian changes in synaptic strength build up slowly over time in ways that reflect time-averaged correlations in the patterns of input activity. It is important to see how far

such simple assumptions can go in explaining cortical development.

12.4.2 Competitive Hebbian Models

When a stimulus is presented to a sheet of cells connected by a Mexican hat pattern of lateral connections, the activity patterns that develop tend to consist of isolated patches of high activity with a size that matches the extent of the lateral excitation in the network. This suggests the following way of simplifying the computationally time-consuming calculation of activity patterns in response to a stimulus: For any particular stimulus, find the cortical point that gives the largest initial response (ignoring lateral interactions); assume nearby cells will likewise be active (because of the lateral connections); and then modify the connections by a Hebbian rule. This will have the effect of making the "winning" cortical point, and its neighbors, more responsive to the stimulus in question. The application of a neighborhood rule enforces continuity in the mapping; i.e., it ensures that nearby cortical locations will develop similar receptive field profiles. The competitive element has the opposite effect and ensures that the map represents diverse features. Thus, even if a stimulus evokes only a very weak response in the cortex initially, that response will still evoke a modification that will strengthen the response, and if the stimulus is presented sufficiently often, it will gain a representation in the map.

This method was first proposed by Kohonen (1982) and is often termed the self-organizing feature map algorithm. It can be expressed mathematically as follows: First, compute the response c_i of each cortical point i to the stimulus r—i.e., $c_i = \sum_j s_{i,j} r_j$, where the summation is over all points, indexed by j, in the input—and find the winning cortical point i^* for which c is a maximum. Then change the connection strengths according to the following rule:

$$s_{i,j}(t+1) = \alpha(t)[s_{i,j}(t) + \varepsilon h(i, i^*)r_j], \qquad (12.2)$$

where $\alpha(t)$ is a normalization factor chosen to keep the sum of the synaptic strengths at each cortical point (or the sum of their squares) a constant; and h is a neighborhood function, which is typically a Gaussian function of the distance between cortical points i and i^*. Input patterns r are chosen according to the mapping problem being studied. For example, when Obermayer et al. (1990) modeled the formation of orientation columns and the retinotopic map, points r_j were randomly positioned in a single 2-D input layer and the activity patterns were elliptical Gaussian blobs of varying position and orientation. Goodhill (1993) modeled the formation of ocular dominance columns and the retinotopic map, assuming short-range within-eye correlations and positive between-eye correlations. This was done to study the effects of changing the interocular correlation, given that this is likely to be changed by visual experience or by manipulations such as strabismus (a condition in which the two eyes point in different directions). Both models produce realistic patterns of orientation preference, or ocular dominance stripes, in which there are interesting accompanying variations in the retinotopic map. For the orientation column model, periodic fluctuations in retinal magnification factor develop, and these correlate with the orientation gradient, i.e., the rate at which preferred orientation changes with distance in the map. Specifically, there is a negative gradient correlation, so that in regions where orientation changes rapidly with position, retinal positions change slowly, and vice versa. This negative gradient correlation has been observed in other models (see section 12.4.4) that implement related developmental principles. In the model of ocular dominance column formation, z-folds developed in the retinotopic map. These would appear to be a re-

alistic feature given the evidence for this type of folding in the macaque monkey (see section 12.1.6).

Periodicity in this model appears to be determined by a variety of factors, including the size of the cortical neighborhood function and, in the case of ocular dominance, by the amount of interocular correlation. Goodhill (1993) showed that if this correlation is low or absent, then the stripes have a larger spacing than if the correlation is high. Since strabismus can be expected to reduce or abolish interocular correlations, Goodhill made the experimentally testable prediction that animals made artificially strabismic during the period when ocular dominance columns are developing should have larger than normal ocular dominance columns. Although initial tests in cats appeared to confirm this (Löwel, 1994; Tieman and Tumosa, 1997), more recent studies have not replicated the effect in cats (Sengpiel et al., 1998) or been able to demonstrate it in monkeys (Crawford, 1998; Murphy et al., 1998). A possible reason for this is that the periodicity of ocular dominance columns becomes established too early for strabismus to change it. Better tests of the prediction are likely to involve manipulations that can alter the correlations present in spontaneously occurring, prenatal patterns of activity.

The learning rule used in Eq. (12.2) differs from that used in the linear correlation models described in section 12.4.1. Here, learning is not strictly Hebbian, because although the rate of change is proportional to the level of presynaptic activity r_j, it is conditional on the synapse in question being close to a region of cortex that is responding strongly to the stimulus, rather than simply being the product of pre- and post-synaptic activities. Some physiological evidence points toward mechanisms similar to this. In rat visual cortex, it has been observed that when a connection between an afferent and a neuron is strengthened by the correlated stimulation of both cells, the connections from a

nearby but unstimulated afferent become strengthened as well (Kossel et al., 1990). This suggests that
synaptic potentiation is accompanied by a signal that
travels through tissue and potentiates nearby synapses.
This mechanism has been termed volume learning
(Montague and Sejnowski, 1994). Possible mechanisms for the spread include glial involvement and
release of nitric oxide, arachidonic acid, carbon mon-
oxide, hydrogen peroxide, or neurotrophins (Thoenen, 1995).

Piepenbrock and Obermayer (2000) have introduced a model that is a blend of linear Hebbian learning (see section 12.4.1) and the nonlinear Kohonen
mechanism (section 12.4.2). This is done by normalizing the net response of the cortex to each stimulus
and by introducing a nonlinearity (parameterized by a
constant β) in the cortical response. Low values of β
approximate the linear case, where development is
driven by the second-order statistics of the input patterns (i.e., the correlation functions C^{LL} etc.), while
large values approximate the competitive case, where
only a single small region of cortex responds to any
stimulus. In this case, learning is essentially feature
based, i.e., driven by higher-order statistics in the input patterns.

12.4.3 Lateral Interaction Models

Most, if not all, models describe the emergence of
pattern in the cortical map as the result of processes
that involve lateral interactions. The origins of these
interactions are varied in the models, just as they are
likely to be in the real brain. They include spatial correlations in the inputs, lateral intracortical interactions,
the release of diffusible substances, and factors generally subsumed under the ambit of normalization—
regulatory mechanisms exerted within individual
axons, and mechanisms regulating the total number
and strength of connections each cell receives.

A considerable simplification can be achieved by
making the following assumptions: (1) all of the
effects occur on a time scale that is short compared
with the time scale of map development; (2) the
effects add linearly; and (3) they are translationally
invariant; i.e., the net effect of each type of interaction
is a function of the distance between points and does
not vary with absolute location in the cortex. It is
then possible to lump all the interactions together and
write down an equation for growth (e.g., of one type
of connection) in terms of convolution with kernels
that describe the lateral interactions within and between the pattern elements (Swindale, 1980, 1982).
For left and right eye synapses, whose densities are,
respectively, given by n_L and n_R as functions of position on the cortical surface, we can write

$$\frac{dn_L}{dt} = (n_L * w_{LL} + n_R * w_{RL})f(n_L)$$

$$\frac{dn_R}{dt} = (n_R * w_{RR} + n_L * w_{LR})f(n_R),$$

(12.3)

where w_{LL} and w_{RR} describe within-eye interactions; w_{RL} and w_{LR} describe, respectively, the effects
of right-eye on left-eye, and left-eye on right-eye
connections; and the asterisk denotes convolution.

The function $f(n)$ is used to terminate growth as it
reaches some upper or lower limiting density; a suitable form is $f(n) = n(N - n)$, where N is the upper
limiting density and the lower limit is assumed to be
zero. If the within-eye interactions are described by
an upright Mexican hat function, and the between-
eye interactions by an inverted Mexican hat function,
then an initial state in which left and right eye synapses have random densities > 0 and $< N$ evolves
into a branching periodic pattern of stripes with the
morphological features of ocular dominance columns
(figure 12.5a).

a

b

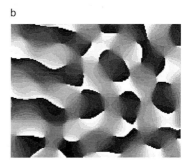

Figure 12.5

(*a*) Simulated pattern of ocular dominance stripes produced by a lateral interaction model [Eq. (12.3)]. (Reproduced from Swindale, 1980.) (*b*) Orientation preference map simulated using Eq. (12.5). For gray scale, see figure 12.3.

A comparison can be made between this type of model and reaction-diffusion models of pattern formation, which were initially developed by Turing (1952) and later applied by others to various aspects of pattern formation, particularly animal coat patterns (Meinhardt, 1982; Murray, 1989; see also chapters 1, 2, and 3). As pointed out earlier, these often bear an interesting resemblance to ocular dominance stripe patterns. In Turing's formulation, substances called morphogens react with each other and diffuse laterally through the substrate (usually assumed to be two-dimensional). For two substances with linear first-order reaction kinetics, the concentrations evolve according to the following differential equations:

$$\frac{\partial X}{\partial t} = aX + bY + D_1 \frac{\partial^2 X}{\partial r^2}$$

$$\frac{\partial Y}{\partial t} = cX + dY + D_2 \frac{\partial^2 Y}{\partial r^2}, \tag{12.4}$$

where X and Y are the morphogen concentrations; a, b, c, and d are rate constants; r is spatial position; and D_1 and D_2 are diffusion coefficients. Turing showed that with suitably chosen rate constants, initially nearly uniform concentrations of X and Y would develop into spatially periodic patterns. He supposed that these patterns, or prepatterns, would then trigger the differentiation of tissues into the observed pattern. Simulations of these, or related, systems of equations show that periodic spots or stripes of morphogen concentration are produced, although stripes are generally formed on narrow cylinders (Bard, 1981; Murray, 1981; Lyons and Harrison, 1991). A comparison of Eq. (12.3) with Eq. (12.4) makes the differences between the two mechanisms clear. Turing assumed that the actions of X and Y on themselves and each other were strictly local, which is appropriate for chemical reactions, while Eq. (12.3) assumes "action at a distance" as subsumed by the lateral interaction terms. Lateral interactions in the Turing model are mediated by the actual movement, via diffusion, of X and Y through the tissue, whereas in Eq. (12.3) lateral movement of synapses does not occur.

The lateral interaction model can also be implemented as a cellular automaton. This is a class of model in which pattern elements have discrete states at discrete times and simple neighborhood rules are used to determine state transitions at each time step (Wolfram, 1984). The stripe-forming behavior of Eq. (12.3) can be adequately approximated by the following procedure: Let n take only values of $+1$ or -1 on a discrete 2-D lattice indexed by (i, j). At each time step, calculate $a_{i,j} = \sum_{k,l} n_{i+k,j+l} w_{k,l}$, where w is also Mexican hat in form (e.g., $w_{k,l} = 1$ for $0 \leq |k,l| \leq d_E$; $w_{k,l} = -1$ for $d_E < |k,l| \leq d_I$, where the distances d_E and d_I define the spread of excitation and inhibition, respectively). The values of $n_{i,j}$ are then updated according to the sign of $a_{i,j}$. That is, if $a_{i,j} > 0$, $n_{i,j}$ is set to 1; if $a_{i,j} < 0$, $n_{i,j}$ is set to -1. It may also be noted that this system is isomorphic with a suitably connected Hopfield net (Hopfield, 1982). Thus ocular dominance stripe patterns are stable (ground) states of a Hopfield net with short-range excitatory and long-range inhibitory connections.

Orientation Columns

The approach of the previous section can be extended to orientation columns (Swindale, 1982). Here we assume that the quantity that is emerging in the map is an orientation, represented by a vector $\mathbf{z} = (a, b)$. In order to ensure that angles differing by 180 degrees are equivalent, we adopt the convention that the orientation represented by $\mathbf{z}$ is $\theta = 0.5 \operatorname{atan}(b/a)$. It is intuitive to regard $|\mathbf{z}|$ as a measure of the strength of orientation tuning; i.e., regions of cortex with narrow orientation tuning will have large values of $|\mathbf{z}|$, while regions of weak or disorganized selectivity will have small values of $|\mathbf{z}|$. As before, the change in $\mathbf{z}$ is assumed to be determined by many different processes, the net outcome of which is to make nearby regions develop similar preferences and to make regions further away develop dissimilar preferences. Thus, we write

$$\frac{d\mathbf{z}}{dt} = \mathbf{z} * w_{\mathbf{z}} f(\mathbf{z}), \qquad (12.5)$$

where $w_{\mathbf{z}}$ is a Mexican hat function describing the lateral interactions and $f(\mathbf{z})$ is used to keep $\mathbf{z}$ within bounds, e.g., $f(\mathbf{z}) = (1 - \mathbf{z})$. Solutions to Eq. (12.5),

with values of $\mathbf{z}$ initially small and randomly distributed, are a good match to real orientation column patterns with respect to periodicity and singularity distribution (figure 12.5b).

A combined model for ocular dominance and orientation columns has been proposed (Swindale, 1992) based on the idea of competition for feature selectivity. It is supposed that development of one feature might be slowed down in regions where the other is emerging most rapidly and vice versa. The resulting model is able to reproduce the orthogonal pattern of the intersection of orientation domains with ocular dominance column borders as well as the tendency of singularities to lie in the centers of ocular dominance stripes (Erwin et al., 1995; Swindale, 1996).

12.4.4   *Low-Dimensional Feature Map Models*

In this section, we retain the simplifying idea of formulating a model in terms of low-level features, rather than the patterns of neural activity that correspond to them, and we return to the framework of the competitive Hebbian models discussed in section 12.4.2. We assume that the input to the cortex is a feature vector $\mathbf{v}$, which is a point in a feature space $\mathbf{S}^N$, as discussed in section 12.1.2. For a combined map of retinotopy, ocular dominance, and orientation, we might let $\mathbf{v} = \{x, y, n, a, b\}$, where x and y represent position in retinal space, n represents ocular dominance, and a and b represent the two components of the orientation vector (using the conventions described in the section on orientation columns). The cortex is represented by a 2-D sheet of points indexed by i, and $\mathbf{w}_i$ represents the feature vector currently mapped to point i. We can picture the sheet as folded inside $\mathbf{S}$ in the manner suggested by figure 12.1. Remember that the Kohonen algorithm (see section 12.4.2) worked by taking a stimulus, finding the most responsive cortical point, and then modifying its connections and those of its neighbors in such a way as to make it more responsive to the stimulus in question. The low-dimensional version of the Kohonen algorithm works in the same way if we assume that the closer $\mathbf{w}_i$ is to $\mathbf{v}$, the stronger is the response of point i to $\mathbf{v}$. As before, i^* denotes the most responsive point, which is given by $\min|\mathbf{w}_i - \mathbf{v}| \; \forall i$. Points in the cortex are then moved toward $\mathbf{v}$ by an amount given by

$$\mathbf{w}_i(t+1) = \mathbf{w}_i(t) + \varepsilon[\mathbf{v} - \mathbf{w}_i(t)]h(i^*, i). \qquad (12.6)$$

As before, ε is a growth rate and $h(i^*, i)$ is a neighborhood function that equals 1 for $i = i^*$ and that falls smoothly to zero with increasing distance between i and i^*. Initial values of $\mathbf{w}(t = 0)$ are typically assumed to be small and random, with the exception of retinotopic space, where a linear mapping with some specified amount of random scatter is usually assumed. At each time step, a new stimulus, typically chosen at random from a defined manifold within $\mathbf{S}$, is presented and the procedure repeated until a stable, or nearly stable, mapping has been obtained.

Obermayer et al. (1991, 1992) used this algorithm as a model for visual cortex map formation. Despite the simplification involved, and the need to define a suitable scaling and metric for determining distances between points in $\mathbf{S}$, the types of mappings obtained in the low- and high-dimensional instances are similar. For the complete case of retinotopy, ocular dominance, and orientation, the resulting maps capture the main features observed in the monkey, including periodic ocular dominance stripes, periodic iso-orientation domains, orientation singularities in the centers of ocular dominance stripes, and orthogonal crossings of iso-orientation domains and ocular dominance stripe borders (figure 12.6). The low-dimensional Kohonen algorithm has been applied to maps of direction preference (Swindale and Bauer,

1998). Mitchison and Swindale (1999) have studied the effects of making the learning rule in Eq. (12.6) more strictly Hebbian by making modification contingent on the receptive field of any cortical unit (not just the winning one) being sufficiently close to $\mathbf{v}$.

It can be seen that the mappings produced by the Kohonen algorithm will tend to satisfy the continuity and completeness constraints discussed in section 12.1.2. That is, neighboring cortical points will tend to be close together in $\mathbf{S}$, maximizing continuity, while for each $\mathbf{v}$ that is presented, there will generally be a $\mathbf{w}_i$ that is close to $\mathbf{v}$, satisfying the completeness (or coverage) requirement. (This is really an empirical observation because there is no analytical proof that the Kohonen algorithm maximizes any combination of these properties.) If the set of stimuli that is used is finite and less than the number of cortical points, then of course a solution to the mapping can always be found where there is a matching point i for every $\mathbf{v}$, i.e., for which $|\mathbf{w}_i - \mathbf{v}| = 0$. Note that finding the solution that minimizes the distances in $\mathbf{S}$ between adjacent cortical points (i.e., that maximizes continuity in the mapping) is the same as solving the traveling salesman problem. In this problem, a route must be found in which neighboring points on the route (cities) are close together, minimizing the total distance traveled, and the route must pass through every city. The only difference is that solutions to the conventional traveling salesman problem are mappings from a 1-D route to a 2-D surface, whereas in the cortex the mapping is from a 2-D surface to an N-dimensional space. This means that any algorithm that can be shown to produce good solutions to the traveling salesman problem can be applied to the problem of cortical map formation, although of course not all algorithms may be equally suitable and the interpretation of the algorithms' behavior in biological terms may be difficult.

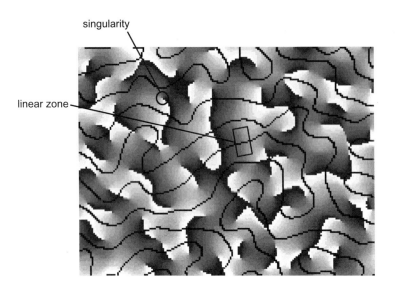

Figure 12.6
Combined feature map of orientation and ocular dominance produced by the low-dimensional Kohonen algorithm [Eq. (12.6)]. Dark lines mark the boundaries of the ocular dominance columns. Each gray level represents a unique orientation preference (for gray scale, see figure 12.3). An example of a singularity (circle) and a linear zone are shown (rectangle). Note the resemblance to the biological map shown in figure 12.3. (Figure courtesy of K. Obermayer; simulation details are given in Blasdel and Obermayer, 1994.)

One such algorithm, the elastic net algorithm (Durbin and Willshaw, 1987; see also chapter 11), does have a plausible biological basis (Willshaw and Von der Malsburg, 1979) and has been applied successfully to cortical map formation (Durbin and Mitchison, 1990; Goodhill and Willshaw, 1990; Erwin et al., 1995; Goodhill and Cimponeriu, 2000). In this model (figure 12.7), a finite number of stimuli $\mathbf{v}_j$ ($j = 1 \ldots M$) exert attractive "forces" and pull nearby cortical receptive fields $\mathbf{w}_i$ toward them. This part of the model can be considered to be Hebbian, inasmuch as Hebbian rules have the effect of making the receptive fields of cells change so that they are more responsive to those input patterns to which they are already most responsive. Units that are neighbors in the cortex are also connected by "elastic" and are

thereby subjected to forces that tend to enforce continuity in the mapping. The learning rule is

$$\mathbf{w}_i(t+1) = \mathbf{w}_i(t) + \alpha \sum_j F_{i,j}(\mathbf{v}_j - \mathbf{w}_i)$$

$$+ \beta K \sum_{k \in N_i} (\mathbf{w}_k - \mathbf{w}_i), \qquad (12.7)$$

where α and β are constants scaling the Hebbian and elastic forces, respectively, and the summation in the second term is over the nearest neighbors k of point i. The "force" $F_{i,j}$ exerted by stimulus j on cortical point i is a Gaussian function of the distance between $\mathbf{w}_i$ and $\mathbf{v}_j$ (i.e., the response, assuming Gaussian receptive fields) normalized by the sum of the responses from all other cortical units, i.e.,

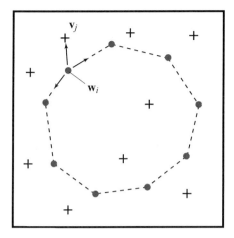

Figure 12.7
Illustration of how the elastic net algorithm works. For simplicity, a 1-D cortex (circles) is showed mapping to a 2-D stimulus space. The diagram shows the "forces" acting on one cortical point $\mathbf{w}_i$. Stimuli (crosses) exert attractive forces whose effects fall off as a Gaussian function of distance. The stimulus that has the strongest effect on the movement of $\mathbf{w}_i$ is thus the one closest to it, $\mathbf{v}_j$; other, more distant stimuli exert weaker attractive forces. Elastic forces are proportional to the distance between neighboring cortical points; in this case, they oppose the motion of $\mathbf{w}_i$ toward $\mathbf{v}_j$.

$$F_{i,j} = \frac{\exp(-|\mathbf{v}_j - \mathbf{w}_i|^2/2K^2)}{\sum_p \exp(-|\mathbf{v}_j - \mathbf{w}_p|^2/2K^2)}. \qquad (12.8)$$

Normalization ensures that stimuli that are far away from any cortical point do not get ignored and exert forces that are as large as those exerted by stimuli that are closer to cortical units. The parameter K scales the receptive field sizes, i.e., the distance in $\mathbf{S}$ over which stimuli exert attractive forces. This distance may be large initially, and it is typically reduced in size (annealing) in order to make individual cortical points approach specific stimuli.

This algorithm provides good solutions to the traveling salesman problem (Durbin and Willshaw, 1987)

and, like the low-dimensional Kohonen algorithm, produces realistic maps of orientation and ocular dominance columns.

To sum up this section, the advantages of feature-based algorithms are that they start with very general principles (continuity and completeness) and show how these lead to detailed predictions about the layout of cortical maps. They are computationally simple, which means that large areas of cortex can be modeled and that it is easy to run many simulations. Finally, the models work well, judged by results. The disadvantages are that it is difficult to reformulate the algorithms in terms of nuts-and-bolts models of neural development, which makes it hard to extend them by incorporating new biological details. Nor, arguably, do the models give insights into the significance of many of the biological mechanisms known to be involved. Setting up the models requires the definition of a suitable stimulus manifold, which generally has to be done in an ad hoc way. In addition, lateral interactions, such as the cortical neighborhood function, cannot be justified by reference to known biological interactions.

12.4.5   Models Based on Sprouting and Neurotrophic Interactions

Many models impose rigid constraints on the types of growth that afferent connections can exhibit. Growth may be restricted to within a predetermined arbor function; synapses may not be allowed to reappear once a connection strength has gone to zero in any area; and connections are generally assumed to be formed in response to scalar (i.e., the local concentration of a trophic molecule) rather than gradient (i.e., movement of a growth cone up or down a concentration gradient) growth cues. Real axons, of course, behave in more complex ways, with new connections formed by sprouting, often in response to chemical gradient cues, and by neurotrophic factors released by target neurons. Competition for neurotrophic support is thought to be one of the main mechanisms governing the selective elimination of connections during development in many parts of the nervous system (Purves, 1988; Van Ooyen and Willshaw, 1999; Van Ooyen, 2001; see also chapter 10), and there is evidence that this is true in the visual cortex as well. It is obviously of interest to explore models that explicitly represent such features. Elliott et al. (1996) have incorporated sprouting and retraction in a model of ocular dominance column formation with the specific intention of avoiding normalization rules. They do this by first assuming that two sheets of LGN neurons project to a cortex, with every LGN cell making connections within a retinotopically defined square region of cortex. Denoting the activity of the ith or jth LGN connection by σ_i or $\sigma_j \in \{1, -1\}$, the following energy function is defined:

$$E = -\frac{1}{2} \sum_{i,j} \sigma_i \sigma_j h(i, j), \qquad (12.9)$$

where $h(i, j)$ is a coupling, or cortical neighborhood, function that has the value 1 when connections i and j are on the same cortical cell or are on nearest-neighbor cells, and that is zero otherwise. The LGN activity patterns are assumed to be randomly positioned circles of activity confined to one or the other sheet. Connections are assumed to either exist, in which case they have a nominal strength of 1, or to not exist, which is signaled by a strength value of zero and their absence from the summation in Eq. (12.9). They can appear or disappear from within an arbor region, with changes that decrease the value of E generally being favored over changes that increase it. Elliott et al. tested a "relocation" model in which fixed upper and lower limits were imposed on the

number of connections per cell, and a connection was allowed to move to a new location within its parent arbor, provided the number of connections per cell remained within the limits. An "interchange" model was also tested; this required that each cell received a fixed number of connections and that a pair of axons was allowed to exchange connections provided they were within each other's arbor regions. In both types of model, change was accepted with a probability $1/(1 + e^{\Delta E/T})$, where T is a temperature, and updates were repeated many times for many different LGN activation patterns. Examples in which $T = 0$, and in which T was slowly reduced during development, were studied.

The advantage of this model is that it is formally simple and has obvious links with physical systems such as spin glasses (Elliott and Shadbolt, 1998a). An interpretation is possible in which the contribution of a pair of connections to E is inversely related to the level of neurotrophic support (low E means high neurotrophin levels and vice versa), while the neighborhood function h represents, in a crude way, the release and diffusion of neurotrophin molecules by postsynaptic cells through the tissue.

More complex models, which explicitly model activity-dependent release of neurotrophins, have recently been proposed (Harris et al., 1997; Elliott and Shadbolt, 1998b, 1999). These models avoid the use of synaptic weight normalization rules, are able to explain the effects of neurotrophin injections (section 12.2.2), and can explain segregation in the presence of positive interocular correlations. In the model by Elliott and Shadbolt (1998b, 1999), the general anatomical framework is similar to that just described. Neurotrophins are released by postsynaptic neurons in amounts proportional to their activity, diffuse through the tissue, and are taken up by afferent axons in amounts that are proportional to their activity and the number of synapses present in their arbor. Afferents are able to make or lose connections in proportion to a recent time average of the neurotrophin uptake. The LGN activity patterns are either correlated random noise (section 12.3.1) (Elliott and Shadbolt, 1998b) or simulated retinal waves (Elliott and Shadbolt, 1999). Like Goodhill (1993), Elliott and Shadbolt predict that a decrease in interocular correlations should increase column spacing (although, as discussed in section 12.4.2, the evidence for this effect is ambiguous). In addition, the model predicts that changes in the spatial extent of within-eye correlations should affect periodicity. This prediction may be testable by pharmacologically manipulating retinal activity at early stages of development in ferrets.

12.4.6 The BCM Modification Rule

Many of the models discussed in this chapter use a simple Hebbian rule for strengthening connections, while weakening occurs in a nonspecific way as a result of normalizing synaptic strengths onto neurons. Nevertheless, it has been realized for a long time that there is a logical complement to Hebb's principle, namely, the postulate that when a presynaptic axon is active but fails to cause the postsynaptic cell to fire, its connection strength is weakened (Stent, 1973). Such behavior has been demonstrated physiologically and is known as long-term depression. Bienenstock, Cooper, and Munro (1982) developed this idea mathematically into what is now generally known as the BCM rule for synaptic modification. Considering only the inputs to a single cell, the learning rule for the jth connection carrying an input v_j to a cortical cell with an activity c is

$$s_j(t+1) = (1 - \varepsilon)s_j(t) + \varphi[c(t)]v_j, \qquad (12.10)$$

where $\varphi(c)$ is a function that is negative when the postsynaptic response c is below a modification threshold θ_M and positive when it is greater than θ_M. The small constant ε produces a constant decay in synaptic strength in the absence of any input or output activity; its effects can be ignored in the present context. In the absence of any kind of normalization, this rule has the undesirable feature that there can easily be situations in which all the inputs either increase or decrease without limit, leading to a loss of any kind of feature selectivity. This problem can be avoided by the use of a sliding modification threshold in which θ_M varies as a function of the recent average activity of the cell, $\bar{c}$. The time period over which this average is taken is not critical. The dependence of θ_M on $\bar{c}$ is important, however, and it can be shown that for stable feature selectivity to be guaranteed, θ_M must increase and decrease more rapidly, relative to a fixed value c_0, than does $\bar{c}$. This is achieved if

$$\theta_M(\bar{c}) = (\bar{c}/c_0)^p \bar{c}, \qquad (12.11)$$

where $p > 1$. With this rule, emergence of feature selectivity, for example, to oriented patterns of activity in a 2-D array of inputs, is guaranteed whatever the initial values of s_j.

Although a simulation of the development of a 1-D layout of orientation selectivity was presented by Bienenstock et al. (1982), the BCM rule does not appear to have been incorporated into any 2-D model of visual cortex map formation. This is a pity because it has the clear advantage of avoiding the complex normalization rules employed in many models. There is also a significant body of evidence supporting the idea of a synaptic modification threshold (Bear and Rittenhouse, 1999). For example, the finding that monocular deprivation produced by lid suture produces a bigger shift in ocular dominance than does silencing one eye's inputs by TTX injections (Rittenhouse et al., 1999) can be explained by a BCM mechanism (Blais et al., 1999).

12.5   Discussion

This chapter has given an overview of a variety of models of visual cortex development (additional coverage of many of the topics discussed here can be found in Erwin et al., 1995; Miller, 1995; Swindale, 1996). Almost all of these models assume as a framework a two-layer (LGN + cortex) feedforward net, spatially correlated patterns of activity in the input layer, and Hebbian modification. Lateral cortical interactions, mediated either by neural connections or by diffusion of plasticity-modifying substances, are universally assumed. Competition among inputs is enforced more variably, e.g., by the use of subtractive or divisive normalization rules, by competition for trophic support, or by the use of learning rules that allow for synaptic weakening as well as strengthening to occur (this is implicit in the low-dimensional Kohonen and elastic net models and explicit in the BCM learning rule).

Some of the models include retinotopic refinement as part of the mechanism, although in others a rigid retinotopy is built in. All of the ocular dominance column models are able to explain the formation of the basic striped pattern of ocular dominance, while individual models are able to account for the effects of experimental manipulations such as monocular deprivation, silencing of retinal inputs, and neurotrophin injections. Of the models for orientation columns, and those for the joint formation of ocular dominance and orientation columns, feature-based and competitive Hebbian models appear to perform better than linear correlation-based models. Despite these undoubted successes, there is probably no single model that is able to account for all of the phenomena listed in sections 12.2.2 and 12.2.3.

Future Modeling Studies

Several recent experimental findings seem relevant to the further development of new models. Studies in the hippocampus (Bi and Poo, 1998) and cortex (Markram et al., 1997) show that the relative timing of pre- and postsynaptic spikes is a critical factor controlling connection strengths. If an action potential follows a synaptic input within about 20 ms, the input is potentiated; if the action potential precedes the presynaptic input by up to 20 ms, the input is weakened. The implications of this finding from a modeling perspective have only just begun to be explored. For example, timing, since it allows for weakening, can be used in place of normalization to mediate competition among inputs (Song et al., 2000) and stabilize postsynaptic firing rates (Kempter et al., 2001). Future models of visual cortex development will probably have to take into account the possibility that two inputs may be positively correlated at one temporal interval but negatively correlated at another. Feedback connections from cortex to LGN may alter this correlation structure (Weliky and Katz, 1999) and therefore may need to be incorporated.

It is possible that models of the formation of ocular dominance columns may need even more radical revision, since recent work has shown that segregation will occur in the ferret even after removal of both retinas (Crowley and Katz, 1999) and is present almost as soon as thalamic afferents have grown into layer IV (Crowley and Katz, 2000) (see also chapter 10). While it is possible that spontaneous activity in deafferented LGN layers might drive segregation, it is conceivable that it might be driven instead by eye-specific chemical labels (perhaps through mechanisms of the type discussed in section 12.4.3). While it is hard to see how chemical diffusion mechanisms might be extended to account for the formation of more complex properties such as orientation selectivity, different kinds of mechanisms might be involved in the formation of the two types of column.

Future Experimental Studies

Although the problems being addressed here are developmental, much remains to be learned about the organization of visual cortical maps in adult animals, and these details are likely to be crucial in constraining developmental models. Current techniques (e.g., optical imaging and functional magnetic resonance imaging) average the signals from large numbers of neurons and across different layers, and it would be useful to have much more detailed information, at the single neuron level, about what receptive field properties are mapped, how they are mapped, and in which layers. Simultaneous extracellular recording of single-cell receptive field properties in groups of neurons whose spatial locations are precisely known (micromapping), relative to each other and to a coarser-scale map determined by optical imaging from the same region of tissue, may be the best approach to this problem. Knowing what happens at the very earliest, prenatal, stages of development is also likely to be crucial. The lack of knowledge about the spatio-temporal patterns of neural activity in the retina, LGN, and cortex during the periods when ocular dominance and orientation columns are forming is probably the weakest component of all models. Finally, little is known about the fine-scale structure of the retinotopic map in the adult visual cortex, or about the mechanisms that establish and refine retinotopy in young animals. Answers to these questions would be of great value to modelers.

References

Albus, K. (1975). A quantitative study of the projection area of the central and paracentral visual field in area 17 of the cat. I. The precision of the topography. *Exp. Brain Res.* 24: 159–179.

Albus, K., and Wolf, W. (1984). Early post-natal development of neuronal function in the kitten's visual cortex: A laminar analysis. *J. Physiol. (London)* 348: 153–185.

Barlow, H. B., and Pettigrew, J. D. (1971). Lack of specificity of neurones in the visual cortex of young kittens. *J. Physiol. (London)* 218: 98P–100P.

Bard, J. (1981). A model for generating aspects of zebra and other mammalian coat patterns. *J. Theor. Biol.* 93: 363–385.

Bartfeld, E., and Grinvald, A. (1992). Relationships between orientation-preference pinwheels, cytochrome oxidase blobs, and ocular dominance columns in primate striate cortex. *Proc. Natl. Acad. Sci. U.S.A.* 89: 11905–11909.

Bear, M. F., and Rittenhouse, C. D. (1999). Molecular basis for induction of ocular dominance plasticity. *J. Neurobiol.* 41: 83–91.

Bi, G., and Poo, M.-M. (1998). Synaptic modifications in cultured hippocampal neurons: Dependence on spike timing, synaptic strength, and postsynaptic cell type. *J. Neurosci.* 18: 10464–10472.

Bienenstock, E. L., Cooper, L. N., and Munro, P. W. (1982). Theory for the development of neuron selectivity: Orientation specificity and binocular interaction in visual cortex. *J. Neurosci.* 2: 32–48.

Blais, B. S., Shouval, H. Z., and Cooper, L. N. (1999). The role of presynaptic activity in monocular deprivation: Comparison of homosynaptic and heterosynaptic mechanisms. *Proc. Natl. Acad. Sci. U.S.A.* 96: 1083–1087.

Blakemore, C., and Cooper, G. F. (1970). Development of the brain depends on the visual environment. *Nature* 228: 477–478.

Blasdel, G. G. (1992). Orientation selectivity, preference, and continuity in monkey striate cortex. *J. Neurosci.* 12: 3139–3161.

Blasdel, G. G., and Fitzpatrick, D. (1984). Physiological organization of layer 4 in macaque striate cortex. *J. Neurosci.* 4: 880–895.

Blasdel, G. G., and Lund, J. S. (1983). Termination of afferent axons in macaque striate cortex. *J. Neurosci.* 3: 1389–1413.

Blasdel, G., and Obermayer, K. (1994). Putative strategies of scene segmentation in monkey visual cortex. *Neur. Net.* 7: 865–881.

Blasdel, G. G., Obermayer, K., and Kiorpes, L. (1995). Organization of ocular dominance and orientation columns in the striate cortex of neonatal macaque monkeys. *Vis. Neurosci.* 12: 589–603.

Bonhoeffer, T., and Grinvald, A. (1991). Orientation columns in cat are organized in pin-wheel like patterns. *Nature* 353: 429–431.

Bosking, W. H., Crowley, J. C., and Fitzpatrick, D. (1997). Fine structure of the map of visual space in the tree shrew striate cortex revealed by optical imaging. *Soc. Neurosci. Abstr.* 23: 1945.

Braastad, B. O., and Heggelund, P. (1985). Development of spatial receptive-field organization and orientation selectivity in kitten striate cortex. *J. Neurophysiol.* 53: 1158–1178.

Burgi, P.-Y., and Grzywacz, N. M. (1994). Model for the pharmacological basis of spontaneous synchronous activity in developing retinas. *J. Neurosci.* 14: 7426–7439.

Cabelli, R. J., Hohn, A., and Shatz, C. J. (1995). Inhibition of ocular dominance column formation by infusion of NT-4/5 or BDNF. *Science* 267: 1662–1666.

Chapman, B., and Gödecke, I. (2000). Cortical cell orientation selectivity fails to develop in the absence of ON-center retinal ganglion cell activity. *J. Neurosci.* 20: 1922–1930.

Crair, M. C., Gillespie, D. C., and Stryker, M. P. (1998). The role of visual experience in the development of columns in cat visual cortex. *Science* 279: 566–570.

Crawford, M. L. J. (1998). Column spacing in normal and visually deprived monkeys. *Exp. Brain Res.* 123: 282–288.

Crowley, J. C., and Katz, L. C. (1999). Development of ocular dominance columns in the absence of retinal input. *Nat. Neurosci.* 2: 1125–1130.

Crowley, J. C., and Katz, L. C. (2000). Early development of ocular dominance columns. *Science* 290: 1321–1324.

Das, A., and Gilbert, C. D. (1997). Distortions of visuotopic map match orientation singularities in primary visual cortex. *Nature* 387: 594–598.

Durbin, R., and Mitchison, G. (1990). A dimension reduction framework for understanding cortical maps. *Nature* 343: 644–647.

Durbin, R., and Willshaw, D. J. (1987). An analogue approach to the travelling salesman problem using an elastic net method. *Nature* 326: 698–691.

Elliott, T., and Shadbolt, N. R. (1998a). Competition for neurotrophic factors: Mathematical analysis. *Neur. Comput.* 10: 1939–1981

Elliott, T., and Shadbolt, N. R. (1998b). Competition for neurotrophic factors: Ocular dominance columns. *J. Neurosci.* 18: 5850–5858.

Elliott, T., and Shadbolt, N. R. (1999). A neurotrophic model of the development of the retinogeniculocortical pathway induced by spontaneous retinal waves. *J. Neurosci.* 19: 7951–7970.

Elliott, T., Howarth, C. I., and Shadbolt, N. R. (1996). Axonal processes and neural plasticity. I: Ocular dominance columns. *Cereb. Cortex* 6: 781–788.

Erwin, E., Obermayer, K., and Schulten, K. (1995). Models of orientation and ocular dominance columns in the visual cortex: A critical comparison. *Neur. Comput.* 7: 425–468.

Erwin, E., and Miller, K. D. (1998). Correlation-based development of ocularly matched orientation and ocular dominance maps: Determination of required input activities. *J. Neurosci.* 18: 9870–9895.

Feller, M. B., Butts, D. A., Aaron, H. L., Rokhsar, D. S., and Shatz, C. J. (1997). Dynamic processes shape spatiotemporal properties of retinal waves. *Neuron* 19: 293–306.

Florence, S. L., and Kaas, J. H. (1992). Ocular dominance columns in area 17 of Old World macaque and talapoin monkeys: Complete reconstructions and quantitative analyses. *Vis. Neurosci.* 8: 449–462.

Galli, L., and Maffei, L. (1988). Spontaneous impulse activity of rat retinal ganglion cells in prenatal life. *Science* 242: 90–91.

Goodhill, G. J. (1993). Topography and ocular dominance: A model exploring positive correlations. *Biol. Cybern.* 69: 109–118.

Goodhill, G. J., and Barrow, H. G. (1994). The role of weight normalization in competitive learning. *Neur. Comput.* 6: 255–269.

Goodhill, G. J., and Cimponeriu, A. (2000). Analysis of the elastic net model applied to the formation of ocular dominance and orientation columns. *Network* 11: 153–168.

Goodhill, G. J., and Willshaw, D. J. (1990). Application of the elastic net algorithm to the formation of ocular dominance stripes. *Network* 1: 41–59.

Harris, A. E., Ermentrout, G. B., and Small, S. L. (1997). A model of ocular dominance column development by competition for trophic factor. *Proc. Natl. Acad. Sci. U.S.A.* 94: 9944–9949.

Hirsch, H. V. B., and Spinelli, D. N. (1970). Visual experience modifies distribution of horizontal and vertical oriented receptive fields in cats. *Science* 168: 869–871.

Hopfield, J. J. (1982). Neural networks and physical systems with emergent collective computational abilities. *Proc. Natl. Acad. Sci. U.S.A.* 79: 2554–2558.

Horton, J. C., and Hocking, D. R. (1996). An adult-like pattern of ocular dominance columns in striate cortex of newborn monkeys prior to visual experience. *J. Neurosci.* 16: 1791–1807.

Hubel, D. H., and Wiesel, T. N. (1962). Receptive fields, binocular interaction, and functional architecture of cat striate cortex. *J. Physiol. (London)* 160: 106–154.

Hubel, D. H., and Wiesel, T. N. (1974). Sequence regularity and geometry of orientation columns in the monkey striate cortex. *J. Comp. Neurol.* 158: 267–294.

Hubel, D. H., Wiesel, T. N., and LeVay, S. (1974). Visual field of representation in layer IVc of monkey striate cortex. *Soc. Neurosci. Abstr.* 4th Annual Meeting, p. 264.

Hubel, D. H., Wiesel, T. N., and LeVay, S. (1977). Plasticity of ocular dominance columns in monkey striate cortex. *Phil. Trans. Roy. Soc. London B.* 278: 131–163.

Hübener, M., Shoham, D., Grinvald, A., and Bonhoeffer, T. (1997). Spatial relationships among three columnar systems in cat area 17. *J. Neurosci.* 17: 9270–9284.

Kandler, K., and Katz, L. C. (1998). Coordination of neuronal activity in developing visual cortex by gap junction-mediated biochemical communication. *J. Neurosci.* 18: 1419–1427.

Katz, L. C., and Shatz, C. J. (1996). Synaptic activity and the construction of cortical circuits. *Science* 274: 1133–1138.

Kempter, R., Gerstner, W., and van Hemmen, J. L. (2001). Intrinsic stabilization of output rates by spike-based Hebbian learning. *Neural Comput.* 13: 2709–2742.

Kohonen, T. (1982). Self-organized formation of topologically correct feature maps. *Biol. Cybern.* 43: 59–69.

Kossel, A., Bonhoeffer, T., and Bolz, J. (1990). Non-Hebbian synapses in rat visual cortex. *NeuroReport* 1: 115–118.

LeVay, S., Wiesel, T. N., and Hubel, D. H. (1980). The development of ocular dominance columns in normal and visually deprived monkeys. *J. Comp. Neurol.* 191: 1–51.

Löwel, S. (1994). Ocular dominance column development: Strabismus changes the spacing of adjacent columns in cat visual cortex. *J. Neurosci.* 14: 7451–7468.

Lyons, M. J., and Harrison, L. G. (1991). A class of reaction-diffusion mechanisms which preferentially select striped patterns. *Chem. Phys. Lett.* 183: 158–164.

Markram, H., Lübke, J., Frotscher, M., and Sakmann, B. (1997). Regulation of synaptic efficacy by coincidence of postsynaptic APs and EPSPs. *Science* 275: 213–215.

Meinhardt, H. (1982). *Models of Biological Pattern Formation*. London: Academic Press.

Miller, K. D. (1994). A model for the development of simple cell receptive fields and the ordered arrangement of orientation columns through the activity-dependent competition between ON- and OFF-center inputs. *J. Neurosci.* 14: 409–441.

Miller, K. D. (1995). Receptive fields and maps in the visual cortex: Models of ocular dominance and orientation columns. In *Models of Neural Networks III*, E. Domany, J. L. van Hemmen, and K. Schulten, eds. New York: Springer-Verlag.

Miller, K. D., and MacKay, D. J. C. (1994). The role of constraints in Hebbian Learning. *Neur. Comput.* 6: 100–126.

Miller, K. D., Keller, J. B., and Stryker, M. P. (1989). Ocular dominance column development: Analysis and simulation. *Science* 245: 605–615.

Mitchison, G., and Swindale, N. V. (1999). Can Hebbian volume learning explain discontinuities in cortical maps? *Neur. Comput.* 11: 1519–1526.

Miyashita, M., and Tanaka, S. (1992). A mathematical model for the self-organization of orientation columns in visual cortex. *NeuroReport* 3: 69–72.

Montague, P. R., and Sejnowski, T. J. (1994). The predictive brain: Temporal coincidence and temporal order in synaptic learning mechanisms. *Learning Memory* 1: 1–33.

Mountcastle, V. B. (1957). Modality and topographic properties of single neurons of cat's somatic sensory cortex. *J. Neurophys.* 20: 408–434.

Murphy, K. M., Jones, D. G., Fenstemaker, S. B., Pegado, V. D., Kiorpes, L., and Movshon, J. A. (1998). Spacing of cytochrome oxidase blobs in visual cortex of normal and strabismic monkeys. *Cereb. Cortex* 8: 237–244.

Murray, J. D. (1981). On pattern formation mechanisms for lepidopteran wing patterns and mammalian coat markings. *Phil. Trans. Roy. Soc. London B.* 295: 473–496.

Murray, J. D. (1989). *Mathematical Biology.* New York: Springer-Verlag.

Obermayer, K., and Blasdel, G. G. (1993). Geometry of orientation and ocular dominance columns in monkey striate cortex. *J. Neurosci.* 13: 4114–4129.

Obermayer, K., and Blasdel, G. G. (1997). Singularities in primate orientation maps. *Neur. Comput.* 9: 555–575.

Obermayer, K., Ritter, H., and Schulten, K. (1990). A principle for the formation of the spatial structure of cortical feature maps. *Proc. Natl. Acad. Sci. U.S.A.* 87: 8345–8349.

Obermayer, K., Blasdel, G. G., and Schulten, K. (1991). A neural network model for the formation and for the spatial structure of retinotopic maps, orientation- and ocular dominance columns. In *Artificial Neural Networks*, T. Kohonen, K. Mäkisara, O. Simula, and J. Kangas, eds. pp. 505–511. Amsterdam: Elsevier.

Obermayer, K., Ritter, H., and Schulten, K. J. (1992). A model for the development of the spatial structure of retinotopic maps and orientation columns. *IEICE Trans. Fundamentals* E75-A: 537–545.

Piepenbrock, C., and Obermayer, K. (2000). The effect of intracortical competition on the formation of topographic maps in models of Hebbian learning. *Biol. Cybern.* 82: 345–353.

Purves, D. (1988). *Body and Brain: A Trophic Theory of Neural Connections.* Cambridge, Mass.: Harvard University Press.

Reiter, H. O., and Stryker, M. P. (1988). Neural plasticity without postsynaptic action potentials: Less active inputs become dominant when kitten visual cells are pharmacologically inhibited. *Proc. Natl. Acad. Sci. U.S.A.* 85: 3623–3627.

Rittenhouse, C. D., Shouval, H. Z., Paradiso, M. A., and Bear, M. F. (1999). Monocular deprivation induces homosynaptic long-term depression in visual cortex. *Nature* 397: 347–350.

Sengpiel, F., Gödecke, I., Stawinski, P., Hübener, M., Löwel, S., and Bonhoeffer, T. (1998). Intrinsic and environmental factors in the development of functional maps in cat visual cortex. *Neuropharmacology* 37: 607–621.

Sengpiel, F., Stawinski, P., and Bonhoeffer, T. (1999). Influence of experience on orientation maps in cat visual cortex. *Nat. Neurosci.* 2: 727–732.

Shatz, C. J., and Stryker, M. P. (1978). Ocular dominance columns in layer IV of the cat's visual cortex and the effects of monocular deprivation. *J. Physiol. (London)* 281: 267–283.

Song, S., Miller, K. D., and Abbott, L. F. (2000). Competitive Hebbian learning through spike-timing-dependent synaptic plasticity. *Nat. Neurosci.* 3: 919–926.

Stent, G. (1973). A physiological mechanism for Hebb's postulate of learning. *Proc. Natl. Acad. Sci. U.S.A.* 70: 997–1001.

Stryker, M. P., and Harris, W. A. (1986). Binocular impulse blockade prevents the formation of ocular dominance columns in cat visual cortex. *J. Neurosci.* 6: 2117–2133.

Swindale, N. V. (1980). A model for the formation of ocular dominance stripes. *Proc. Roy. Soc. London B* 208: 243–264.

Swindale, N. V. (1982). A model for the formation of orientation columns. *Proc. Roy. Soc. London B.* 215: 211–230.

Swindale, N. V. (1991). Coverage and the design of striate cortex. *Biol. Cybern.* 65: 415–424.

Swindale, N. V. (1992). A model for the coordinated development of columnar systems in primate striate cortex. *Biol. Cybern.* 66: 217–230.

Swindale, N. V. (1996). The development of topography in the visual cortex: A review of models. *Network: Comput. Neural Syst.* 7: 161–247.

Swindale, N. V. (2000). How many maps are there in visual cortex? *Cereb. Cortex* 10: 633–643.

Swindale, N. V., and Bauer, H.-U. (1998). Application of Kohonen's self-organising feature map algorithm to cortical maps of orientation and direction preference. *Proc. Roy. Soc. London B.* 265: 827–838.

Swindale, N. V., Vital-Durand, F., and Blakemore, C. B. (1981). Recovery from monocular deprivation in the monkey: 3. Reversal of anatomical effects in the visual cortex. *Proc. Roy. Soc. London B* 213: 435–450.

Swindale, N. V., Matsubara, J. A., and Cynader, M. S. (1987). Surface organization of orientation and direction selectivity in cat area 18. *J. Neurosci.* 7: 1414–1427.

Swindale, N. V., Shoham, D., Grinvald, A., Bonhoeffer, T., and Hübener, M. (2000). Visual cortex maps are optimized for uniform coverage. *Nat. Neurosci.* 3: 822–826.

Thoenen, H. (1995). Neurotrophins and neuronal plasticity. *Science* 270: 593–598.

Tieman, S. B., and Tumosa, N. (1997) Alternating monocular exposure increases the spacing of ocularity domains in area 17 of cats. *Vis. Neurosci.* 14: 929–938.

Turing, A. M. (1952). The chemical basis of morphogenesis. *Phil. Trans. Roy. Soc. London B* 237: 37–72.

Van Ooyen, A. (2001) Competition in the development of nerve connections: A review of models. *Network: Comput. Neur. Syst.* 12: R1–R47.

Van Ooyen, A., and Willshaw, D. J. (1999). Competition for neurotrophic factor in the development of nerve connections. *Proc. Roy. Soc. London B.* 266: 883–892.

Von der Malsburg, C. (1973). Self-organization of orientation sensitive cells in the striate cortex. *Kybernetik* 14: 85–100.

Von der Malsburg, C., and Willshaw, D. J. (1976). A mechanism for producing continuous neural mappings: Ocularity dominance stripes and ordered retino-tectal projections. *Exp. Brain Res.* Suppl. 1: 463–469.

Weliky, M., and Katz, L. C. (1999). Correlational structure of spontaneous neuronal activity in the developing lateral geniculate nucleus in vivo. *Science* 285: 599–604.

Wiesel, T. N., and Hubel, D. H. (1974). Ordered arrangement of orientation columns in monkeys lacking visual experience. *J. Comp. Neurol.* 158: 307–318.

Willshaw, D. J., and Von der Malsburg, C. (1976). How patterned neural connections can be set up by self-organisation. *Proc. Roy. Soc. London B.* 194: 431–445.

Willshaw, D. J., and Von der Malsburg, C. (1979). A marker induction mechanism for the establishment of ordered neural mappings: Its application to the retinotectal problem. *Phil. Trans. Roy. Soc. London B* 287: 203–243.

Wiskott, L., and Sejnowski, T. J. (1998). Constrained optimization for neural map formation: A unifying framework for weight growth and normalization. *Neur. Comput.* 10: 671–716.

Wolfram, S. (1984). Cellular automata as models of complexity. *Nature* 311: 419–424.

Wong, R. O. (1999). Retinal waves and visual system development. *Annu. Rev. Neurosci.* 22: 29–47.

Structural Plasticity at the Axodendritic Interface: Some Functional Implications

13

Bartlett W. Mel

Models of development and learning have often focused on the establishment or modification of synaptic connections between classical "point" neurons. However, the notion that a neuron's internal state can be characterized by a single value is challenged by recent evidence indicating that dendritic trees can provide a significant degree of internal compartmentalization of their electrical signals. Recent experiments have shown, for example, that synaptically evoked dendritic spikes mediated by NMDA, Na^+, and Ca^{2+} channels can remain confined to individual thin dendrites or even small portions of dendrites, suggesting that nonlinear thresholding of synaptic inputs could take place on a compartment-by-compartment basis. The potential importance of neuronal compartmentalization is magnified when coupled with evidence for continuous structural remodeling at the interface between axons and dendrites, which may continue throughout life. This combination of effects—subunitized neurons and fine-grained structural plasticity—raises the possibility that activity-dependent mechanisms may control the targeting of synaptic contacts, not just onto neurons as undifferentiated wholes, but onto specific dendritic subregions. We show how this shift in model granularity fundamentally alters the way new information is incorporated into neural tissue, and leads to much higher storage capacities than have been conventionally assumed.

13.1 Introduction

Models of neural development have often focused on the mechanisms that establish or modify the point-to-

point wiring diagram within some area of the brain. A critical assumption of such models is that the neuron is the appropriate level of granularity at which to study development, and when operating at this level, the outcome of development is usually expressed in terms of the learned strengths of connections w_{AB} between point neurons A and B. The development of retinotopy, and the formation of orientation and ocular dominance maps, are canonical examples of developmental modeling at neuron-level granularity (see chapters 11 and 12).

Other developmental levels must be considered, however. For example, activity-dependent rules appear to modulate the density and spatial distribution of the ion channels governing a cell's basic electrical behavior (see chapter 8) and can modulate neurite outgrowth and branching (see chapter 6). Both involve types of changes in the neural substrate that are not naturally described in terms of changes in neuron-to-neuron connection strengths.

In this chapter we focus on a level between that of the ion channel and the whole neuron: development at the level of the dendritic subunit and its inputs. We examine the hypothesis that both the development of neural tissue and learning in the mature brain depend critically on a continual process of structural remodeling at the axodendritic interface. In a shift from the usual perspective, we view development as the correlation-based sorting of synaptic contacts onto the many separate dendrites of a developing neuron rather than onto the many separate neurons of a developing neural map. This shift in the granularity of our analysis is justified by the assumption that individual dendrites, or parts of dendritic trees, act as separately thresholded neuronlike subunits in a manner analogous to the point neurons that populate coarser-grained developmental models. In our discussion, we focus on the main projection neurons of cortical tissue—pyramidal cells—though our conclusions are likely to apply to other types of cells as well.

We first consider recent anatomical and physiological data that support the dendritic subunit hypothesis, and review the basic biological mechanisms that are thought to shape the three-dimensional interface between axons and dendrites in cortical tissue during development and learning in the adult brain. We then consider the functional significance of fine-grained structural plasticity at the axodendritic interface, with an emphasis on the memory-related functions of cortical neurons.

13.2   Neurobiological Background

13.2.1   *Active Responses and Dendritic Compartmentalization*

The dendrites of pyramidal cells contain a large number and variety of voltage-dependent channels, which are likely to profoundly affect their integrative behavior. These include NMDA channels and voltage-dependent Na^+ and Ca^{2+} conductances capable of amplifying synaptic inputs (Thomson et al., 1988; Fox et al., 1990; Cauller and Connors, 1993; Schwindt and Crill, 1995; Lipowsky et al., 1996; Seamans et al., 1997; Margulis and Tang, 1998; Schiller et al., 2000) (although see Urban et al., 1998; Cash and Yuste, 1999) and of generating regenerative responses, including full-blown fast and slow dendritic spikes both in vitro (Spencer and Kandel, 1961; Wong et al., 1979; Poolos and Kocsis, 1990; Jaffe et al., 1992, Wong and Stewart, 1992; Amitai et al., 1993; Kim and Connors, 1993; Stuart and Sakmann, 1994;

Spruston et al., 1995; Magee and Johnston, 1997; Larkum et al., 1999) and in vivo (Pockberger, 1991; Hirsch et al., 1995; Svoboda et al., 1997; Kamondi et al., 1998; Zhu and Connors, 1999). (For a review, see Hausser et al., 2000.)

A variety of evidence from intracellular recordings and imaging studies suggests that active spikelike responses can be localized within the dendritic arbor, i.e., can occur independently of the main axonal spike-generating mechanism (e.g., Benardo et al., 1982; Schwindt and Crill, 1997; Schiller et al., 1997; Golding and Spruston, 1998; Schiller et al., 2000; Wei et al., 2001). For example, Golding and Spruston (1998) used dual intracellular recordings in hippocampal slices to show that synaptically evoked dendritic spikes could occur with or without accompanying somatic spikes.

More recently, Schiller et al. (2000) used ultraviolet (UV)-laser uncaging of glutamate to study synaptic responses in thin basal dendrites of neocortical pyramidal neurons. They found that up to a point, a steadily increasing synaptic stimulus led to steadily increasing somatic excitatory postsynaptic potentials (EPSPs), after which large, slow, all-or-none responses were elicited, which the authors dubbed NMDA spikes. Similar findings were reported for thin branches in the apical dendritic trees of CA1 pyramidal cells (Wei et al., 2001). Of particular interest, slow NMDA-dependent responses imaged by calcium fluorescence were often confined to a small region within the stimulated branch, failing to propagate to other branches within the basal arbor. This finding of a nonlinear thresholding response evoked by synaptic input and confined to a single branch provides the first direct support for the idea that individual thin branches can act as surrogate "neurons" capable of separately summing and thresholding their synaptic inputs. In light of such evidence, we must take seriously the possibility that the relevant postsynaptic processing unit in models of learning and development may be the individual dendritic branch rather than the cell as a whole.

13.2.2 Neuronal Form, Neuronal Function

Anatomical hints are similarly supportive of such a possibility. If advantages accrue to a cell that maintains multiple integrative subregions within its dendritic tree, one might expect pyramidal cell morphologies to maximize the number of subunits available for independent synaptic processing. This is subject, of course, to practical constraints such as that the cell remain of manageable size, the dendritic tree retain cytoplasmic continuity, the cell body give rise to a bounded number of primary neurites, nonlinear voltage-dependent synaptic interactions remain confined to individual subunits, all participating subunits communicate effectively with the cell body, and so on.

Recent quantitative anatomical studies shed new light on this issue. Pace et al. (2000) found that in layer 4 stellate cells of the cat striate cortex, the number of long, thin terminal sections is nearly constant around forty and is independent of the number of primary dendritic branches emanating from the cell body, which ranged from four to eleven in their population (see figure 13.1). Thus, cells with many primary dendrites exhibited fewer stages of branching, while cells with few primary dendrites exhibited more stages of branching. In combination with the fact that most of the synapses onto basal dendrites lie on the long, thin unbranched terminal sections, while the shorter proximal branches contain a much lower density of synaptic contacts (Beaulieu and Colonnier, 1985; Elston and Rosa, 1997; Pace et al., 2000; Megías et al., 2001; Jacobs et al., 2001), the data of Pace et al. (2000) suggest a developmental program that tightly regulates the production of mutually isolated dendritic subunits and then arranges for synapses

to be formed primarily there. Moreover, while the morphologies of basal dendritic trees at different stages of cortical processing differ greatly in the size of their dendritic trees and in their spine densities—which increase nearly 20-fold from the primary visual cortex to the prefrontal cortex (Elston, 2000)—one stable and highly salient morphological feature of these cells is the tendency to show several quick bifurcations near the cell body, where spine densities are low (see also Megías et al., 2001), culminating in a moderately large number of long, thin, unbranched, spine-dense terminal sections (see also chapter 4), which appear biophysically optimized for independent integrative processing. It is interesting to note that this short-to-long branching pattern is by no means universal among dendritic structures and in fact is the opposite of that found in most terrestrial trees.

13.2.3 Structural Plasticity at the Axodendritic Interface

The possibility that neurons contain multiple, separately thresholded dendritic subunits has profound implications for the mechanisms governing the formation and remodeling of the interface between axons and dendrites, an issue central to the focus of this chapter. Axons, dendrites, and spines are strikingly dynamic structures (Greenough and Bailey, 1988; Goodman and Shatz, 1993; Cline, 1999; Woolley, 1999; Harris, 1999; McAllister et al., 1999; Klintsova and Greenough, 1999; Lüscher et al., 2000; Segal et al., 2000). New dendritic spines or filopodia can emerge within minutes in vitro (Dailey and Smith, 1996; Engert and Bonhoeffer, 2000; Maletic-Savatic et al., 1999; Toni et al., 1999) or in vivo (O'Rourke and Fraser, 1990; Lendvai et al., 2000), while large-scale growth and remodeling of axonal and dendritic arbors and/or proliferation of new spinous synapses can occur in the adult brain within

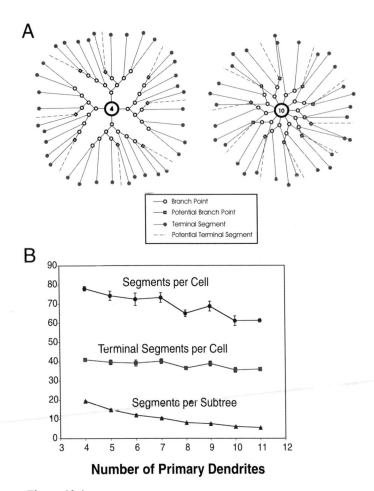

Figure 13.1
Evidence for regulation of number of functional dendritic subunits. (*A*) Schematics of spiny stellate cell morphology in cat visual cortex derived from 3-D reconstructions, shown for cells with four or ten primary dendrites. (*B*) The number of unbranched terminal segments per cell is nearly constant (about forty) for cells with widely varying numbers of primary dendrites (middle curve). (Adapted with permission from Pace et al., 2000.)

days (Greenough et al., 1985; Woolley et al., 1990; Darian-Smith and Gilbert, 1994). This structural lability at the axodendritic interface is correlated with the high concentration of actin found in dendrites and spines, underlying their motility (Crick, 1982; Matus, 1999; see also chapters 3, 4, and 6).

One conception regarding the role of structural plasticity in learning and development involves correlation-based sorting of synaptic contacts on their postsynaptic targets (Shatz, 1990; Cline, 1999). According to this idea, (1) synapses are initially formed between axons and dendrites in a random, activity-independent fashion; (2) newly formed synapses begin their life cycle in a probationary, or "silent," phase (i.e., containing only NMDA channels), which leaves them unable to unilaterally activate their postsynaptic targets (Liao et al., 1995; Isaac et al., 1995; Durand et al., 1996); and (3) silent synapses that are frequently coactivated with mature (nonsilent) synapses within the same postsynaptic compartment are structurally stabilized and thus retained, perhaps via the insertion of AMPA receptors (Lynch and Baudry, 1984), while those that are poorly correlated with their neighbors are eliminated (Cline et al., 1997; Lüscher et al., 2000; Segal et al., 2000).

Later in this chapter we take up an extension of this idea in which synaptic contacts are sorted onto the many separate dendrites of a developing neuron rather than, or in addition to, the many separate whole neurons of a developing neural map. Before considering this scenario in detail, we first review some relevant concepts relating to the theory of dendritic computation.

13.3 In Search of the Single Neuron Model

The neuronlike unit most often used in models of brain function for the past 50 years is the thresh-olded linear point neuron, in which synaptic inputs are summed (with weights) and then passed through a single global nonlinearity representing the output spiking mechanism of the cell (McCullough and Pitts, 1943; Rosenblatt, 1962; Rumelhart et al., 1986). An accumulation of evidence, however, including results discussed earlier from anatomical and physiological studies, suggests that the venerated point neuron may be a poor model of synaptic integration in many cell types, including pyramidal cells of the cerebral cortex.

Cable theory (Jack et al., 1975; Koch, 1999) informs us that the dendritic morphology of a typical CNS neuron, consisting of many thin-branched subtrees radiating outward from the cell body and/or main dendritic trunks, is ideally suited for isolating voltage responses within individual branches of the dendritic arbor. Thus, the impedance mismatch between a thin branch and a main trunk or soma is expected to produce a significant attenuation of distally generated voltage signals when measured at the trunk, with the most profound suppression expected for rapid voltage transients associated with fast synaptic potentials or fast dendritic spikes.

In an early theoretical look at the compartmentalization of synaptic interactions in a passive dendritic tree, Koch et al. (1982) defined a dendritic subunit as a region within which the steady-state voltage attenuation is small between any pair of sites within the subunit, but large between every site within the subunit and the soma. Under reasonable biophysical assumptions, Koch et al. (1982) found that large retinal ganglion cells have a considerable capacity for subunit-specific synaptic integration. The possibility that dendritic trees could support complex multisite nonlinear operations, including logiclike operations, was further considered in a number of subsequent modeling studies (Koch et al., 1986; Shepherd and Brayton, 1987; Rall and Segev, 1987; Mel, 1992a,b, 1993).

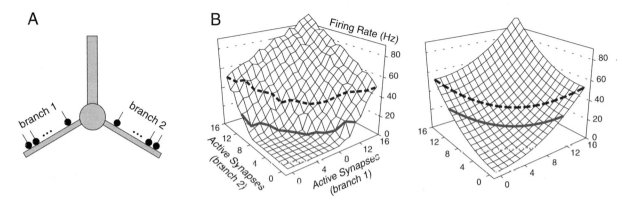

Figure 13.2
Summation of synaptic inputs delivered to two branches of a ball-and-stick model with active dendrites. (*A*) Schematic of a model cell whose thin basal branches contained AMPA- and NMDA-type synapses and low concentrations of voltage-dependent sodium and potassium channels (see Archie and Mel, 2000). The simulations were carried out in the NEURON simulation environment. (*B*) The plot shows the cell's mean firing rate (averaged over sixty-four 1-sec simulations) as the number of active (100 Hz) synapses was increased on the two branches. Within each branch, synapse locations were random. Solid and dashed lines show two iso-synapse-count contours (solid = twelve total synapses; dashed = sixteen total synapses); dips in these contours reflect compartmentalization of the cell (a point neuron would exhibit a constant response along iso-synapse-count contours). (*C*) An abstract model cell whose output is proportional to $n_1^2 + n_2^2$, where n_1 and n_2 are the numbers of active synapses on branches 1 and 2, respectively. Iso-synapse-count contours are similar to those shown in (*B*).

13.3.1 *The Nonlinear Sum-of-Subunits Hypothesis*

One recent study utilized a simplified model pyramidal cell whose dendrites contained AMPA/NMDA synapses and low concentrations of voltage-dependent Na$^+$/K$^+$ channels capable of generating dendritic spikes. When total synaptic drive to the cell was held constant but distributed in varying spatial patterns to two dendritic branches, the average firing rate of the cell was approximated by a simple sum-of-squares model (figure 13.2). The finding is intriguing in that it suggests a possible connection to the quadratic "energy" models used to describe a variety of visual receptive field (RF) nonlinearities (Pollen and Ronner, 1983; Adelson and Bergen, 1985; Heeger, 1992; Ohzawa et al., 1997; for further discussion, see Mel, 1999). This compartmental modeling study pro-

vided the first direct test of the two-layer sum-of-subunits model for synaptic integration in which (1) the thin dendritic branches, which receive the bulk of the cell's synaptic input, act like separately thresholded neuronlike subunits; and (2) the outputs of these thin-branched subunits are summed linearly via the main trunks and cell body prior to global thresholding:

$$y(\mathbf{x}) = g\left(\sum_{i=1}^{m} \alpha_i b\left[\sum_{j=1}^{k} w_{ij} x_{ij} \right] \right), \tag{13.1}$$

where y is the global output, m is the number of subunits, k is the number of synapses per subunit, w_{ij} is the weight and x_{ij} is the activity of the jth input to the ith subunit, $b(\ )$ is the subunit nonlinearity, α_i is the coupling of the ith subunit to the cell body, and $g(\ )$ is a global output nonlinearity.

In more recent work, we developed a biophysically detailed CA1 pyramidal cell model calibrated with a broad spectrum of in vitro data and used it to more closely examine the form of the thin-branched input-output nonlinearity $b(\ )$. Using Eq. (13.1) to predict the time-averaged response of the compartmental model, we found that the assumption of a sigmoidal subunit nonlinearity led to the most accurate predictions of the cell's firing rate, surpassing the performance of any simple accelerating curve tested (e.g., x^2, x^3, etc.) (Poirazi et al., 2003b).

In summary, evidence from physiological, anatomical, and modeling studies suggests that neurons may be capable of providing a moderately large number of separately thresholded integrative subunits within their dendritic trees, perhaps 50 to 100 per cell. If correct, this view implies that the key postsynaptic processing unit in cortical tissue is the thin dendritic branch rather than the neuron as a whole. In the remainder of this chapter we consider the consequences of this shift in processing granularity for the neural basis of learning and memory.

13.4 Impact of Subunit-Containing Neurons on Neural Learning and Development

A critical issue for the interpretation of data relating to structural plasticity in neural tissue involves the definition of the postsynaptic unit. In line with standard concepts from the neural network literature, the formation of new synapses and the elimination of old ones in the course of learning could be viewed as a means to dynamically regulate the overall connection strength w_{AB} between two neurons A and B. However, if postsynaptic neuron B contains separately thresholded dendritic subunits, analogous to hidden units in a multilayer neural network, then the overall connection strength between two neurons is no

longer well defined. Rather, the shift in granularity leads to a higher-dimensional parameterization of the learning system in terms of weights w_{ABk}, i.e., the weight from neuron A to the kth subunit of neuron B.

In the limiting case where each postsynaptic subunit contains enough sites to accommodate only a small fraction of the presynaptic axon population, and the modifiable synaptic weights are all positive and of low resolution (e.g., binary valued), then it may be conceptually parsimonious to parameterize the interface between axons and dendrites in terms of addresses rather than weights, i.e., the set of dendritic subunits with which each presynaptic axon makes contact (for discussion, see Poirazi and Mel, 2000, 2001).

The relevance of this address-based parameterization can be seen from the perspective of axon i in the process of "choosing" which subunit $s \in \{1 \ldots k\}$ to enervate on postsynaptic neuron j during learning or development. The subunit function b, which generates nonlinear interactions among the set of inputs to each subunit, ensures that i's effectiveness in driving cell j depends, not just on its own activity x_i and associated weights w_{ijs}, but also on the activity and weights of the other axons providing input to the same subunit(s).

Thus, given compartmentalized neurons, the "receptive field" of the neuron changes, in general, when any single axon withdraws a synaptic contact from one subunit and forms a new contact on another, even when the change of address involves two branches of the same postsynaptic cell (figure 13.3). By contrast, models operating at neuron-level granularity, which encode only the overall connection strength between neurons, lack the parameters needed to represent such changes. This highlights the danger in counting sheer numbers of synaptic contacts formed between two neurons as a measure of the outcome of learning.

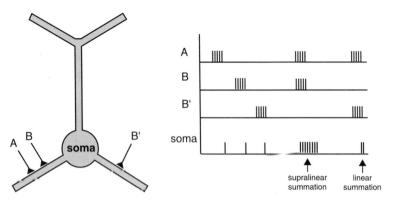

Figure 13.3

The existence of subunit (branch) nonlinearity means that the effectiveness of a synapse can depend on its address on the post-synaptic cell. At left, an afferent axon faces a choice between two subunits for synaptic contact (location B versus B'). At right, a conceptual experiment shows how coactivation of synapses A and B within the same subunit leads to supralinear summation at the cell body, while coactivation of synapses in different subunits (A and B') leads to linear summation. In this scenario, the effectiveness of synapse B, and hence its "weight," is in part determined by the ongoing activity of other synapses.

13.4.1 Partnership Combinatorics at the Axodendritic Interface

The possibility that learning-related mechanisms could orchestrate the correlation-based sorting of synaptic contacts, not just onto whole neurons, but one level down onto specific dendritic subunits, raises the question as to whether the physical interface between axons and dendrites in cortical tissue is amenable to this type of fine-scaled structural plasticity. In particular, when a neuron contains subunits, we have found that its capacity to absorb learned information is closely tied to the addressing flexibility of the tissue, i.e., the flexibility to establish arbitrary partnerships between presynaptic axons and postsynaptic subunits (Poirazi and Mel, 2001).

A serious practical difficulty arising from the need to establish on-demand partnerships between arbitrary pairs of axons and dendrites is that of physical proximity, or the lack thereof. It is unreasonable to expect that during the course of learning, particularly in the densely packed neuropil of the mature brain, axons or dendrites should be regularly required to advance and retract over long distances in search of appropriate partnerships.

What physical properties of axons and dendrites might enhance their partnership flexibility, minimizing the need for long-distance travel to form arbitrary pairings between presynaptic axons and dendritic subunits? In one feature of the anatomy that could contribute significantly to this process, axonal and dendritic arborizations are heavily interdigitated within the three-dimensional volume of the cortical neuropil, an arrangement that maximizes the probability of a close approach between any given axon and any given compartment of a postsynaptic cell. More generally, treelike structures—including circulatory systems, root systems, terrestrial trees, and so on—are specifically designed to penetrate a volume for the delivery or retrieval of some substance, be it oxygen, nutrients, or in the case of neuronal arborizations, information. By design, then, it is difficult for a den-

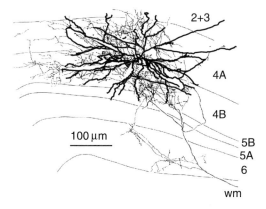

Figure 13.4
Interdigitated axonal and dendritic trees provide an ideal interface for flexible partnering between a presynaptic axon and the several dozen dendritic subunits of a single postsynaptic neuron. The picture was created by superimposing a dendritic arbor from a cat layer 4 spiny stellate cell (courtesy Judith Hirsch) with a thalamocortical afferent taken from Freund et al. (1985).

dritic branch to be far away from every branch of an axon in the region in which the two trees ramify. This qualitative observation is supported by the conceptual montage of two axons shown in figure 13.4, but is not yet backed by hard numbers. One useful experiment would be to measure, over a dense 3-D grid spanning a volume of neural tissue, the distance between each sample point and the point of closest approach to a particular axonal arbor. This kind of measure would serve as an estimate of the distances that would need to be traversed by dynamic axonal and/or dendritic outgrowths in the process of forming correlation-based partnerships.

It is important to note that the three-dimensional interdigitation of axonal and dendritic arbors found in the cerebral cortex, which creates an interface rich in partnership potential, is by no means a universal pattern in the CNS. A striking counterexample is found

at the interface between parallel fibers and Purkinje cells in the cerebellum, an arrangement that virtually predetermines the site of contact between a granule cell axon and the Purkinje cell dendrites to which it has access.

We next consider how correlation-based sorting of synaptic contacts onto the many separate dendrites of a neuron could contribute to long-term memory storage in cortical pyramidal cells. A discussion of the possible contributions of dendritic subunit processing to the image-processing functions of visual cortex is available elsewhere (Mel, 1999).

13.5 Implications of Dendritic Subunits and Structural Plasticity for Long-Term Memory Storage

Ten years ago we found that nonlinear interactions between synapses coactivated on the same branch of an active dendritic tree could break the symmetry among the many subunits of a single postsynaptic cell, thereby providing a location-dependent mode of long-term storage orthogonal to that contained in the overall connection strength between neurons (Mel, 1992a,b, 1993).

We recently set out to quantify the excess trainable capacity contained in the addressing of synaptic contacts onto dendritic subunits, and to characterize how this excess capacity depends on dendritic geometry. We extended a previously developed function-counting approach (Poirazi and Mel, 2000) to compare the capacity of a subunit–containing neuron with m branches (subunits) and k synapses per branch [Eq. (13.1)] with that of a point neuron with the same number of synaptic sites, but with a linear summation rule, i.e., with $b(x) = x$ in Eq. (13.1). The two neuron models were thus identical except for the presence or absence of a fixed subunit nonlinearity.

13.5.1 Quantifying Memory Capacity with and without Dendritic Subunits

Assuming synaptic contacts of unit weight—although any of the d input lines could form multiple connections to the same or different branches—we derived the upper bounds on the capacity of a linear (B_L) versus a nonlinear (B_N) cell:

$$B_L = 2 \log_2 \binom{s + d - 1}{s},$$

$$B_N = 2 \log_2 \left[\binom{k + d - 1}{k}_m + m - 1 \right]. \quad (13.2)$$

The expressions in each case estimate the number of distinct input-output functions that can be expressed by the respective model when assigning $s = m \times k$ synaptic contacts with replacement from d distinct input lines. The combinatorial terms take into account the redundancies associated with the two models, i.e., the changes in synaptic connectivity that have no effect on the cell's receptive field. The linear cell, for example, treats as equivalent every spatial rearrangement of the same set of synaptic contacts, since its summation rule encodes no notion of location on the postsynaptic cell. For this type of cell, then, the capacity to selectively target synaptic contacts to dendritic subunits would not readily translate into additional memory capacity.

In the nonlinear model, a similar type of redundancy, but of much lesser magnitude, arises from an insensitivity to rearrangements of synapses within any given subunit, or with rearrangements of branches at the cell level—such as the swapping of the entire synaptic contents of two branches. More detail regarding the derivation of Eq. (13.2) can be found in Poirazi and Mel (2001).

The expressions for B_L and B_N are plotted in figure 13.5A for a cell with 10,000 synaptic contacts and three values of d. The capacity is shown in bits on the y-axis for a range of cell geometries represented along the x-axis. The values of B_L are shown on the left and right edges of the plot, since the capacity of a point neuron is equivalent to a subunitized neuron in a degenerate state either with a single branch containing 10,000 synapses or with 10,000 branches containing one synapse each. The peak capacity occurs for cells containing approximately 1000 subunits of size 10, where the optimal geometry depends little on d over the order-of-magnitude range tested. Also of interest is that the capacity of the optimally configured nonlinear cell exceeds that of the same-sized linear cell by a factor of 23.

13.5.2 Empirical Testing of Memory Capacity

To validate the analytical model, we trained both linear and nonlinear cells on random two-class classification problems. All target and distractor patterns were drawn from a forty-dimensional spherical Gaussian distribution and randomly assigned labels of 1 or −1. The patterns were then recoded into 400 dimensions through a set of ten nonoverlapping one-dimensional boxcar receptive fields per input dimension, with bins sized to contain equal (10 percent) shares of the probability density along each dimension.

A stochastic gradient descent learning rule patterned after the "clusteron" learning rule described in Mel (1992a) was used to train both linear and nonlinear cells. The learning rule employed two mechanisms known to contribute to neural development: (1) random, activity-independent synapse formation and (2) activity-dependent synapse stabilization or elimination. In each iteration of the learning process,

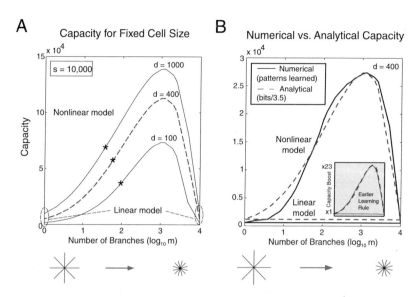

A Capacity for Fixed Cell Size

B Numerical vs. Analytical Capacity

Figure 13.5

Linear versus nonlinear cell capacity as a function of branch geometry. (*A*) The capacity of a nonlinear cell with 10,000 sites for three values of *d*. Branch count *m* grows and branch size *k* correspondingly shrinks moving along the *x*-axis. Cells at both ends of the *x*-axis have a capacity equivalent to that of a linear model. The capacity of the nonlinear model is maximal for cells with 1250 branches containing eight synapses each. The asterisks indicate half-maximum capacity. (*B*) A comparison of memory capacity predicted by analysis with that found empirically in random memorization problems. The dashed lines show the analytical curves for linear and nonlinear cells [the nonlinear capacity curve corresponds to the dashed curve in (*A*)]. The solid curves show capacity measured empirically with a 2 percent error criterion. The analytical curves were scaled down together by a factor of 3.5 to align peak analytical and empirical capacity values for the nonlinear model. The analytical and empirical curves were similar in form. However, the predicted capacity boost for an optimal nonlinear cell was 23 relative to its size-matched linear counterpart, whereas the actual boost realized in empirical trials was nearly 46, corresponding to 27,400 versus 600 patterns learned with a 2 percent error criterion. Using a less effective variant of the learning rule developed earlier, for which the peak empirical capacity boost for a nonlinear cell was (fortuitously) 23, the empirical and analytical capacity curves were nearly identical in form (inset).

a random set T of $n_{\mathcal{T}}$ synapses was targeted for possible replacement. For each synapse in T, a fitness score was computed using a delta rule that measured the degree to which the synapse contributed to the classification performance of the postsynaptic cell. The fitness ϕ_{ij} for the ith synapse on the jth branch was given by a product of four terms:

$$\phi_{ij} = \langle x_{ij} b_j'(\mathbf{x}) g'(\gamma)[t - g(\gamma)]\rangle, \qquad (13.3)$$

where angle brackets indicate the expectation value over the training set, x_{ij} is the presynaptic activity, $b_j'(\mathbf{x})$ is the derivative of the postsynaptic branch response, the sigmoid $g(\gamma) = 1/[1 + \exp(-\gamma/0.05)]$ is a global output nonlinearity with $g'(\gamma)$ its derivative, and $t = \{0, 1\}$ is an external (i.e., network-derived) supervisory signal instructing the neuron whether to respond to a given input pattern. For the linear model, the value of $b_j'(\mathbf{x})$ was replaced by 1, so that

Eq. (13.3) reverted to the standard single-layer delta rule (see Bishop, 1995). Rules of this form are sometimes called Hebbian because the change in a weight depends on the product of the presynaptic activity and some measure of the postsynaptic activity. According to Eq. (13.3), the synapse that most merits potentiation (stabilization) is one that fires strongly whenever (1) its branch is also strongly activated, (2) the cell-wide supervisory signal is strongly activated, and (3) the cell as a whole is firing somewhere in the middle of its dynamic range.

The poorest-performing (minimum ϕ) existing synapse in T was tagged for replacement with the best-performing (maximum ϕ) synapse in a pool of randomly chosen replacement candidates R consisting of $n_{\mathscr{R}}$ of the d input lines. The replacement set was analogous to a pool of silent synapses that resided on the branch in an immature state, unable to contribute to the branch activity in the absence of other synaptic input. In the event a silent synapse proved itself to be strongly correlated with other synapses on the branch, however, it was eligible to be "upgraded" to mature status with an increased measure of structural stability. We found that memory performance varied significantly for different choices of $n_{\mathscr{T}}$ and $n_{\mathscr{R}}$, which controlled the degree of randomness in the gradient descent process. In the experiments discussed here, $n_{\mathscr{T}} = n_{\mathscr{R}} = 25$.

Memory capacity was measured for cells of different geometry by determining how large the training set could be until the cell's error rate reached 2 percent.

13.5.3 Comparison of Analytical and Empirical Capacity Measures

A comparison of analytical and empirical capacities for both linear and nonlinear cells is shown in figure 13.5B. Since the analytical and empirical capacities are reported in different units (bits versus patterns learnable with a 2 percent error), the analytical curves (dashed lines) for both linear and nonlinear models were scaled down by a factor of 3.5 to match the peak empirical capacity (solid lines) of the nonlinear model. As shown in figure 13.5B, the analytical and empirical capacity curves are remarkably similar in shape, with peak capacity again occurring for cells of very similar shape, and with similar costs for deviations from the optimal geometry. However, in a departure from the predictions of Eq. (13.2), the optimal nonlinear cell with 10,000 synapses outperformed its size-matched linear counterpart by a factor of 46 (rather than 23), learning 27,400 rather than 600 patterns at the 2 percent error criterion. This mismatch and the slight difference in form of the analytical and empirical capacity curves are discussed elsewhere (Poirazi and Mel, 2001).

In further experiments with a population of independently trained cells, which could occur within a cortical minicolumn, we found that the excess storage capacity available to a structural learning rule could easily approach two orders of magnitude.

13.6 Discussion

Several types of evidence call into question the classical point neuron as a model for a pyramidal cell or other large dendritic neuron of the CNS. We presented an alternative model, supported by physiological, anatomical, and modeling studies, in which the output of the cell represents the sum of a moderately large set of separately thresholded dendritic subunits—a formulation that looks remarkably like a conventional two-layer neural network. While the validity of this subunitized view of a neuron remains to be proven empirically, we have nonetheless gone about exploring some of its major consequences for development and learning.

Most important, if we adopt a functionally compartmentalized model of the individual neuron, the formation of new synapses and the elimination of old ones during learning or development can no longer be viewed simply as a means to increase or decrease the overall connection strength between two neurons—a common interpretation of the process of new synapse or spine formation. Indeed, the concept of an overall connection strength between two neurons is no longer well defined, in the sense that the interaction between two neurons can no longer be captured by a single positive or negative coefficient. The granularity has changed; we must now worry about the role of learning-related mechanisms in tuning the connection strengths between a many-fingered presynaptic fiber and the multiple dendritic subunits of a given postsynaptic cell.

From the perspective of learning theory, this change in granularity brings with it a large increase in the number of modifiable parameters available to the tissue. This is not a purely theoretical construct. We have found in simulation studies that these extra parameters can translate directly into additional long-term storage capacity hidden in the fine structure of the axodendritic interface.

Future Modeling Studies

Future modeling studies will be needed to assess the impact of subunitized neurons on higher-level developmental processes such as map formation. In the visual realm, a number of models based on arrays of point neurons have been devised to account for the development of retinotopy and/or formation of maps of orientation and ocular dominance among the simple cell population in the visual cortex (see chapters 11 and 12). In contrast, few studies have explored the development of higher-order receptive field properties. In the visual cortex, Foldiak (1991) showed how "complex cells" (following the original terminology of Hubel and Wiesel) could develop from "simple cells" given structured spatiotemporal input, while Hyvarinen and Hoyer (2001) showed how the principles of sparseness and independence and the statistics of natural images could lead to the formation of both simple and complex cell receptive fields. Although complex cell receptive fields have been almost universally assumed to arise from the pooling of simple cell outputs, they provide an interesting new area for developmental modeling in light of our previous work showing that the subunit structure of complex cell receptive fields could arise from the contributions of individual dendrites acting as simple cell surrogates (Mel et al., 1998; Archie and Mel, 2000).

A biologically detailed treatment of this type of multilayered developmental process will be complicated, since it will likely require (1) modeling of the intracellular mechanisms that give rise to functional compartmentalization within the dendritic tree; (2) finer-grained anatomical assumptions regarding the targeting of axons to specific dendritic subregions (e.g., apical versus basal trees) or the cell body; (3) finer-grained biophysical assumptions regarding the rules for synaptic modification within individual cells; and as always, (4) external network effects mediated by various forms of lateral excitation and inhibition. In short, a host of additional assumptions must be made in the construction of such a developmental model, once the inadequacy of the point neuron abstraction has been accepted as a premise. The difficulty of this type of research program should, however, be considered along with the potential payoff: the high likelihood that such explorations will lead to the discovery of altogether new principles of neural development.

Future Experimental Studies

Modeling studies aside, experiments must ultimately decide which are the relevant postsynaptic integrative units operating in neural tissue—whether portions of dendritic branches or entire cells—and they must specify the changes that occur to these integrative units in the course of development or learning. Experimental predictions arising from the ideas discussed in this chapter lie along two main tracks.

The first track involves predictions regarding the form of synaptic integration in the dendritic trees of pyramidal cells of the hippocampus and neocortex [Eq. (13.1)]. In its simplest form with only two stimulus sites involved, the dendritic subunit hypothesis predicts that for weak inputs, the cell will respond more strongly when two stimuli are delivered to the same subunit rather than split between two different subunits. Several variants of this paired-input scenario were tested in a recent compartmental modeling study, leading to the expected results (Poirazi et al., 2003a) (see also Mel, 1992b, 1993; Mel et al., 1998; Archie and Mel, 2000; Poirazi et al., 2003b). Experiments of this type can be done and could be carried out with relatively minor modifications to methods used in published in vitro studies (Cash and Yuste, 1999; Schiller et al., 2000).

The second experimental track addresses the proposal that in the course of neural development, and later as a manifestation of long-term memory formation, synaptic contacts sort themselves into like-activated cohorts across the population of postsynaptic dendritic subunits. Specifically, groups of afferent axons that fire together should be more likely to form synaptic contacts, not just onto the same postsynaptic cells, but onto the same dendritic compartments. This prediction represents only a modest departure from the widely accepted principle of neural development,

which holds that axodendritic connections are initially formed at random and then stabilized or eliminated based on the correlation of pre- and postsynaptic signals (Shatz, 1990; Cline, 1999). Our proposed elaboration of this principle simply requires that the relevant postsynaptic compartment be the dendritic branch or subtree rather than the cell as a whole.

An experimental approach to this issue has been described recently (Poirazi and Mel, 2001). The basic idea is to arrange through artificial means for strong correlations between the firing of two groups of axons, say, groups A and B, that interpenetrate the same target dendritic field. If the activity correlations remain strong during a period of synaptogenesis, then synapses arising from axons in group A are expected to be more frequently cocompartmentalized with those of group B (i.e., lying on the same dendritic branch or within the same minor subtree) and to be less frequently cocompartmentalized with the terminals of an uncorrelated (but equally active) control group C. Experiments of this general kind could be carried out in slice cultures or in vivo using triple labeling to visualize and quantify conjunctions of A and B (or A and C) terminals with identified postsynaptic dendritic branches in the target population.

In summary, clever use of coordinated experimental and modeling studies will be needed to help direct our attention to the proper levels of analysis as we seek to more fully understand the many-faceted mechanisms of neural development.

Acknowledgments

Thanks to Judith Hirsch for providing the stellate cell pictured in figure 13.4. This work was funded by the National Science Foundation and the U.S. Office of Naval Research.

References

Adelson, E., and Bergen, J. (1985). Spatiotemporal energy models for the perception of motion. *J. Opt. Soc. Am. A* 2: 284–299.

Amitai, Y., Friedman, A., Conners, B., and Gutnick, M. (1993). Regenerative electrical activity in apical dendrites of pyramidal cells in neocortex. *Cereb. Cortex* 3: 26–38.

Archie, K. A., and Mel, B. W. (2000). An intradendritic model for computation of binocular disparity. *Nat. Neurosci.* 3: 54–63.

Beaulieu, C., and Colonnier, M. (1985). A laminar analysis of the number of round-asymmetrical and flat-symmetrical synapses on spines, dendritic trunks, and cell bodies in area 17 of the cat. *J. Comp. Neurol.* 231: 180–189.

Benardo, L., Masukawa, L., and Prince, D. (1982). Electrophysiology of isolated hippocampal pyramidal dendrites. *J. Neurosci.* 2: 1614–1622.

Bishop, C. (1995). *Neural Networks for Pattern Recognition.* Oxford: Oxford University Press.

Cash, S., and Yuste, R. (1999). Linear summation of excitatory inputs by CA1 pyramidal neurons. *Neuron* 22: 383–394.

Cauller, L. J., and Connors, B. W. (1993). Synaptic physiology of horizontal afferents to layer I in slices of rat SI neocortex. *J. Neurosci.* 14: 751–762.

Cline, H. T. (1999). Development of dendrites. In *Dendrites*, G. Stuart, N. Spruston, and M. Häusser, eds. pp. 35–67. Oxford: Oxford University Press.

Cline, H. T., Wu, G.-Y., and Malinow, R. (1997). *In vivo* development of neuronal structure and function. *Cold Spring Harbor Symp. Quant. Biol.* 61: 95–104.

Crick, F. (1982). Do dendritic spines twitch? *Trends Neurosci.* 5: 44–46.

Dailey, M., and Smith, S. (1996). The dynamics of dendritic structure in developing hippocampal slices. *J. Neurosci.* 16: 2983–2994.

Darian-Smith, C., and Gilbert, C. (1994). Axonal sprouting accompanies functional reorganization in adult cat striate cortex. *Nature* 368: 737–740.

Durand, G. M., Kovalchuk, Y., and Konnerth, A. (1996). Long-term potentiation and functional synapse induction in developing hippocampus. *Nature* 381: 71–75.

Elston, G. N. (2000). Pyramidal cells of the frontal lobe: All the more spinous to think with. *J. Neurosci.* 20RC95: 1–4.

Elston, G. N., and Rosa, M. G. (1997). The occipitoparietal pathway of the macaque monkey: Comparison of pyramidal cell morphology in layer III of functionally related cortical visual areas. *Cereb. Cortex* 7: 432–452.

Engert, F., and Bonhoeffer, T. (2000). Dendritic spine changes associated with hippocampal long-term synaptic plasticity. *Nature* 399: 66–70.

Foldiak, P. (1991). Learning invariance from transformation sequences. *Percep. Psychophys.* 3: 194–200.

Fox, K., Sato, H., and Daw, N. (1990). The effect of varying stimulus intensity on NMDA-receptor activity in cat visual cortex. *J. Neurophysiol.* 64: 1413–1428.

Freund, T. F., Martin, K. A., and Whitteridge, D. (1985). Innervation of cat visual areas 17 and 18 by physiologically identified X- and Y-type thalamic afferents. I. Arborization patterns and quantitative distribution of postsynaptic elements. *J. Comp. Neurol.* 242: 263–274.

Golding, N. L., and Spruston, N. (1998). Dendritic sodium spikes are variable triggers of axonal action potentials in hippocampal CA1 pyramidal neurons. *Neuron* 21: 1189–1200.

Goodman, C. S., and Shatz, C. J. (1993). Developing mechanisms that generate precise patterns of neuronal connectivity. *Cell* (Suppl.) 72: 77–98.

Greenough, W., and Bailey, C. (1988). The anatomy of a memory: Convergence of results across a diversity of tests. *Trends Neurosci.* 11: 142–147.

Greenough, W. T., Larson, J. R., and Withers, G. S. (1985). Effects of unilateral and bilateral training in a reading task on dendritic branching of neurons in the rat sensory-motor forelimb cortex. *Behav. Neur. Biol.* 44: 301–314.

Harris, K. M. (1999). Structure, development, and plasticity of dendritic spines. *Curr. Opin. Neurobiol.* 9: 343–348.

Hausser, M., Spruston, N., and Stuart, G. J. (2000). Diversity and dynamics of dendritic signaling. *Science* 290: 739–744.

Heeger, D. (1992). Half-squaring in responses of cat striate cells. *Vis. Neurosci.* 9: 427–443.

Hirsch, J., Alonso, J., and Reid, R. (1995). Visually evoked calcium action-potentials in cat striate cortex. *Nature* 378: 612–616.

Hyvarinen, A., and Hoyer, P. O. (2001). A two-layer sparse coding model learns simple and complex cell receptive fields and topography from natural images. *Vision Res.* 41: 2413–2423.

Isaac, J. T. R., Nicoll, R. A., and Malenka, R. C. (1995). Evidence for silent synapses. Implications for the expression of LTD. *Neuron* 15: 427–434.

Jack, J., Noble, D., and Tsien, R. (1975). *Electric Current Flow in Excitable Cells.* Oxford: Oxford University Press.

Jacobs, B., Schall, M., Prather, M., Kapler, E., Driscoll, L., Baca, S., Jacobs, J., Ford, K., Wainwright, M., and Treml, M. (2001). Regional dendritic and spine variation in human cerebral cortex: A quantitative Golgi study. *Cereb. Cortex* 11: 558–571.

Jaffe, D., Johnston, D., Lasser-Ross, N., Lisman, J., Miyakawa, H., and Ross, W. (1992). The spread of Na$^+$ spikes determines the pattern of dendritic Ca^{2+} entry into hippocampal neurons. *Nature* 357: 244–246.

Kamondi, A., Acsády, L., and Buzsáki, G. (1998). Dendritic spikes are enhanced by cooperative network activity in the intact hippocampus. *J. Neurosci.* 18: 3919–3928.

Kim, H. G., and Connors, B. W. (1993). Apical dendrites of the neocortex: Correlation between sodium- and calcium-dependent spiking and pyramidal cell morphology. *J. Neurosci.* 13: 5301–5311.

Klintsova, A. Y., and Greenough, W. T. (1999). Synaptic plasticity in cortical systems. *Curr. Opin. Neurobiol.* 9: 203–208.

Koch, C. (1999). *Biophysics of Computation.* Oxford: Oxford University Press.

Koch, C., Poggio, T., and Torre, V. (1982). Retinal ganglion cells: A functional interpretation of dendritic morphology. *Phil. Trans. Roy. Soc. London B* 298: 227–264.

Koch, C., Poggio, T., and Torre, V. (1986). Computations in the vertebrate retina: Gain enhancement, differentiation and motion discrimination. *Trends Neurosci.* 9: 204–211.

Larkum, M. E., Zhu, J. J., and Sakmann, B. (1999). A new cellular mechanism for coupling inputs arriving at different cortical layers. *Nature* 398: 338–341.

Lendvai, B., Stern, E. A., Chen, B., and Svoboda, K. (2000). Experience-dependent plasticity of dendritic spines in the developing rat barrel cortex *in vivo. Nature* 404: 876–880.

Liao, D. Z., Hessler, N. A., and Malinow, R. (1995). Activation of postsynaptically silent synapses during pairing induced LTP in CA1 region of hippocampal slice. *Nature* 375: 400–404.

Lipowsky, R., Gillessen, T., and Alzheimer, C. (1996). Dendritic Na$^+$ channels amplify EPSPs in hippocampal CA1 pyramidal cells. *J. Neurophysiol.* 76: 2181–2191.

Lüscher, C., Nicoll, R. A., Malenka, R. C., and Muller, D. (2000). Synaptic plasticity and dynamic modulation of the postsynaptic membrane. *Nat. Neurosci.* 3: 545–550.

Lynch, G., and Baudry, M. (1984). The biochemistry of memory: A new and specific hypothesis. *Science* 224: 1057–1063.

Magee, J., and Johnston, D. (1997). A synaptically controlled, associative signal for Hebbian plasticity in hippocampal neurons. *Science* 275: 209–213.

Maletic-Savatic, M., Malinow, R., and Svoboda, K. (1999). Rapid dendritic morphogenesis in CA1 hippocampal dendrites induced by synaptic activity. *Science* 283: 1923–1927.

Margulis, M., and Tang, C.-M. (1998). Temporal integration can readily switch between sublinear and supralinear summation. *J. Neurophysiol.* 79: 2809–2813.

Matus, A. (1999). Postsynaptic actin and neuronal plasticity. *Curr. Opin. Neurobiol.* 9: 561–565.

McAllister, A. K., Katz, L. C., and Lo, D. C. (1999). Neurotrophins and synaptic plasticity. *Annu. Rev. Neurosci.* 22: 295–318.

McCullough, W., and Pitts, W. (1943). A logical calculus of the ideas immanent in nervous activity. *Bull. Math. Biophys.* 5: 115–133.

Megías, M., Emri, Z., Freund, T. F., and Gulyás, A. I. (2001). Total number and distribution of inhibitory and excitatory synapses on hippocampal CA1 pyramidal cells. *Neuroscience* 102: 527–540.

Mel, B. W. (1992a). The clusteron: Toward a simple abstraction for a complex neuron. In *Advances in Neural Information Processing Systems*, Vol. 4, J. Moody, S. Hanson, and R. Lippmann, eds. pp. 35–42. San Mateo, Calif.: Morgan Kaufmann.

Mel, B. W. (1992b). NMDA-based pattern discrimination in a modeled cortical neuron. *Neur. Comput.* 4: 502–516.

Mel, B. W. (1993). Synaptic integration in an excitable dendritic tree. *J. Neurophysiol.* 70: 1086–1101.

Mel, B. W. (1999). Why have dendrites? A computational perspective. In *Dendrites*, G. Stuart, N. Spruston, and M. Häusser, eds. pp. 271–289. Oxford: Oxford University Press.

Mel, B. W., Ruderman, D. L., and Archie, K. A. (1998). Translation-invariant orientation tuning in visual "complex" cells could derive from intradendritic computations. *J. Neurosci.* 17: 4325–4334.

Ohzawa, I., DeAngelis, G., and Freeman, R. (1997). Encoding of binocular disparity by complex cells in the cat's visual cortex. *J. Neurophysiol.* 77: 2879–2909.

O'Rourke, N., and Fraser, S. (1990). Dynamic changes in optic fiber terminal arbors lead to retinotopic map formation: An *in vivo* confocal microscopic study. *Neuron* 5: 159–171.

Pace, C. J., Tieman, D. G., and Tieman, S. B. (2000). Neuronal form: Patterns of dendritic branching in layer 4 stellate cells. *Soc. Neurosci. Abstr.* 2(794.2), 489.

Pockberger, H. (1991). Electrophysiological and morphological properties of rat motor cortex neurons *in vivo*. *Brain Res.* 539: 181–190.

Poirazi, Y., and Mel, B. W. (2000). Choice and value flexibility jointly contribute to the capacity of a subsampled quadratic classifier. *Neur. Comput.* 12: 1189–1205.

Poirazi, Y., and Mel, B. W. (2001). Impact of active dendrites and structural plasticity on the memory capacity of neural tissue. *Neuron* 29: 779–796.

Poirazi, P., Brannon, T. M., and Mel, B. W. (2003a). Arithmetic of subthreshold synaptic summation in a model CA1 pyramidal cell. *Neuron*, in press.

Poirazi, P., Brannon, T. M., and Mel, B. W. (2003b). Pyramidal neuron as 2-layer neural network. *Neuron*, in press.

Pollen, D., and Ronner, S. (1983). Visual cortical neurons as localized spatial frequency filters. *IEEE Trans. Sys. Man. Cybern.* 13: 907–916.

Poolos, N., and Kocsis, J. (1990). Dendritic action potentials activated by NMDA receptor-mediated EPSPs in CA1 hippocampal pyramidal cells. *Brain Res.* 524: 342–346.

Rall, W., and Segev, I. (1987). Functional possibilities for synapses on dendrites and on dendritic spines. In *Synaptic Function*, G. Edelman, W. Gall, and W. Cowan, eds. pp. 605–636. New York: Wiley.

Rosenblatt, F. (1962). *Principles of Neurodynamics*. New York: Spartan.

Rumelhart, D., Hinton, G., and McClelland, J. (1986). A general framework for parallel distributed processing. In *Parallel Distributed Processing: Explorations in the Microstructure of Cognition*, Vol. 1, D. Rumelhart and J. McClelland, eds. pp. 45–76. Cambridge, Mass.: Bradford.

Schiller, J., Schiller, Y., Stuart, G., and Sakmann, B. (1997). Calcium action potentials restricted to distal apical dendrites of rat neocortical pyramidal neurons. *J. Neurophysiol.* 505: 605–616.

Schiller, J., Major, G., Koester, H. J., and Schiller, Y. (2000). NMDA spikes in basal dendrites of cortical pyramidal neurons. *Nature* 404: 285–289.

Schwindt, P. C., and Crill, W. E. (1995). Amplification of synaptic current by persistent sodium conductance in apical dendrite of neocortical neurons. *J. Neurophysiol.* 74: 2220–2224.

Schwindt, P. C., and Crill, W. E. (1997). Local and propagated dendritic action potentials evoked by glutamate iontophoresis on rat neocortical pyramidal neurons. *J. Neurophysiol.* 77: 2466–2483.

Seamans, J. K., Gorelova, N. A., and Yang, C. R. (1997). Contributions of voltage-gated Ca^{2+} channels in the proximal versus distal dendrites to synaptic integration in prefrontal cortical neurons. *J. Neurosci.* 17: 5936–5948.

Segal, M., Korkotian, E., and Murphy, D. D. (2000). Dendritic spine formation and pruning: Common cellular mechanisms? *Trends Neurosci.* 23: 53–57.

Shatz, C. (1990). Impulse activity and the patterning of connections during CNS development. *Neuron* 5: 745–756.

Shepherd, G., and Brayton, R. (1987). Logic operations are properties of computer-simulated interactions between excitable dendritic spines. *Neuroscience* 21: 151–166.

Spencer, W., and Kandel, E. (1961). Electrophysiology of hippocampal neurons. IV. Fast prepotentials. *J. Neurophysiol.* 24: 272–285.

Spruston, N., Schiller, Y., Stuart, G., and Sakmann, B. (1995). Activity-dependent action potential invasion and calcium influx into hippocampal CA1 dendrites. *Science* 286: 297–300.

Stuart, G., and Sakmann, B. (1994). Active propagation of somatic action potentials into neocortical pyramidal cell dendrites. *Nature* 367: 69–72.

Svoboda, K., Denk, W., Kleinfeld, D., and Tank, D. (1997). In-vivo dendritic calcium dynamics in neocortical pyramidal neurons. *Nature* 385: 161–165.

Thomson, A., Girdlestone, D., and West, D. (1988). Voltage-dependent currents prolong single-axon post-synaptic potentials in layer III pyramidal neurons in rat neocortical slices. *J. Neurophysiol.* 60: 1896–1907.

Toni, N., Buchs, P. A., Nikonenko, I., Bron, C. R., and Muller, D. (1999). LTP promotes formation of multiple spine synapses between a single axon terminal and a dendrite. *Nature* 402: 421–425.

Urban, N. N., Henze, D. A., and Barrionuevo, G. (1998). Amplification of perforant-path EPSPs in CA3 pyramidal cells by LVA calcium and sodium channels. *J. Neurophysiol.* 80: 1558–1561.

Wei, D. S., Mei, Y. A., Bagal, A., Kao, J. P., Thompson, S. M., and Tang, C. M. (2001). Compartmentalized and binary behavior of terminal dendrites in hippocampal pyramidal neurons. *Science* 293: 2272–2275.

Wong, R., and Stewart, M. (1992). Different firing patterns generated in dendrites and somata of CA1 pyramidal neurones in guinea-pig hippocampus. *J. Physiol.* 457: 675–687.

Wong, R., Prince, D., and Busbaum, A. (1979). Intra-dendritic recordings from hippocampal neurons. *Proc. Natl. Acad. Sci. U.S.A.* 76: 986–990.

Woolley, C. S. (1999). Structural plasticity of dendrites. In *Dendrites*, G. Stuart, N. Spruston, and M. Häusser, eds. pp. 339–364. Oxford: Oxford University Press.

Woolley, C., Gould, E., Frankfurt, M., and McEwen, B. (1990). Naturally occurring fluctuation in dendritic spine density on adult hippocampal pyramidal neurons. *J. Neurosci.* 10: 4035–4039.

Zhu, J. J., and Connors, B. W. (1999). Intrinsic firing patterns and whisker-evoked synaptic responses of neurons in the rat barrel cortex. *J. Neurophysiol.* 81: 1171–1183.

Modeling the Neural Basis of Cognitive Development

Steven R. Quartz

14

Until recently, developmental modeling efforts at the systems level were frustrated by the lack of experimental data that would constrain the model. A number of new in vivo and in vitro probes now exist (two-photon laser scanning microscopy and magnetic resonance imaging, for example) that can provide critical data on the dynamics of development. These new experimental technologies now make it possible to begin to construct computational models of brain development at the network and systems level. My aim in this chapter is to sketch some of the fundamental issues and challenges that arise for computational modeling efforts at this level in light of the progress stemming from these new technologies. In particular, I focus on questions regarding the link between developmental processes at the cellular level and the systems level, with reference to human development where possible. As I will explore, one of the most intriguing issues that arises at this level concerns the use of computational modeling in understanding how structural brain development relates to functional, or cognitive, development.

14.1 Introduction

As illustrated by the chapters in this volume, computational approaches to development are important tools for investigating the collective properties of dynamic systems. In particular, such models aid in understanding how multiple interactions at one level of organization give rise to phenomena at higher levels of organization. Because brain function takes place

across a number of organizational levels—from the molecular to the systems level—one goal of modeling neural development is to construct simulation frameworks that provide insight into how developmental processes at one level give rise to developmental phenomena at another level, linking molecules to networks and systems. Since the systems level is the appropriate level for investigating many cognitive phenomena, linking multiple levels of organization through modeling efforts suggests that computational frameworks will offer a means for probing the neural basis of cognitive development.

Computational models of development depend upon the availability of developmental data at multiple levels of organization both to construct suitably constrained simulation frameworks and to test their predictions. A major hurdle in constructing such frameworks, however, has been the lack of requisite developmental data. In part, this has been the result of a dearth of investigative probes at a number of spatial and temporal scales.

At the cellular level, the lack of in vivo and in vitro probes that can monitor developmental processes across extended time periods has meant that a great deal about development had to be inferred from cross-sectional histological preparations. Although this provided a wealth of structural information, it did not provide much knowledge regarding the dynamics of developmental processes. Because it is generally believed that cellular processes underlying development are largely conserved across mammalian species, some insight into the dynamics of developmental processes has been obtained by using animal models to investigate the cellular mechanisms of development, which in turn may be used to constrain computational models. With regard to human brain development, the situation is more tenuous at the network and systems level, since good animal models either do not exist or are infeasible from a practical point of view. For this reason, the neural basis of human development at the systems level in nonclinical populations has been largely inferred either from clinical populations or from a limited number of postmortem studies.

In the past few years, a number of new investigative probes have been developed that fill large temporal and spatial gaps, providing important new constraints at the cellular level and making it possible to begin to construct computational models at increasingly complex levels of organization. At the subcellular and cellular levels, two-photon laser scanning microscopy can be used both in vivo and in vitro to monitor developmental processes across extended time periods; at the level of regional and whole brain development, MRI allows both structural and functional assays of human brain development in nonclinical populations. Optical intrinsic signal imaging and near-infrared spectroscopy also show promise in a developmental context.

As other chapters in this volume illustrate, computational models have been particularly important in unraveling the relation between neuronal activity and structural development, which generally falls under the rubric of self-organization. Since such processes of activity-dependent development correspond to cognitive processes of learning at the network and systems level, computational modeling holds great promise in explicating the role of learning in brain development. In recent years, the burgeoning field of developmental cognitive neuroscience has emerged with the explicit aim of integrating cognitive and neural perspectives on brain development.

14.2   Learning and the Computational Perspective

A fundamental issue in computational modeling involves a characterization of the abstract problem a system of interest is attempting to solve. The most thorough discussion of this kind of analysis stems from David Marr's work on computational vision (Marr, 1982). According to Marr, there are three explanatory levels at which any machine carrying out an information-processing task must be understood. The highest, computational, level involves a theory of the computational goal the system instantiates, why it is appropriate, and the logic of the strategy by which it can be carried out. The second level involves an algorithmic description of the computation, while the third level involves an implementational description of the physical substrates of the computation. Although the details of Marr's analysis have rightly been criticized from a variety of perspectives, there remains much value in articulating an abstract characterization of the computational goal a system is attempting to achieve. Chief among the reasons for such an analysis is that it is impossible, and undesirable, to incorporate every parameter in a computational model. The question of what parameters must be incorporated depends on the explanatory role of the model, and this depends on the computational goal of the natural system in question.

To cite two examples, Montague et al. (1995) constructed a highly illuminating model of bee foraging that centered on an idealized model of diffuse ascending systems. Although the model was highly idealized, its theoretical basis, which stemmed from work in reinforcement learning and temporal differences learning (for a review, see Sutton and Barto, 1998), demarcated what level of biological detail was necessary for the model to have explanatory value. In another area of research, the originally highly empirical exploration of neural network supervised learning algorithms has increasingly been replaced by a theoretically motivated one, such as work on support vector machines (Vapnik, 1998), which provides numerous theoretical insights into the nature of supervised learning.

In this light, an often-overlooked value of the computational perspective lies in how it characterizes what I will refer to as "the problem of brain development" and how this characterization conceptualizes the interplay between the processes of learning and those of biological maturation. At its most general, development is the process by which an organism or artificial system constructs those representations that both guide adaptive behavior and facilitate alterations in representational structures through experience in the mature state.[1] The process of fitting representations to ecological niches involves both phylogenetic and ontogenetic strategies, which at a high level of abstraction instantiate similar strategies of error correction, although on different time scales (see Koza, 1992). A great deal of debate concerns the relative contribution of these two processes to cognitive development, as manifest in the innateness debate (Elman et al., 1996).

A number of theoretical approaches have attempted to characterize the problem of development as a learning problem. In the context of language identification "in the limit," early on Gold (1967) attempted to establish upper bounds or worst-case scenario results by asking what a general learner could learn when presented with example sentences of some language. Gold supposed that the learner's task was to create a hypothesis regarding the grammar that might generate that language. The learner was said to identify the language in the limit if it eventually chose a grammar that was consistent with every string. The major implication of Gold's work was that uncon-

strained learning was prohibitive. Simple counting arguments show that the probability of a learner searching through a fixed hypothesis space to successfully learn a concept chosen at random is exponentially small (reviewed in Dietterich, 1990). For this reason, the hypothesis space must be an exponentially small subset of possible concepts (see Blumer et al., 1988).

There were two limitations of Gold's model: the concern for convergence in the limit and its requirement that the learner precisely identify the target concept (no mistakes allowed). Valiant (1984) introduced a probabilistic model of learning that remedied these two limitations and which accordingly became the standard model of inductive inference in the field (see Dietterich, 1990, in the case of machine learning). Rather than disallowing any mistakes, Valiant's learner could make a hypothesis that was only a good approximation with a high probability. This framework was thus dubbed the probably approximately correct (PAC) model of learning. It also addressed the question of convergence time because it distinguished between feasible and infeasible learning by classifying problems according to whether they could be learned in polynomial time. Valiant's model thus shifted the main emphasis of the learning problem from what is in principle learnable to what is learnable from some representation class in feasible time.

Like Gold's, Valiant's model viewed learning as a process of selective induction, i.e., learning as a search through a hypothesis space that the learner posits. A key result stemming from the Gold paradigm was that the child must come equipped with a highly restricted set of hypotheses regarding the target function—in the case of language, a universal grammar. This conclusion derives from the view of learning as essentially a search problem in a hypothesis space (e.g., searching through the grammars) to the target concept. To make this a feasible search, the space must be restricted by building in an inductive bias, roughly the system's background knowledge. One of the key virtues of Valiant's model was that it quantified the relation between inductive bias and learning performance from within a complexity-based account (e.g., Haussler, 1989). The results with Valiant's model thus showed how difficult it was to learn some problem with various inductive biases, or background knowledge. The Valiant model thus demonstrated what could not be fully characterized in the earlier limit-based formal learning theory: learning systems face severe learning-theoretical pressures and can be successful in some domain only if they have solved this difficult prior problem involving representation. That is, from the perspective of the PAC model of learning, the fundamental problems of learning are not those involving statistical inference; instead they center around how to find appropriate representations to support efficient learning (reviewed in Geman et al., 1992). This problem precedes the treatment of learning as statistical inference because a learner's choice of representation class (background knowledge) largely determines the success of learning as statistical inference.

In artificial systems, it is often the designer's task to build in inductive bias to make learning feasible. In natural systems, however, this is a developmental problem. Indeed, this is the "problem of development" I alluded to earlier. In the late 1980s, as neural network research grew in popularity, there was a growing suspicion that this problem might not be so severe. In particular, the claim that algorithms such as back-propagation learned internal representations as a function of exposure to some domain attracted a great deal of attention. Since the networks in question typically began with a randomized set of weights, a popular interpretation was that such networks did not require domain-specific constraints to successfully learn (for an evaluation of this claim, see Quartz, 1993). Part of the promise surrounding the

rise of neural network research concerned the potential of neural network learning algorithms to minimize the requirement for a richly structured initial state.

Many of these insights into the learning properties of neural network algorithms were based on experimental investigations. As more theoretic work was pursued, it became apparent that these early claims were not accurate (see Geman et al., 1992). Like other learning algorithms, neural network algorithms confronted a basic tradeoff between the two contributors to error: bias and variance. Bias is a measure of how closely the learner's best concept in its representation space approximates the target function (the thing to be learned). Variance refers to the actual distance between what the learner has learned so far and the target function.

To make this a bit more concrete, a small neural network will be highly biased in that the class of functions allowed by weight adjustments is very small. If the target function is poorly approximated by this class of functions, then the bias will contribute to error. By making a network large, hence flexible in terms of what it can represent (by decreasing bias), the contribution of variance to error is typically increased. That is, the network has many more possible states, and so is likely to be distant from the function of interest. This means that very large training sets will be required because many examples will be needed in order to rule out all the possible functions. As Geman et al. (1992) state it, this results in a dilemma: highly biased learners will work only if they have been carefully chosen for the particular problem at hand, whereas flexible learners seem to place too high a demand on training time and resources. Geman et al. (1992, p. 22) state, "learning complex tasks is essentially impossible without the *a priori* introduction of carefully designed biases into the machine's architecture."

From this theoretical perspective, then, the problem of development is a formidable one. Indeed, although the bulk of research in neural network modeling has centered around learning as statistical inference, there is a more fundamental problem, namely, the construction of an efficient set of representations that make statistical inference possible at all. In natural systems, this latter issue is the critical problem of development.

14.3 Development as Construction of Representation: Contrasting Selective Induction and Constructive Induction

In order to investigate the nature of the construction of representations that underlie behavior and facilitate efficient learning in the mature state, it is necessary to consider the following questions:

1. What structural measures correspond to representational complexity in the brain, and how do such measures change across the developmental time course?

2. Is this change dependent on environmental interaction, and, if so, at what level of specificity?

As Quartz and Sejnowski (1997) indicated, there are three nonexclusive measures that relate structural and functional complexity: axonal arborization, synaptic numbers, and dendritic arborization. Since its beginnings, developmental neurobiology has been embroiled in debate over whether these neural structures are added progressively over the developmental time course, or whether developmental processes are analogous to those seen in population biology, where an initial overproduction of structures or individuals is acted on by selective mechanisms (see Purves et al., 1996, for a summary of this debate). In developmental neurobiology, the most programmatic statements of

such a view are known as selectionist models. In these models, the initial production of neural structures is regulated by intrinsic mechanisms whose main purpose is to create a diversity of representations for a later, activity-dependent process of selective elimination to act on, eliminating those structures that do not reflect the informational structure of the environment appropriately (Changeux and Danchin, 1976; Edelman, 1987).

The abstract characterization of the computational goal of a developing system provides insight into the selectionist strategy. From a learning-theoretic perspective, these exuberant structures can be regarded as encoding a hypothesis space that contains the target function as a proper subset. The initial construction of this space is insensitive to information originating from the environment; instead the space is constructed through intrinsic genetic and epigenetic processes. The role of environmentally derived activity is limited to a process of error correction in which representations that do not appropriately reflect the information structure of the system's environment are eliminated.

The selectionist model thus implements learning as a process of selective induction in which the essential feature is a search through a fixed hypothesis space. There is an alternative framework, which treats learning as a process of constructive induction (for a review, see Quartz, 1999). According to this framework, a system begins with an initially restricted hypothesis space and constructs a more complex one as some function of exposure to a problem domain. Constructive induction provides a markedly different answer to the question of the source of the representations underlying acquisition. Rather than presupposing a set of fixed representations in the initial state, constructive induction regards these representations as unfolding dynamically across the developmental time course.

A number of analyses have demonstrated the intriguing properties of constructive induction. From a theoretical perspective, White (1990) demonstrated that a network that adds units at an appropriate rate relative to its experience is a consistent nonparametric estimator. This asymptotic property means that it can learn essentially any arbitrary mapping. The intuition behind this result, which plays a central role in characterizing constructive learning, follows a general nonparametric strategy: slowly increase representational capacity by reducing bias at a rate that also reduces variance. Since network bias depends on the number of units, as a network grows its approximation capacities increase. The secret is regulating the rate of growth so that the contribution of variance to error does not increase. Encouraging bounds on the rate of convergence have been obtained (Barron, 1994).

White's demonstration of the power of neural networks depends on allowing the network to grow as it learns. In fact, many of the limitations encountered by neural networks are due to a fixed architecture. Judd (1988), for example, demonstrated that learning the weights in a neural network is a nondeterministic polynomial time (NP)–complete problem, and therefore computationally intractable, a result that extended to architectures of just three nodes (Blum and Rivest, 1988). These results suggest that severe problems may be lurking behind the early success of network learning. As Blum and Rivest (1988) note, however, these results stem from the fixed architecture property of the networks under consideration.

In contrast, the loading problem becomes polynomial (feasible) if the network is allowed to add hidden units. This suggests fundamentally different learning properties for networks that can add structure during learning. This has been confirmed by studies such as that of Redding et al. (1993), who presented a constructivist neural network algorithm that can learn very general problems in polynomial time by building its architecture to suit the demands of the specific problem. Since the construction of the learner's hypothesis space is sensitive to the problem domain facing the learner, this is a way of tailoring the hypothesis space to suit the demands of the problem at hand. This allows the particular structure of the problem domain to determine the connectivity and complexity of the network. Since the network has the capacity to respond to the structure of the environment in this way, the original high bias is reduced through increases in network complexity, which allows the network to represent more complex functions. Hence, the need to find a good representation beforehand is replaced by the flexibility of a system that can respond to the structure of some task by building its representation class as it samples that structure to learn any polynomially learnable class of concepts. Research on constructive algorithms has become increasingly sophisticated (reviewed in Quinlan, 1998).

Given that constructive induction has intriguing learning properties, it is important to consider whether there is evidence for such processes in brain development. Were the brain to implement such a strategy, then one would expect to find measures of representational complexity that increase across the developmental time course as some function of learning. As I indicate in the following discussion, recent advances in experimental techniques provide important new evidence for this kind of learning as construction of representation.

14.4   Rethinking Synaptic Numbers

Many influential studies in developmental neurobiology focused on changes in synaptic numbers across development. To be more accurate, some of these studies focused on changes in synaptic density and made estimates of area to determine absolute numbers. Since determining areal boundaries across samples is not a trivial task, much of this work is limited to the primary visual cortex. Alternatively, many of these studies relied on the assumption that changes in area could be discounted, thereby making density estimates meaningful. Even granting that reliable estimates of area are possible, a number of assumptions are involved in the claim that synaptic numbers are a good measure of representational complexity. Principal among these is the traditional view of synaptic integration in which dendrites are passive cables that linearly sum a set of inputs. Such a model has led to one of the central assumptions of developmental neurobiology and much of cognitive and computational neuroscience: that the synapse is the basic computational unit of the brain. For this reason, developmental neurobiologists have been concerned with changes in synaptic numbers over the developmental time course (e.g., Rakic et al., 1986). Likewise, many computational models employ a fixed architecture and regard long-term changes in connection strengths as the main modifiable parameter, reflecting a central tenet of brain function that information is stored in the patterning of synaptic weight values.

The central result of the cellular basis of learning—synaptic plasticity—has been the discovery of long-term potentiation (Bliss and Lomo, 1973), which has contributed to the view that information is stored in the pattern of synaptic weights. According to this view, learning is mediated by correlated patterns of firing that induce long-term increases in the strength of connections between neurons. The rise of connectionist network modeling has contributed to the view that learning involves the long-term modification of connection strengths between neurons, as computational studies demonstrated the power of computing by adjusting connection strengths in a fixed architecture. Theoretical analyses of these learning algorithms also demonstrated the powerful statistical methods they implemented. The combination of experimental, computational, and theoretical analyses has made the view that connection strengths are the main modifiable parameters extremely widespread, thereby presenting learning as involving a modification of this encoding of information.

As Poirazi and Mel (2001) note, however, much new evidence weakens the link between synaptic weights and information representation and processing in the brain (see also chapter 13). Perhaps the most important source of evidence, and one central to developmental issues, concerns the nonlinear summation properties of many dendrites. The dendrites of pyramidal cells, for example, contain numerous voltage-dependent channels that play an important role in determining the cell's information-processing function. Most important, a variety of channels, including NMDA, Na^+, and Ca^{2+} channels, are capable of amplifying synaptic inputs and generating fast and slow dendritic spikes. Numerous laboratories have localized active nonlinear responses to synaptic inputs within the dendritic arbor (Bernardo et al., 1982; Golding and Spruston, 1998) and even within a single thin dendritic branch (Schiller et al., 2000). These nonlinear properties complicate the notion of a connection strength, since the weight of a given synaptic contact will be dependent on the activity of neighboring synapses.

These considerations suggest that the dendritic arbor itself is the basic unit of computation in the cortex (for an extended discussion, see Quartz and Sejnowski, 1997).[2] Structural alterations in the axodendritic interface, particularly the growth and retraction of dendritic arbors, would be expected to have a significant impact on information processing. Indeed, rather than the learning of internal representations being primarily a process of weight adjustment, structural alterations in the axodendritic interface could be a primary mechanism by which the representational properties of the cortex are constructed (see Poirazi and Mel, 2001, for extended discussion; see also chapter 13). From a developmental perspective, an important issue thus concerns the processes that regulate the growth of the axodendritic interface and their relation to cognitive processes of learning.

In earlier work (Quartz and Sejnowski, 1997), I hypothesized from histological preparations that dendritic development follows a neural constructivist theme. In recent years, the development of such probes as two-photon laser scanning microscopy has provided important new insights into the processes regulating dendritic development and their dynamics. Recent cellular work provides the strongest evidence to date for the role of patterned activity in the development of neural structures. Specifically, recent advances in microscopy that allow the continuous monitoring of cellular components at high resolution (Maletic-Savatic et al., 1999; Engert and Bonhoeffer, 1999; reviewed in Wong and Wong, 2000) have revealed a highly dynamic view of development at the cellular level and provide strong evidence for the instructive role of activity in neural development (see also chapters 6 and 8). These results indicate that dendritic filopodial formation may be tightly and locally regulated by the activity of presynaptic axons. Previously, it had been difficult to determine the level of specificity that activity played in the construction of neural structures. Specifically, it was difficult to differentiate between a permissive role for activity, in which its mere presence is sufficient to induce growth, and an instructive role, in which activity regulates growth according to learning rules at specific sites. These new results demonstrate that activity is not simply permissive in its regulation of development. Rather, as Maletic-Savatic et al. (1999) demonstrated, temporally correlated activity between pre- and post-

synaptic elements that induces long-term potentiation results in the highly spatially defined local sprouting of dendritic elements, which is in agreement with Hebb's original postulate in its developmental context (Hebb, 1949).

Similar results have been obtained in vivo. Recently, Lendvai et al. (2000) used two-photon laser scanning microscopy to characterize the experience-dependent plasticity of dendritic spines in the developing rat barrel cortex in vivo. They found that sensory deprivation markedly reduced protrusive motility ($\sim$40 percent) in deprived regions of the barrel cortex during a critical period around postnatal days 11–13, which resulted in a degraded tuning of the receptive fields of layer 2/3 cells.

This research, along with much other evidence I considered elsewhere (Quartz and Sejnowski, 1997; Quartz, 1999), strongly suggests that cortical development is not exclusively mediated by mechanisms of selective elimination operating on transient, exuberant structures. Rather, neural development during the acquisition of major cognitive skills is best characterized as a progressive construction of neural structures in which environmentally derived activity plays a role in the construction of neural circuits. Selective processes do indeed play an important role, but they are the consequence of stochastic sampling mechanisms and should be seen as complementary to constructive mechanisms. This revised view of the role of activity in the construction of neural circuits forms the basis for neural constructivism, which examines how representational structures are progressively elaborated during development through activity-dependent growth mechanisms, in interaction with intrinsic developmental programs.

From the perspective of cognitive development, the far-reaching interaction between neural growth and environmentally derived neural activity blurs the distinction between biological maturation and learning. In place of this dichotomy, "constructive learning" thus appears to be an important theme in development, which from a learning-theoretical perspective appears to possess more powerful acquisition properties than traditional accounts of cognitive development assumed.

As I mentioned earlier, constructive learning is aimed at a prior problem facing natural systems that is in many ways more fundamental than the problem of learning as statistical inference. Development can be divided into two distinct phases. The first involves the development of representations, while the second involves using those representations to learn efficiently, which continues in the mature state.

A study in the owl (Knudsen, 1998) highlights the role of early experience as a constructor of the representations that facilitate learning later in life. The optic tectum of barn owls contains a multimodal map of space. In particular, auditory visual neurons in the optic tectum associate values of auditory spatial cues with locations in the visual field. This is done by matching the tuning of tectal neurons for interaural time differences with their visual receptive fields. During development, but not adulthood, there is considerable plasticity in this system, allowing a wide range of associations to be learned. When juvenile animals were fitted with goggles that shifted the visual field, the resulting abnormal associations were learned. Knudsen (1998) demonstrated that the range of associations that adult owls could learn was greatly expanded in those animals that had learned abnormal associations during development.

14.5 From Cellular Processes to Networks

Recent research on the cellular basis of developmental processes demonstrates that dendritic structures are highly dynamic and that their growth is modulated by patterned activity at a high level of specificity. These results provide strong evidence that at the cellular level the process of constructing representations is one of constructive rather than selective induction. However, it is necessary to consider how these cellular-level processes are integrated into network-level processes to determine the processes and strategies underlying the construction of representations. In particular, it is important to consider the nature of the constraints that make this constructive induction paradigm efficient.

The abstract characterization of the computational goals of development highlighted the need for constraints on a developing system. It appears prohibitive for a developing system to efficiently construct representations without substantial constraints that limit the possible forms of such representations. It is important to note that most considerations of the form such constraints might take have centered on static restrictions on a fixed hypothesis space. Rather than conceiving of constraints in terms of restricting a fixed hypothesis space, it appears necessary to consider time-dependent constraints on the developing cortex. Such constraints include the interaction among activity-dependent growth mechanisms, intrinsic developmental pathways, an initially small hypothesis space, and the contribution of generic initial cortical circuitry, conduction velocities, subcortical organization, learning rates, and hierarchical development.

To understand how the constraints I have enumerated here operate, it is necessary to characterize the dynamics of development and its time-dependent properties. That is, it is necessary to understand how these constraints shape the developmental path. For example, what are the implications of a limited initial architecture for the acquisition properties of a learning system? Whereas traditional accounts suggested that these limitations weakened the learning system, neural network modeling casts these limitations in a new, advantageous, light (Elman, 1993). An initially restricted network must pass through a phase of limited representational power during early exposure to some problem and then build successively more powerful representational structures. Thus, these early limitations may actually help the system first learn the lower-order structure of some problem domain and subsequently use what it has learned to bootstrap itself into more complex knowledge of that domain (reviewed in Plunkett et al., 1997).

Such considerations suggest that the developmental path that is determined by the dynamics of the constraints enumerated above play an important but neglected role in constraining development. Thus, understanding the process of developmental change, rather than simply its initial state or final outcome, is paramount in developmental science. It is here that the study of self-organizing systems provides a number of important insights. For example, self-organizing systems have helped to explicate the developmental function of spontaneous neural activity, which is known to play a role in constructing neural circuits (reviewed in Wong, 1999). As Linsker demonstrated (reviewed in Linsker, 1990), randomly generated activity, which essentially appears as noise, can create feature filters, given the functional properties that neural circuits possess in combination with their geometric properties (e.g., interaction functions through which nearby activity is excitatory but becomes inhibitory with increasing distance). Ordered structure is thus an emergent property of the dynamics and geometric organization of such systems. In the following section, I consider in more detail how order may be generated from the dynamic interaction of brain systems.

14.6 Hierarchical Development

Given the evidence for constructive processes at the cellular level, it is important to consider how these are related to larger levels of organization in the brain. One of the most intriguing time-dependent developmental constraints is hierarchical development. The core idea that development involves the expansion of hierarchically organized sequential operations, beginning with perceptual and sensorimotor functions and becoming more combinatorially complex, remains popular. For example, Luciana and Nelson (1998) recently examined the developmental emergence of functions involved in prefrontally guided working memory systems in 4–8-year-old children. The development of these memory systems, which is thought to involve particularly the dorsolateral region of the prefrontal cortex (PFC), in 4–8-year-old children, appears to proceed dimensionally, beginning with the refinement of basic perceptual and sensorimotor functions and culminating with the emergence of distributed networks that integrate complex processing demands. This is a paradigmatic case of time-dependent development in which increasingly complex representations are built as a function of exposure to problem domains.

As recently as a few years ago, however, it was unclear whether cortical development proceeded in a manner that was consistent with hierarchical development. According to the influential results of Rakic et al. (1986), cortical development followed a pattern of concurrent synaptogenesis. This influential view was based on electron microscope studies of synaptogenesis in the rhesus monkey. This view suggested that the entire cerebral cortex develops as a whole and that the establishment of cell-to-cell communication may be orchestrated by a single genetic or humoral signal. As Rackic et al. (1986) pointed out, this view ruled out a hierarchical view of cortical development, i.e., a developmental ordering from the sensory periphery to higher associational areas. This theory of concurrent synaptogenesis was difficult to reconcile with other structural measures, including patterns of myelination and dendritic arborization, which showed a regional, or heterochronic, pattern of development.

More recent work has indicated that synatogenesis in human development is not concurrent across different regions of the cortex (Huttenlocher and Dabholkar, 1997), but rather follows a regional pattern. According to this finding, human cortical synaptogenesis occurs regionally and in accord with the hierarchical developmental schedule observed for axonal growth, dendritic growth, and myelination. Assimilating the developmental schedule for these various measures, it appears that primary sensory and motor cortical areas are both closer to their mature measures at birth and reach those measures earlier than do areas of association in temporal and parietal regions and the PFC.

Recently, MRI studies have shed light on this issue. For example, Thompson et al. (2000) found that regions of the cortex develop at different rates. These studies suggest that the brain develops hierarchically, with early sensory regions developing before more complex representations in association areas. This regional pattern of cortical development, proceeding from the sensory periphery to higher association areas, is particularly intriguing given that cortical representations are arranged hierarchically in a way that matches this regional hierarchy. According to Fuster (1997, p. 451), "the cortical substrate of memory, and of knowledge in general, can be viewed as the upward expansion of a hierarchy of neural structures." Although the existence of extensive feedback connections suggests that the notion of a strict hierarchy must be qualified, cortical areas closer to the sensory periphery encode lower-order, or more elementary, representations than do areas further removed, which involve more distributed networks that do not have the topographical organization of lower areas.

All three sensory modalities—vision, touch, and audition—involve what Fuster (1997, p. 455) refers to as a "hierarchical stacking of perceptual memory categories in progressively higher and more widely distributed networks." All three modalities then converge on the polysensory association cortex and the limbic structures of the temporal lobe, particularly the hippocampus. This hierarchical organization of representations, combined with its hierarchical developmental pattern, lends support to the view of development as a cascade of increasingly complex representational structures, in which construction in some regions depends on the prior development of others.

Given the importance of dendritic morphology in information processing (see chapter 13), it is particularly intriguing to note that dendritic complexity is not uniform across the cortex. Rather, regional dendritic variation is extensive, with far-reaching implications for cortical processing. In a series of studies in monkeys, Elston and Rosa (Elston et al., 1996; Elston and Rosa, 1998a,b) documented a caudal-to-rostral progression in dendritic field size and spine number. This suggests a more extensive sampling of input in dendritic systems at higher levels of visual processing.

Recently Jacobs et al. (2001) explored regional differences in dendritic and spine extent across several human cortical layers. Specifically, they utilized the cortical hierarchy scheme proposed by Benson (1993), in which the cerebral cortex is classified into four subdivisions. Each subdivision corresponds to a progressively more complex level of neural processing. The primary cortex is involved in the initial processing of sensory impulses, or the final output stage for motor functions; unimodal regions discriminate, categorize, and integrate information within a single modality; the heteromodal cortex is involved in cross-modal integration; and supramodal association regions are involved in executive control and cognitive networks. It should be borne in mind that these divisions are somewhat heuristic because the existence of reciprocal connections makes it infeasible to impose a strict hierarchy on cortical regions. Jacobs et al. (2001) examined dendritic and spine extent in four areas corresponding to each of Benson's subdivisions and found a progressive increase among hierarchically arranged cortical regions of the human brain. These differences were substantial. The increase in total dendritic length from primary cortical regions to supramodal regions was approximately 30 percent, and total spine number was about 60 percent.

These results suggest that the processing demands placed on dendritic systems may substantially influence their mature form, as Ramón y Cajal hypothesized. Given the hierarchical pattern of regional development I reviewed earlier and its correspondence with regional variation in dendritic complexity, it will be important for future research to investigate how dendritic processing may be determined in a stagelike fashion as increasingly complex dendritic structures are constructed across the developmental time course.

14.7   Supervised, Unsupervised, and Self-Supervised Learning

A largely open question at the network level concerns how multiple brain systems interact during development. It is likely that multiple brain systems instantiate multiple acquisition strategies. The hierarchical organization of these systems suggests that some structures may constrain the development of others by directing their development. Just as a structure such as the pri-

mary visual cortex can be constrained by the nature of the sensory modality innervating it, so too some neural structures can be constrained by the pattern of input from other neural structures. In the case of the primary visual cortex, it is generally believed that incoming sensory information, reflected in patterns of activity, is utilized in an unsupervised mode. Unsupervised learning, or self-organization in its developmental context, involves developing an efficient internal model of the salient statistical structure of the environment. For example, Hebbian learning can be understood in the context of principal component analysis, which is a method of efficiently representing the correlational structure of the environment.

Over the past few years, significant progress has been made in exploring unsupervised learning algorithms for neural network models (Hinton and Sejnowski, 1999). Unlike the earlier supervised learning algorithms, which required a detailed teacher to provide feedback on performance, the goal of unsupervised learning is to extract an efficient internal representation of the statistical structure implicit in the stream of inputs. Infants in the womb are bombarded by sensory inputs, and their environment after birth is filled with latent information about the environment. During development, unsupervised learning could shape circuits in the early stages of sensory processing to represent the environment more efficiently; in the adult brain, similar forms of implicit learning could provide cues to help guide behavior.

While computational neurobiologists investigate unsupervised learning in analyses of the developing visual system (see chapters 11 and 12), cognitive scientists typically investigate supervised algorithms in connectionist-style architectures (Shultz et al., 1994; for a review, see Plunkett et al., 1997). There is an additional class of learning algorithms—reinforcement, or self-supervised, algorithms—that may be utilized by one neural region to direct the development of another. As Piaget stressed, a central theme of development involves the developing system's active exploration of its environment in which learning is mediated through the consequences of the system's actions on that environment. This places a premium on the presence of reward systems that both engage a developing system in its environment and drive learning through the patterns of reward (and punishment) that such engagement brings about. Elsewhere (Quartz, in press), I have suggested that evolutionary considerations support a behavioral systems model of the brain that regards the brain as a hierarchical control structure in which reward plays a central computational role and in which this hierarchical organization is evident both developmentally and evolutionarily.

A key source of evidence regarding this view is that despite the apparent diversity of nervous systems, most share a deep structure, or common design principles. Even the simplest motile organisms require control structures to regulate the goal-directed behavior necessary for survival in a variable environment (for discussion, see Allman, 1999). For example, although the bacterium *Escherichia coli* does not possess a nervous system, it does possess control structures for sensory responses, memory, and motility that underlie its capacity to alter behavior in response to environmental conditions. The ability to approach nutritive stimuli and avoid aversive stimuli in the maintenance of life functions is the hallmark of behavioral systems across phyla. Whereas chemotaxis in bacteria involves a single step from sensory transduction to motor behavior, some multicellular organisms contain control structures that involve intercellular communication via hormonal signaling, while others possess nervous systems with control structures that add layers of mediating control between sensory transduction and motor behavior.

There are several alternative design possibilities for biological control structures. One is to make a closed system, in the sense of linking fixed behavioral patterns between internal goal states and their environmental targets. Although there are many examples of this strategy (Gallistel, 1990), there are more powerful and flexible control structures. One such strategy involves leaving the path from the internal goal state to the target state open and discoverable via learning. Principal among this latter design strategy are reinforcement-based systems that are capable of learning an environment's reward structure.[3]

A variety of experimental techniques, ranging from psychopharmacology to neural imaging, have demonstrated the striking ubiquity and conservation of reward structures across species. At virtually all levels of the human nervous system, for example, reward systems can be found that play a central role in goal-directed behavior (Schultz, 2000). Here I focus on one such system, the midbrain dopamine system (figure 14.1). This system projects principally from the ventral tegmental area to the nucleus accumbens and the temporal and frontal cortex. Studies utilizing self-stimulation paradigms revealed that activation of this system was highly reinforcing; often laboratory animals preferred to self-stimulate this system than to eat or copulate with a receptive partner (reviewed in Wise, 1996). Most addictive substances involve this system, giving rise to the hedonic theory of dopamine as the signal underlying pleasure (although see Garris et al., 1999).

Given what I have previously stated regarding the possibility that control structures are highly conserved, it is interesting to note, as figure 14.1 illustrates, the striking homology between the dopamine system in humans and a reward system in the honeybee. The honeybee subesophogeal ganglion contains an identified neuron, VUMmx1, which delivers information about reward during classical conditioning

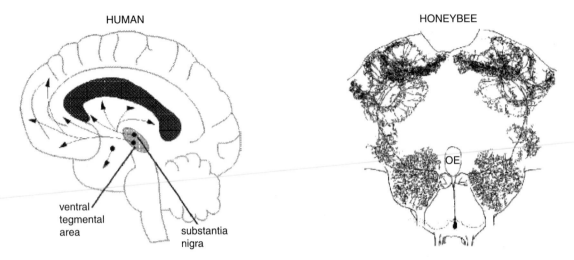

HUMAN HONEYBEE

ventral tegmental area

substantia nigra

OE

Figure 14.1
Neuromodulatory (octopamine) neurons in the bee brain and dopamine projections in the human brain play homologous roles. Neural activity in these neurons distributes information about expected reward.

experiments via the neurotransmitter octopamine, which is similar in molecular structure to dopamine (Hammer, 1993).

Both experimental and computational work on the role of VUMmx1 in honeybee foraging has provided important insights into the signal carried by octopamine and the system's functional significance (Real, 1991; Montague et al., 1995). Rather than simply carrying information regarding reward, it appears that octopamine signals information about prediction errors. Whereas reward is traditionally a behavioral notion, prediction is a computational notion. The difference between certain rewarding outcomes and the predictions of these outcomes can be used to guide adaptive behavior. A system that learns through prediction learning need not have the path from goal to reward specified, in contrast to fixed behavioral patterns, such as stimulus-response learning. Instead, the path from goals to rewards may be left open and discoverable via learning, resulting in flexible action.

Evolution, then, may shape the pattern of the basic rewards that animals are motivated to obtain, but the behavioral path is left open to discovery, as are more complex relations among predictors. In this sense, brains are prediction machines that use information gathered from past experience to predict future events important for survival (reviewed in Montague and Quartz, 1999).

Experiments utilizing neurophysiological recording in behaving monkeys by Schultz and colleagues demonstrate that the midbrain dopamine system plays an important role in prediction learning in the mammalian brain (Schultz et al., 1993). When these monkeys were presented with various appetitive stimuli, dopaminergic neurons responded with short, phasic activations, which typically lasted for only a few repeated presentations. In an important finding, however, Schultz and colleagues showed that when the rewarding stimuli were preceded by an auditory or visual cue, dopamine neurons changed their time of activation to just after the time of cue onset. In contrast, when the reward did not follow the conditioned stimulus, dopamine neurons were depressed below their basal firing rate exactly at the time the reward should have been presented. These results indicate that the dopamine signal encodes expectations about the delivery of reward. That is, the dopamine neurons code for an error between the actual reward received and predictions of the time and magnitude of reward. Like the octopamine signal in the honeybee, the dopamine signal codes a prediction error that can be used in learning and in selecting action. This mode of action is equivalent to temporal difference learning, a thoroughly examined form of reinforcement learning (Sutton and Barto, 1998) that learns the predictive structure of an environment. Simulations demonstrate that despite the apparent simplicity of this model, it is a very powerful learner, capable of learning master-level backgammon, for example (Tesauro, 1995).

14.8 The Developmental Relation between the Midbrain Dopamine System and the Prefrontal Cortex

It is deeply intriguing to note where the midbrain dopamine system projects to in the human brain. In particular, what is most intriguing is the fact that it projects to the dorsolateral prefrontal, premotor, and parietal cortices, which are structures believed to mediate goal representations; and the orbitofrontal cortex, which is believed to mediate the representation of relative reward value and reward expectation (for a review, see Schultz, 2000). A great deal of attention has centered on dorsolateral and orbitofrontal prefrontal cortex as structures implicated in crucial components of human cognition, particularly social cognition and theory of mind (Stone et al., 1998), symbolic learning (Deacon, 1997), representa-

tions of self (Craik et al., 1999), and executive function and behavioral inhibition (Norman and Shallice, 1986).

It is important to ask what the functional significance is of the fact that a phylogenetically old part of the brain projects to a phylogenetically new part. According to neural constructivism, these structures constitute a hierarchically organized control structure in which additional layers of control have been added to the evolutionarily conserved dopamine system and in which this hierarchical organization is evident developmentally as well. To see this, it is important to examine the developmental links between these components, which I explore in more detail later.

Diamond and colleagues (reviewed in Diamond, 1998) have demonstrated that a functional midbrain dopaminergic system is necessary for normal development of prefrontal functions. The most compelling evidence for this developmental dependence stems from studies of phenylketonuria (PKU). Patients suffering from PKU do not naturally produce a particular enzyme, phenylalanine hydroxylase, which converts the essential amino acid phenylalanine to another amino acid, tyrosine, the precursor of dopamine. When untreated, PKU leads to severe mental retardation. Diamond and colleagues found that lowered levels of tyrosine uniquely affect the cognitive functions that are dependent on the prefrontal cortex because of the special sensitivity of prefrontally projecting dopamine neurons to small decreases in tyrosine. In a 4-year longitudinal study, they found that PKU children performed worse than matched controls, their own siblings, and children from the general population on tasks that required the working memory and inhibitory control abilities that are dependent on dorsolateral prefrontal cortex. In contrast, these PKU children performed well on control tasks that were not mediated by the prefrontal cortex (Diamond et al., 1997).

The hierarchical organization of the control structures that constitute the human cognitive architecture is apparent developmentally. In contrast to the early functional involvement of midbrain dopamine systems, prefrontal structures develop relatively late and exhibit a protracted development that continues into adolescence. Thus, behavior and cognition increasingly come under the mediation of frontal structures from subcortical structures, a process sometimes referred to as frontalization of behavior (Rubia et al., 2000). For example, executive function is a control mechanism that guides, coordinates, and updates behavior in a flexible fashion, particularly in novel or complex tasks (Norman and Shallice, 1986). This requires that information related to behavioral goals be actively represented and maintained so that these representations may guide behavior toward goal-directed activities. In humans, executive function follows a special developmental path, reflecting an evolutionary reorganization of prefrontal structures and their development. Between 7.5 and 12 months of age, infants show developmental progress on on A-not-B (Diamond, 1985), delayed-response (Diamond and Doar, 1989), and object retrieval tasks (Diamond, 1988). There is substantial evidence that these tasks are mediated by dorsolateral prefrontal cortex and rely on working memory, neural representations of goal-related information, and behavioral inhibition (Goldman-Rakic, 1990; Petrides, 1995). Furthermore, various sources of evidence indicate that dopamine is necessary for successful performance on these tasks (Sawaguchi and Goldman-Rakic, 1994).

Although there is strong evidence that an intact dopamine system is necessary for the developmental emergence of prefrontal functions, a largely unresolved question concerns the specific nature of this developmental link. One particularly intriguing possibility is that the dopamine signal serves as a learning signal that guides the construction of prefrontal structures during development. Computational work on the midbrain dopamine system suggests such a learning role, with strong analogies to temporal difference learning, a form of reinforcement learning (Sutton and Barto, 1998).

A key notion underlying reinforcement learning is that of learning through interacting with one's environment. For example, a major source of knowledge stems from an infant's interaction with its environment, which produces a wealth of information about cause and effect, about the consequences of actions, and about what to do in order to achieve goals, all without the need for an explicit teacher. Of course, Piaget also emphasized the central importance of the developing child's agency and active exploration with its environment in his constructivist theory of cognitive development.

Learning through interacting with one's environment requires structures that direct the system to its environment. According to the view I have been outlining here, this is mediated in part by the midbrain dopamine system. One clue for this role derives from studies of the neurobiology of personality, which view personality as deriving from motivational systems. From this perspective, the midbrain dopamine system constitutes a behavioral facilitation system that underlies fundamental properties of personality, specifically extraversion, and positive emotionality (Depue and Collins, 1999). From a developmental perspective, this behavioral facilitation system appears to be operative at an early age and most likely underlies major dimensions of temperament, along with other diffuse ascending systems, such as noradrenergic and serotonergic systems. Thus, given this system's computational properties and its role as a behavioral facilitation system early in postnatal development, it is ideally situated to be involved in the reinforcement or self-supervised construction of prefrontal structures underlying complex behavioral control.

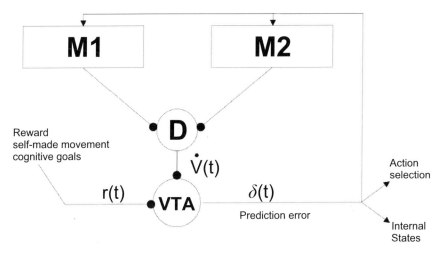

Figure 14.2

Architecture of prediction learning. M1 and M2 represent two different cortical modalities whose output is assumed to arrive at the ventral tegmental area (VTA) in the form of a temporal derivative $\dot{V}(t)$. The structure D is a placeholder in this model to represent the highly convergent input into the VTA (in biological systems, D is likely to be part of the VTA itself). Information about reward $r(t)$ also converges on the VTA. The VTA output is taken as a simple linear sum $\delta(t) = r(t) + \dot{V}(t)$, which is taken to be a measure of prediction error. The output connections of the VTA make the prediction error $\delta(t)$ simultaneously available to structures constructing the predictions.

This computational role can be illustrated by comparing reinforcement models of learning with models of self-organization, or unsupervised learning. The best-known account of unsupervised learning is Hebbian learning, which in its simplest form is

$$w_{ki}(t + 1) = w_{ki}(t) + \eta \gamma_k(t)x_i(t), \qquad (14.1)$$

where $w_{ki}(t)$ is the synaptic weight between neurons i and k at time t, $x_i(t)$ and $\gamma_k(t)$ are, respectively, the presynaptic and the postsynaptic activity, and η is a positive constant that determines the rate of learning. Algorithms such as Eq. (14.1), and a variety of modifications, essentially find efficient representations of salient environmental information by implementing such data reduction strategies as principal component analysis. Such algorithms can be modified to become reinforcement learning algorithms by making weight updates dependent on the Hebbian correlation of a prediction error and the presynaptic activity at the previous time step. This takes the following form:

$$w_i(t) = w_i(t - 1) + \eta x_i(t - 1)\delta(t), \qquad (14.2)$$

where $x_i(t - 1)$ represents presynaptic activity at connection i and time $t - 1$, η is a learning rate, and $w_i(t - 1)$ is the value of the weight at time $t - 1$. The term $\delta(t)$ is a prediction error term (see figure 14.2) and is the difference between a prediction of reward and the actual reward, represented as the output of the dopaminergic projection to the cortex in the simulation framework. The addition of this term changes the Hebbian framework to a predictive Hebbian one (Montague and Sejnowski, 1994) and is the computed differential in the temporal differences method of reinforcement learning (Sutton and Barto, 1998), with close connections to dynamic programming (Bellman, 1957).

The developmental link between the midbrain dopamine system and prefrontal structures suggests that complex developmental skills decompose into developmental precursors, which may often be mediated by structures that are distinct from those mediating the mature state. For example, processing of faces is believed to be mediated by subcortical structures during early postnatal development, but it subsequently shifts to cortical sites (reviewed in Johnson, 1997). The framework I have outlined here suggests a possible way of bootstrapping a system into such complex representations by biasing development through making the system selectively attentive to faces (for a review of the relation between reward structures and selective attention, see Dayan et al., 2000). An economical means of implementing such a strategy would be by making faces, or primitive template representations of them, rewarding to the system, thereby designing a system that preferentially attends to faces. It is clear that human infants possess such behavioral biases (Metzloff and Moore, 1977), which may be implemented through projections to midbrain dopamine systems that constitute unconditioned stimuli.

Although an investigation into how one brain region may direct the development of another through such learning procedures as temporal differences learning is only in early stages, it offers a new framework for analyzing the dynamics of developmental change. Current work in my laboratory involves exploring the relationship between these algorithms and neural outgrowth. It is intriguing to note that dendritic structures in the prefrontal cortex display a protracted development (figure 14.3) and that dopamine may modulate this development by acting trophically (Levitt et al., 1997). These links merit further research.

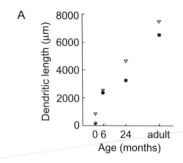

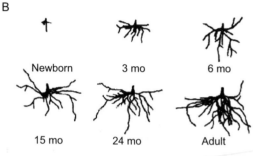

Figure 14.3
Human postnatal dendritic development in dorsolateral prefrontal cortex. (*A*) The total dendritic length of basal dendrites of layer III and V pyramidal cells develops over a protracted period. (*B*) Camera lucida drawings of layer V basal dendrites reveal the extent of this protracted postnatal development. (Modified from Schade and Van Groenigan, 1961.)

14.9 Environmental Structure and Brain Development

In examining the relationship between structured neural activity and neural development, neural constructivism stresses the importance of environmental structure and information in the developmental process. Although proponents of innateness discounted such influences, on the basis of various poverty of the stimulus arguments, there are good reasons to believe this discounting was premature (see Cowie, 1998).

Indeed, as a number of investigators have recently stressed (Tomasello, 1999), it appears that human cognitive development depends on an extremely rich social and cultural interaction. Although prefrontal function has traditionally been most closely associated with purely cognitive functions, its central involvement in social cognition has become increasingly apparent in recent years (Damasio, 1994). Indeed, one potential reason for protracted development lies in the difficulty of developing the social competence necessary for a complex social life. There is now good evidence to indicate that one component of social competence, theory of mind, depends at least in part on the appropriate social exposure for its development since many deaf children show delays on theory of mind tasks (Peterson and Siegal, 1995; Russell et al., 1998). This is believed to be because the parents of deaf children are typically naïve signers, and so household social interactions are limited by communicative ability. Constructive learning, then, may be one particularly powerful route to building complex cognitive and social skills by allowing the structure of the environment to play a central role in cognitive development.

Although developmental models at the network level are still rudimentary, the evidence I have reviewed in this chapter indicates that they will be increasingly important in understanding the complex interplay between the cognitive processes of learning and the biological processes of maturation that underlie the construction of neural representations across the developmental time course.

14.10   Discussion

Computational models of the neural basis of cognitive development are only beginning to be explored, in part because the experimental constraints necessary to build such models are just beginning to emerge from developmental neurobiology and developmental cognitive neuroscience. As I have shown in this chapter, this modeling effort has begun to characterize the nature of learning during development as one that is richer than traditional notions of statistical inference. I have referred to this activity-dependent construction as constructive learning, which from a learning-theoretical perspective appears to possess more powerful acquisition properties than traditional accounts of cognitive development assumed.

Future Modeling Studies

With the advent of structural MRI, it appears that brain development is regional and follows a hierarchical ordering of neural systems (Giedd et al., 1999; Jacobs et al., 2001). It will be an important goal of future theoretical research to link computational models of development at the cellular level with neurobiological evidence on the hierarchical construction of neural systems. It will also be important to integrate this research more closely with behavioral studies of human cognitive development (e.g., Luciana and Nelson, 1998). A further important goal will be to explore self-supervised developmental algorithms because a critical element of development involves the developing system's interaction with its environment and the possibilities for learning that arise as a consequence of this interaction.

Future Experimental Studies

The rapid progress in many developmental sciences suggests that it will become increasingly important to nurture interdisciplinary efforts that can integrate these multiple perspectives on development. Descriptions of behavioral change at the cognitive level are too unconstrained without reference to underlying mechanisms and processes, whereas an understanding of the mechanisms of developmental change is in itself insufficient to explain neural functional or cognitive development. Only by integrating experimental results across a number of levels of organization can a more complete account of development begin to emerge. Computational modeling promises to offer a conceptually powerful framework in which to integrate these perspectives. Future interdisciplinary experimental work will thus play a critical role in constructing this framework by providing results that can suitably constrain it.

Notes

1. The notion of representation is a complex and contentious one in cognitive neuroscience. See O'Reilly and Munakata (2000) for a discussion on the use of representation in cognitive neuroscience.

2. More technically, the main unit would include the axodendritic interface, although the main information-processing functions of interest would be dendritic integration.

3. Although this strategy emphasizes learning, it is important to bear in mind that it requires a primitive set of target states that have intrinsic reward value for the organism (classically known as an unconditioned stimulus).

References

Allman, J. (1999). *Evolving Brains*. New York: Freeman.

Barron, A. R. (1994). Approximation and estimation bounds for artificial neural networks. *Machine Learn.* 14: 115–133.

Bellman, R. E. (1957). *Dynamic Programming*. Princeton, N.J.: Princeton University Press.

Benson, D. F. (1993). Prefrontal abilities. *Behav. Neurol.* 6: 75–81.

Bernardo, L. S., Masukawa, L. M., and Prince, D. A. (1982). Electrophysiology of isolated hippocampal pyramidal dendrites. *J. Neurosci.* 2: 1614–1622.

Bliss, T. V. P., and Lomo, T. (1973). Long-lasting potentiation of synaptic transmission in the dentate area of the anesthetized rabbit following stimulation of the perforant path. *J. Physiol. (London)* 232: 331–356.

Blum, A., and Rivest, R. L. (1988). Training a 3-node neural network is NP-complete. In *Advances in Neural Information Processing Systems*, D. S. Touretzky, ed. San Jose: Morgan Kaufmann, pp. 494–501.

Blumer, A., Ehrenfeucht, A., Haussler, D., and Warmuth, M. (1988). Learnability and the Vapnik-Chervonenkis dimension. Technical Report UCSC-CRL-87-20.

Changeux, J. P., and Danchin, A. (1976). Selective stabilisation of developing synapses as a mechanism for the specification of neuronal networks. *Nature* 264: 705–712.

Cowie, F. (1998) *What's within? Nativism Reconsidered*. Oxford: Oxford University Press.

Craik, F. I. M., Moroz, T. M., Moscovitch, M., Stuss, D. T., Winocur, G., Tulving, E., and Kapur, S. (1999). In search of the self: A positron emission tomography study. *Psychol. Sci.* 10: 26–34.

Damasio, A. R. (1994). Descartes' error: Emotion, reason, and the human brain. New York: G. P. Putnam.

Dayan, P., Kakade, S., and Montague, P. R. (2000). Learning and selective attention. *Nat. Neurosci.* 3: 1218–1223.

Deacon, T. W. (1997). *The Symbolic Species: The Co-evolution of Language and the Brain*. New York: W. W. Norton.

Depue, R. A., and Collins, P. F. (1999). Neurobiology of the structure of personality: Dopamine, facilitation of incentive motivation, and extraversion. *Behav. Brain Sci.* 22: 491–569.

Diamond, A. (1985). Development of the ability to use recall to guide action, as indicated by infants' performance on AB. *Child Dev.* 56: 868–883.

Diamond, A. (1988). Abilities and neural mechanisms underlying AB performance. *Child Dev.* 59: 523–527.

Diamond, A. (1998). Evidence for the importance of dopamine for prefrontal cortex functions early in life. In *The Prefrontal Cortex: Executive and Cognitive Functions*, pp. 144–164. New York: Oxford University Press.

Diamond, A., and Doar, B. (1989). The performance of human infants on a measure of frontal cortex function, the delayed response task. *Dev. Psychobiol.* 22: 271–294.

Diamond, A., Prevor, M. B., Callender, G., and Druin, D. P. (1997). Prefrontal cortex cognitive deficits in children treated early and continuously for PKU. *Monogr. Soc. Res. Child Dev.* 62: 1–205.

Dietterich, T. G. (1990). Machine learning. *Annu. Rev. Comp. Sci.* 4: 255–306.

Edelman, G. (1987). *Neural Darwinism: The Theory of Neuronal Group Selection.* New York: Basic Books.

Elman, J. L. (1993). Learning and development in neural networks: The importance of starting small. *Cognition* 48: 71–99.

Elman, J. L., Bates, E. A., Johnson, M. H., Karmiloff-Smith, A., Parisi, D., and Plunkett, K. (1996). *Rethinking Innateness: A Connectionist Perspective on Development.* Cambridge, Mass.: MIT Press.

Elston, G. N., and Rosa, M. G. (1998a). Morphological variation of layer III pyramidal neurones in the occipito-temporal pathway of the macaque monkey visual cortex. *Cereb. Cortex.* 8: 278–294.

Elston, G. N., and Rosa, M. G. (1998b). Complex dendritic fields of pyramidal cells in the frontal eye field of the macaque monkey: Comparison with parietal areas 7a and LIP. *NeuroReport* 9: 127–131.

Elston, G. N., Rosa, M. G., and Calford, M. B. (1996). Comparison of dendritic fields of layer III pyramidal neurons in striate and extrastriate visual areas of the marmoset: A Lucifer yellow intracellular injection. *Cereb. Cortex.* 6: 807–813.

Engert, F., and Bonhoeffer, T. (1999). Dendritic spine changes associated with hippocampal long-term synaptic plasticity. *Nature* 399: 66–70.

Fuster, J. M. (1997). Network memory. *Trends Neurosci.* 20: 451–459.

Gallistel, C. R. (1990). *The Organization of Learning.* Cambridge, Mass.: MIT Press.

Garris, P. A., Kilpatrick, M., Bunin, M. A., Michael, D., Walker, Q. D., and Wightman, R. M. (1999). Dissociation of dopamine release in the nucleus accumbens from intracranial self-stimulation. *Nature* 398: 67–69.

Geman, S., Bienenstock, E., and Doursat, R. (1992). Neural networks and the bias/variance dilemma. *Neur. Comput.* 4: 1–58.

Giedd, J. N., Blumenthal, J., Jeffries, N. O., Castellanos, F. X., Liu, H., Zijdenbos, A., Paus, T., Evans, A. C., and Rapoport, J. L. (1999). Brain development during childhood and adolescence: A longitudinal MRI study. *Nat. Neurosci.* 2: 861–863.

Gold, E. M. (1967). Language identification in the limit. *Info. Control* 10: 447–474.

Golding, N. L., and Spruston, N. (1998). Dendritic sodium spikes are variable triggers of axonal action potentials in hippocampal CA1 pyramidal neurons. *Neuron* 21: 1189–1200.

Goldman-Rakic, P. S. (1990). Cortical localization of working memory. In *Brain Organization and Memory: Cells, Systems, and Circuits*, pp. 285–298. Oxford: Oxford University Press.

Hammer, M. (1993). An identified neuron mediates the unconditioned stimulus in associative olfactory learning in honeybees. *Nature* 366: 59–63.

Haussler, D. (1989). Quantifying inductive bias: AI learning algorithms and Valiant's learning framework. *Art. Intel.* 36: 177–222.

Hebb, D. O. (1949). *The Organization of Behavior: A Neuropsychological Theory.* New York: Wiley.

Hinton, G. E., and Sejnowski, T. J. (1999). *Unsupervised Learning: Foundations of Neural Computation.* Cambridge, Mass.: MIT Press.

Huttenlocher, P. R., and Dabholkar, A. S. (1997). Regional differences in synaptogenesis in human cerebral cortex. *J. Comp. Neurol.* 387: 167–178.

Jacobs, B., Schall, M., Prather, M., Kapler, E., Driscoll, L., Baca, S., Jacobs, J., Ford, K., Wainwright, M., and Treml,

M. (2001). Regional dendritic and spine variation in human cerebral cortex: A quantitative golgi study. *Cereb. Cortex* 11: 558–571.

Johnson, M. H. (1997). *Developmental Cognitive Neuroscience: An Introduction*. Oxford: Blackwell Science.

Judd, S. (1988). On the complexity of loading shallow neural networks. *J. Complex.* 4: 177–192.

Knudsen, E. I. (1998). Capacity for plasticity in the adult owl auditory system expanded by juvenile experience. *Science* 279: 1531–1533.

Koza, J. (1992). *Genetic Programming: On the Programming of Computers by means of Natural Selection*. Cambridge, Mass.: Bradford Books.

Lendvai, B., Stern, E. A., Chen, B., and Svoboda, K. (2000). Experience-dependent plasticity of dendritic spines in the developing rat barrel cortex in vivo. *Nature* 404: 876–881.

Levitt, P., Harvey, J. A., Friedman, E., Simansky, K., and Murphy, E. H. (1997). New evidence for neurotransmitter influences on brain development. *Trends Neurosci.* 20: 269–274.

Linsker, R. (1990). Perceptual neural organization: Some approaches based on network models and information theory. *Annu. Rev. Neurosci.* 13: 257–281.

Luciana, M., and Nelson, C. A. (1998). The functional emergence of prefrontally guided working memory systems in four- to eight-year-old children. *Neuropsychologia* 36: 273–293.

Maletic-Savatic, M., Malinow, R., and Svoboda, K. (1999). Rapid dendritic morphogenesis in CA1 hippocampal dendrites induced by synaptic activity. *Science* 283: 1923–1927.

Marr, D. (1982). *Vision: A Computational Investigation into the Human Representation and Processing of Visual Information*. San Francisco: W. H. Freeman.

Metzloff, A. N., and Moore, M. K. (1977). Imitation of facial and manual gestures by human neonates. *Science* 298: 75–78.

Montague, P. R., and Quartz, S. R. (1999). Computational approaches to neural reward and development. *Ment. Retard. Dev. Disabil. Res. Rev.* 5: 86–99.

Montague, P. R., and Sejnowski, T. J. (1994). The predictive brain: Temporal coincidence and temporal order in synaptic learning mechanisms. *Learn. Mem.* 1: 1–33.

Montague, P. R., Dayan, P., Person, C., and Sejnowski, T. J. (1995). Bee foraging in uncertain environments using predictive hebbian learning. *Nature* 377: 725–728.

Norman, D. A., and Shallice, T, (1986). Attention to action: Willed and automatic control of behavior. In *Consciousness and Self-Regulation*, R. J. Davidson, G. E. Schwartz, and D. Shapiro, eds. pp. 1–18. New York: Plenum.

O'Reilly, R. C., and Munakata, Y. (2000). *Computational Explorations in Cognitive Neuroscience: Understanding the Mind by Simulating the Brain*. Cambridge, MA: MIT Press.

Peterson, C. C., and Siegal, M. (1995). Deafness, conversation and theory of mind. *J. Child Psych. Psych.* 36: 459–474.

Petrides, M. (1995). Functional organization of the human frontal cortex for mnemonic processing: Evidence from neuroimaging studies. In *Structure and Functions of the Human Prefrontal Cortex*, pp. 85–96. New York: New York Academy of Sciences.

Plunkett, K., Karmiloff-Smith, A., Bates, E., and Elman, J. L. (1997). Connectionism and developmental psychology. *J. Child Psychol. Psychiat., Allied Discipl.* 38: 53–80.

Poirazi, P., and Mel, B. W. (2001). Impact of active dendrites and structural plasticity on the memory capacity of neural tissue. *Neuron* 29: 779–796.

Purves, D., White, L. E., and Riddle, D. R. (1996). Is neural development Darwinian? *Trends Neurosci.* 19: 460–464.

Quartz, S. R. (1993). Nativism, neural networks, and the plausibility of constructivism. *Cognition* 48: 123–144.

Quartz, S. R. (1999). The constructivist brain. *Trends Cognit. Sci.* 3: 48–57.

Quartz, S. R. (in press). Toward a developmental evolutionary psychology: genes, development, and the evolution of the human cognitive architecture. In *Evolutionary Psychology: Alternative Approaches*, S. Scher and M. Rauscher, eds. Dordrecht: Kluwer.

Quartz, S. R., and Sejnowski, T. J. (1997). The neural basis of cognitive development: A constructivist manifesto. *Behav. Brain Sci.* 20: 537–596,

Quinlan, P. T. (1998). Structural change and development in real and artificial neural networks. *Neur. Net.* 11: 577–599.

Rakic, P., Bourgeois, J. P., Eckenhoff, M. F., Zecevic, N., and Goldman-Rakic, P. S. (1986). Concurrent overproduction of synapses in diverse regions of the primate cerebral cortex. *Science* 232: 232–235.

Real, L. A. (1991). Animal choice behavior and the evolution of cognitive architecture. *Science* 253: 980–986.

Redding, N. J., Kowalczyk, A., and Downs, T. (1993). Constructive higher-order network algorithm that is polynomial time. *Neur. Net.* 6: 997–1010.

Rubia, K., Overmeyer, S., Taylor, E., Brammer, M., Williams, S. C. R., Simmons, A., Andrew, C., and Bullmore, E. T. (2000). Functional frontalisation with age: Mapping neurodevelopmental trajectories with fMRI. *Neurosci. Biobehav. Rev.* 24: 13–19.

Russell, P. A., Hosie, J. A., Gray, C. D., Scott, C., Hunter, N., Banks, J. S., and Macaulay, M. C. (1998). The development of theory of mind in deaf children. *J. Child Psychol. Psychiat.* 39: 903–910.

Sawaguchi, T., and Goldman-Rakic, P. S. (1994). The role of D1-dopamine receptor in working memory: Local injections of dopamine antagonists into the prefrontal cortex of rhesus monkeys performing an oculomotor delayed-response task. *J. Neurophysiol.* 71: 515–528.

Schade, J. P., and Van Groenigan, W. B. (1961). Structural organization of the human cerebral cortex: I. Maturation of the middle frontal gyrus. *Acta Anatom.* 47: 72–111.

Schiller, J., Major, G., Koester, H. J., and Schiller, Y. (2000). NMDA spikes in basal dendrites of cortical pyramidal neurons. *Nature* 404: 285–289.

Shultz, T. R., Mareschal, D., and Schmidt, W. C. (1994). Modeling cognitive development on balance scale phenomena. *Machine Learning* 16: 57–86.

Schultz, W. (2000). Multiple reward signals in the brain. *Nat. Rev. Neurosci.* 1: 199–207.

Schultz, W., Apicella, P., and Ljungberg, T. (1993). Responses of monkey dopamine neurons to reward and conditioned stimuli during successive steps of learning a delayed response task. *J. Neurosci.* 13: 900–913.

Stone, V. E., Baron-Cohen, S., and Knight, R. T. (1998). Frontal lobe contributions to theory of mind. *J. Cognit. Neurosci.* 10: 640–656.

Sutton, R. S., and Barto, A. G. (1998). *Reinforcement Learning: An Introduction.* Cambridge, Mass.: MIT Press.

Tesauro, G. (1995). Temporal difference learning and TD-Gammon. *Commun. ACM* 38: 58–68.

Thompson, P. M., Giedd, J. N., Woods, R. P., MacDonald, D., Evans, A. C., and Toga, A. W. (2000). Growth patterns in the developing brain detected by using continuum mechanical tensor maps. *Nature* 404: 190–193.

Tomasello, M. (1999). *The Cultural Origins of Human Cognition.* Cambridge, Mass.: Harvard University Press.

Valiant, L. G. (1984). A theory of the learnable. *Commun. ACM* 27: 1134–1142.

Vapnik, V. N. (1998). *Statistical Learning Theory.* New York: Wiley.

White, H. (1990). Connectionist nonparametric regression: Multilayer feedforward networks can learn arbitrary mappings. *Neur. Net.* 3: 535–549.

Wise, R. A. (1996). Addictive drugs and brain stimulation reward. *Annu. Rev. Neurosci.* 19: 319–340.

Wong, R. O. (1999). Retinal waves and visual system development. *Annu. Rev. Neurosci.* 22: 29–47.

Wong, W. T., and Wong, R. O. (2000). Rapid dendritic movements during synapse formation and rearrangement. *Curr. Opin. Neurobiol.* 10: 118–124.

Contributors

L. F. Abbott
Volen Center and Department of Biology
Brandeis University
Waltham, Massachusetts

Jean-Pierre Changeux
Récepteurs et Cognition
Institut Pasteur
Paris, France

Peter G. H. Clarke
Institut de Biologie Cellulaire et de Morphologie
Université de Lausanne
Lausanne, Switzerland

Michael A. Corner
Graduate School Neurosciences Amsterdam
Netherlands Institute for Brain Research
Amsterdam, The Netherlands

Stephen J. Eglen
Institute for Adaptive and Neural Computation
University of Edinburgh
Edinburgh, Scotland
United Kingdom

Alan Fine
Department of Physiology and Biophysics
Faculty of Medicine
Dalhousie University
Halifax, Nova Scotia
Canada

Lucia Galli-Resta
Istituto di Neuroscienze CNR
Pisa, Italy

Geoffrey J. Goodhill
Department of Neuroscience
Georgetown University Medical Center
Washington, D.C.

Bruce P. Graham
Department of Computing Science and Mathematics
University of Stirling
Stirling, Scotland
United Kingdom

H. G. E. Hentschel
Department of Physics
Emory University
Atlanta, Georgia

Stanley B. Kater
Department of Neurobiology and Anatomy
University of Utah School of Medicine
Salt Lake City, Utah

Michel Kerszberg
Récepteurs et Cognition
Institut Pasteur
Paris, France

Eve Marder
Volen Center and Department of Biology
Brandeis University
Waltham, Massachusetts

George Marnellos
Bioinformatics Group
Sequenom, Inc.
San Diego, California

Bartlett W. Mel
Department of Biomedical Engineering
University of Southern California
Los Angeles, California

Eric D. Mjolsness
Institute for Genomics and Bioinformatics
School of Information and Computer Science
University of California, Irvine

David J. Price
Developmental Biology Laboratory
Department of Biomedical Sciences
University of Edinburgh
Edinburgh, Scotland
United Kingdom

Astrid A. Prinz
Volen Center and Department of Biology
Brandeis University
Waltham, Massachusetts

Steven R. Quartz
Division of Humanities and Social Sciences and
Computation and Neural Systems Program
California Institute of Technology
Pasadena, California

Benjamin E. Reese
Neuroscience Research Institute and Department of
Psychology
University of California
Santa Barbara, California

Richard R. Ribchester
Department of Neuroscience
University of Edinburgh
Edinburgh, Scotland
United Kingdom

N. V. Swindale
Department of Ophthalmology and Visual Sciences
University of British Columbia

Vancouver, British Columbia
Canada

Vatsala Thirumalai
Cold Spring Harbor Laboratory
Cold Spring Harbor, New York

Kurt A. Thoroughman
Department of Biomedical Engineering
Washington University
St. Louis, Missouri

Jeffrey S. Urbach
Department of Physics
Georgetown University
Washington, D.C.

Harry B. M. Uylings
Graduate School Neurosciences Amsterdam
Netherlands Institute for Brain Research and
Department of Anatomy, VU Medical Center
Amsterdam, The Netherlands

Arjen van Ooyen
Graduate School Neurosciences Amsterdam
Netherlands Institute for Brain Research
Amsterdam, The Netherlands

Jaap van Pelt
Graduate School Neurosciences Amsterdam
Netherlands Institute for Brain Research
Amsterdam, The Netherlands

David J. Willshaw
Institute for Adaptive and Neural Computation
School of Informatics
University of Edinburgh
Edinburgh, Scotland
United Kingdom

Index